Annual Editions:
Global Issues, 33/e

Robert Weiner

http://create.mheducation.com

ISBN-10: 1259883302 ISBN-13: 9781259883309

Contents

Detailed Table of Contents

Unit 1: Global Issues in the Twenty-First Century: An Overview

Unit 2: Population, Natural Resources, and Climate Change

Unit 3: The Global Political Economy

Unit 4: Terrorism

ISIS Is Not a Terrorist Group: Why Counterterrorism Won't Stop the Latest Jihadist Threat, Audrey Kurth Cronin, *Foreign Affairs*, 2015

ISIS is not a terrorist organization but rather a pseudostate which controls territory to establish a caliphate in the Middle East. The United States should pursue a policy of containment toward ISIS.

ISIS and the Third Wave of Jihadism, Fawaz A. Gerges, *Current History*, 2014

ISIS emerged as an offshoot of the branch of Al-Qaeda in Iraq. It emerged as a result of the grievances of the Sunnis' repression by the Shia regime of former Prime Minister Maliki as well as the failure of state institutions in Iraq. ISIS continues to focus on sectarian war as its priority in waging war with extreme brutality and violence.

Strategic Amnesia and ISIS, David V. Gioe, *The National Interest*, 2016

In attempting to degrade and destroy ISIS, the United States has enjoyed tactical victories, but the author argues that a strategic victory has eluded it. The United States needs to apply the lessons learned from military history, ranging from the American Revolutionary War to the Vietnam conflict.

Obama and Terrorism: Like It or Not, the War Goes On, Jessica Stern, *Foreign Affairs*, 2015

The author writes that Obama has continued the war against terrorism and violent extremism with targeted killings, aid to allied and indigenous forces, and intensive electronic surveillance. Obama has relied heavily on armed drones to a much greater extent than the Bush administration.

Fixing Fragile States, Dennis Blair et al., *The National Interest*, 2014

Since the 9/11 attacks, the United States has waged major postwar reconstruction campaigns in Iraq and Afghanistan and smaller programs in other countries that harbor Al-Qaeda affiliates. Much of the threat stems from fragile states with weak institutions, higher rates of poverty, and deep ethnic, religious, or tribal divisions.

Unit 5: Conflict and Peace

The Growing Threat of Maritime Conflict, Michael T. Klare, *Current History*, 2013

Prospects for conflict over disputed borders have declined, but conflict over maritime boundaries is growing. A major reason for these conflicts is energy consumers are increasingly reliant on offshore oil and gas deposits.

Afghanistan's Arduous Search for Stability, Thomas Barfield, *Current History*, 2016

The Bush administration supported the Karzai regime, but the Obama administration was critical of the corrupt nature of the regime. Relations between the United States and Afghanistan soured with the advent of the Obama administration, resulting in Karzai's replacement by the dual executive regime of Abdullah Abdullah and Ashraf Ghani which was plagued by disunity, as the reduction of US forces resulted in the resurgence of the Taliban.

Water Wars: A Surprisingly Rare Source of Conflict, Gregory Dunn, *Harvard International Review*, 2013

Competition for access to the increasingly scarce resource of freshwater has surprisingly been mostly resolved through peaceful means via negotiated treaties.

Taiwan's Dire Straits, John J. Mearsheimer, *The National Interest*, 2014

The rise of China in the international system will upset the balance of power in Beijing's favor, with profound implications for Taiwan. China will attempt to dominate Asia as a regional hegemon.

Why 1914 Still Matters, Norman Friedman, The US Naval Institute Proceedings, 2014

There is a similarity in the outbreak of war between the United Kingdom and Germany in 1914 and the possibility of war between the United States and China given that both cases involved a naval arms race that challenged the trading hegemon.

Unit 6: Ethics and Values

The G-Word: The Armenian Massacre and the Politics of Genocide, Thomas de Waal, *Foreign Affairs*, 2015

Year 2015 marks 100 years since the Armenian community in Ottoman Turkey faced efforts on the part of the Turkish government to destroy it. Over one million Armenians perished in the genocide. For strategic reasons, because Turkey is a major US ally, Washington refuses to use the G-word to describe the great catastrophe which befell the Armenians.

Race in the Modern World: The Problem of the Color Line, Kwame Anthony Appiah, *Foreign Affairs*, 2015

The author discusses various efforts that have been made over the years to define race, with a great deal of emphasis placed on the work of W. E. B. Du Bois. Du Bois discussed race as a transnational phenomenon as illustrated by demonstrations in Nigeria protesting the shooting by a police officer of an unarmed black teenager, Michael Brown, in Ferguson, Missouri.

Democracy and Its Discontents, John Shattuck, *The Fletcher Forum of World Affairs*, 2016

Illiberal democracy, which places more emphasis on national identity and centralization of power in the state, marks a regression from the values of pluralism of liberal democracy. East European states such as Poland and Hungary have rejected the values of liberal democracy as represented by the European Union.

Preface

This book engages in an analysis of contemporary global issues based on a careful reading of the major elite newspapers and magazines, as well as the issues that have been emphasized at recent meetings of the International Studies Association and the International Political Science Association. An effort to identify important global issues has also been culled from an analysis of major US governmental reports such as the National intelligence Council, the Pentagon Quadrennial Defense Review, and the US State Department's Quadrennial Diplomatic report.

The ability of the international community to deal with global issues takes place within the framework of the forces of globalization, in a mixed international system, whose structure consists of both state and nonstate actors. Consequently, globalization is occurring in a multipolar system that is characterized by a diffusion of power. Although states, especially the "Great Powers," are still the primary actors in the international system, their power has been eroded somewhat by the phenomenal growth of such nonstate actors as international governmental organizations, nongovernmental organizations, multinational corporations, and terrorist organizations such as Al-Qaeda and the Islamic State.

The identification of current global issues was also facilitated by the editor's attendance of many events and conferences held by the think tanks in Washington, DC.

In publishing *Annual Editions*, we recognize the enormous role played by magazines, newspapers, and journals of the public press in a broad spectrum of areas. A number of articles are drawn from influential journals such as *Foreign Affairs*, *Foreign Policy*, and *The National Interest* as well, which deal with the most important global issues of the day. Many of these articles are appropriate for students, researchers, and professionals seeking accurate, current information to help bridge the gap between theories and the real world. These articles, however, become more useful for study when those of lasting value are carefully collected, organized, indexed, and reproduced in a low-cost format that provides easy and permanent access when the material is needed. That is the role of *Annual Editions*.

A number of learning tools are also included in the book. Each article is followed by a set of *Critical Thinking* questions designed to allow the student to engage in further research and to stimulate classroom discussion, and valuable *Internet References* that provide the reader with more information about the themes addressed in each article. Each article is also preceded by *Learning Outcomes*, which helps the student to focus on the major themes of each article.

I would like to express my thanks to McGraw-Hill's Senior Product Developer, Jill Meloy, without whose guidance this project would not have been completed. Special thanks are also due to Dan Torres, whose research assistance was invaluable in selecting articles that appear in this book.

Robert Weiner
Editor

Editor

Robert Weiner is a Center Associate at the Davis Center for Russian and Eurasian Studies at Harvard University and a Fellow at the Center for Peace, Democracy, and Development at the McCormick Graduate School of Policy and Global Studies, University of Massachusetts/Boston. He has worked as a consultant for Global Integrity, a Washington-based nongovernmental organization that investigates corruption in countries around the world. He is the author of *Romanian Foreign Policy at the United Nations* and *Change in Eastern Europe*. He is also the author of more than 20 book chapters and articles and book reviews. Most recently, he has published chapters entitled "The European Union and Democratization in Moldova" and "The Failure to Prevent and Punish Genocide." He has published articles and book reviews in such journals as *The Slavic Review*, *Sudost-Europa*, *The East European Quarterly*, *The International and Comparative Law Quarterly*, *Orbis*, *The Journal of Cold War Studies*, and *The International Studies Encyclopedia*. Between 2001 and 2011, he was the Graduate Program Director of the Master's program in International Relations at the McCormack Graduate School of Policy and Global Studies at the University of Massachusetts/Boston. He is currently a lecturer in the online BA Program in Global Affairs at the University of Massachusetts/Boston.

Academic Advisory Board Members

Members of the Academic Advisory Board are instrumental in the final selection of articles for *Annual Editions* books. Their review of the articles for content, level, and appropriateness provides critical direction to the editor(s) and staff. We think that you will find their careful consideration reflected here.

Tahereh Alavi Hojjat
Desales University

Chi Anyansi-Archibong
North Carolina A&T State University

Augustine Ayuk
Clayton State University

Dilchoda Berdieva
Miami University of Ohio

Karl Buschmann
Harper College

Steven J. Campbell
University of South Carolina, Lancaster

Jianyue Chen
Northeast Lakeview College

Ravi Dhangria
Sacred Heart University

Charles Fenner
SUNY Canton

Dorith Grant-Wisdom
University of Maryland, College Park

Heather Hawn
Mars Hill University

JohnPatrick Ifedi
Howard University

Richard Katz
Antioch University

Steven L. Lamy
University of Southern California

Allan Mooney
Strayer University | Ashford University

Derek Mosley
Meridian Community College

Vanja Petricevic
Florida Gulf Coast University

Nathan Phelps
Western Kentucky University

Amanda Rees
Columbus State University

Kanishkan Sathasivam
Salem State University

Thomas Schunk
SUNY Orange County Community College

James C. Sperling
University of Akron

Uma Tripathi
St. John's University

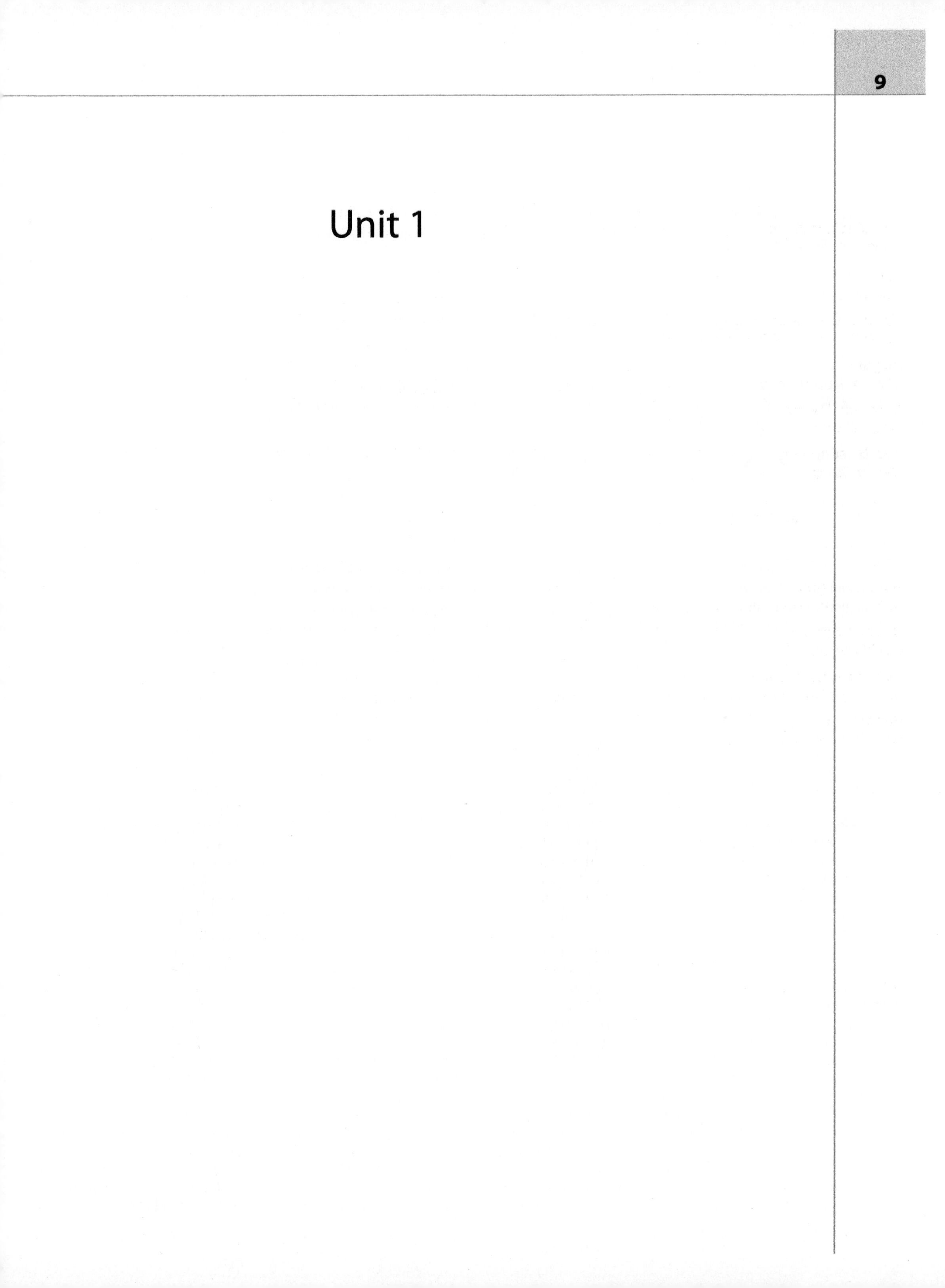

Unit 1

UNIT

Prepared by: Robert Weiner, *University of Massachusetts, Boston*

Global Issues in the Twenty-First Century: An Overview

As the various units that follow indicate, globalization has not necessarily meant the institutionalization of a more effective system of global governance. This was evident in 2014 by the rather slow reaction of the World Health Organization to the outbreak of the Ebola virus in the West African states of Liberia, Sierra Leone, and Guinea. In 2015, the Ebola epidemic was eventually contained but its lingering effects continued to be felt in post-conflict societies like Liberia. In 2016, the potential dangers of a pandemic in a globalized world were underscored again with the outbreak of the Zika virus.

Globalization has also been accompanied by the growth of the world's population to over 7 billion, putting a strain on global natural resources. As the Millennium State of the Future Report for 2014 indicates, humanity is becoming healthier, wealthier, and better educated. However, the growth in the world's population has been accompanied by increased competition for scarce resources, such as freshwater. For example, about 60% of the world's freshwater is located in a small number of states. Central Asian states, such as Kazakhstan, find themselves facing the dilemma of seeking access to safe, clean freshwater.

China imports via the oceans, much of the resources, such as oil and iron ore, which it needs to maintain its economic development. On the other hand, in 2016, there continued to be a glut of energy in the world market, as the benchmark price of Brent crude oil fell around $50 a barrel. Advanced economies such as the United States have developed new energy technologies that allow them to significantly increase the extraction of natural gas and oil domestically. This has profound implications for the geopolitics of energy, as the United States is poised to become the world's leading energy producer.

Globalization has not eliminated conflict. Civil conflicts have resulted in the displacement of millions of people, with the number of migrants, internally displaced persons, and refugees standing at one of the highest marks since the end of the Second World War, straining the capacity of the international community to care for them. New masses of "boat people" fleeing the wars in the Middle East, especially in Syria, have sought sanctuary and safety in Europe, with over 3,000 drowning making the dangerous crossing across the Mediterranean, or putting their lives in the hands of criminal gangs of human traffickers. The European Union, faced with hundreds of thousands of migrants flooding into Eastern and Western Europe, has lacked a coherent policy to deal with this latest crisis. Immigration also emerged as a major issue in the 2016 Presidential campaign. The US presidential debate revolved around immigration reform and the question of the deportation of illegal immigrants and their children, especially to Mexico. The United States also recently became a sanctuary for thousands of children from Central America fleeing poverty and criminal gangs.

Globalization is also occurring in an international system which is marked by a diffusion of power, along with the rise and decline of "Great Powers," and the emergence of new centers of power. Change and power transitions in the international system have resulted in what might be described as an emerging multipolar system. China's rise, for example, has drawn international attention to its Grand Strategy and the dangers of maritime conflict in the South and East China Seas, as Beijing seeks to consolidate its position as a regional hegemon. China seeks to deny access to external powers to what it sees as its sphere of influence, as Beijing emphasizes its sovereignty over various disputed islands in the South China and East China Seas. However, in 2016 the Permanent Court of Arbitration found against China's claim to sovereignty over the disputed islands and reefs in the South China Sea. China rejected the finding of the Permanent Court of Arbitration. The construction of artificial islands in the seas also contributes to a second line of defense for China. China identifies itself as a Eurasian power and seeks to expand its influence through the reconstruction of ancient trade routes through Central Asia. Even though China experienced a slowdown in economic growth in 2016, it continued to focus on the Chinese dream. The Chinese dream was based on several initiatives, such as the "One Belt, One Road Project." The One

Belt, One Road project envisaged a vast infrastructure initiative which would connect China through Central Asia to the Middle East and Europe and also had a maritime component. The Chinese initiative also included an Asian Infrastructure Bank (AIIB) and a regional organization known as the Regional Comprehensive Economic Partnership (RCEP).

The United States, on the other hand, has been categorized by a number of analysts as a declining power. However, China's "peaceful" rise has been countered by US rebalancing toward Asia with the creation of the Trans-Pacific Partnership (TPP), and US acquiescence in Japanese development of more offensive military strategy. As of 2016, however, it remained to be seen whether the TPP would be approved by the next president and Congress in 2017.

The emerging multipolar system has been characterized by the rise of the BRICS (Brazil, Russia, India, China, South Africa) as new centers of power. However, by 2016, most of the BRICS states with the exception of India were experiencing serious economic difficulties. Nevertheless, several African states have undergone impressive amounts of economic growth within the framework of a western dominated international and economic financial system.

The emerging multipolar system has also been marked by regional instability, especially in the Middle East, as ISIL (the Islamic State in Iraq and the Levant) continued to function as a major force in the civil conflicts in Syria and Iraq, against a backdrop of unsuccessful US efforts to extricate itself from the region. However, by 2016, ISIL was being pushed back from the territory which it had gained in Iraq, but responding by orchestrating a set of global attacks in such countries as Belgium. In 2015 the world witnessed the deployment of Russian military assets in Syria, in support of President Assad. The increasing involvement of Russian airpower in Syria raised the possibility of a potential conflict with US air strikes against Assad's forces. Critics argued that the United States needed a Grand Strategy to deal with the regional conflicts in the Middle East, raging in Syria, Iraq, Yemen, and Libya. However, the Obama administration continued to pursue a policy of disengagement from the Middle East, based in part on a perception of growing US energy independence. A nuclear agreement, reached between Iran and the United States in the summer of 2015, also had an effect on the regional security architecture in the Middle East, given the opposition of such states as Saudi Arabia and Israel to the agreement.

The United States in 2016 found it difficult to disentangle itself from Afghanistan as promised by President Obama, given a resurgent Taliban, as well as the inroads being made into Afghanistan by ISIL.

The globalized world of the 21st century has witnessed the growth of new digital technologies, such as "the Internet of things." On the darker digital side of globalization, the revolution in global communications and technology has increased the vulnerability of advanced information societies to cyberthreats, hacking, and cyberwar in cyberspace. This was underscored by allegations of Russian hacking into the computer systems of the Democratic Party in an effort to influence the results of the election. Moreover, as Edward Snowden revealed, information technology has also increased the surveillance capacity of the state both globally and at the domestic level.

Globalization has taken place within the framework of increasing threats to the environment, such as climate warming. Most scientists agree that climate warming is here now, due to the release into the atmosphere of greenhouse gases, which has the result of raising the temperature of the earth. This has a number of effects, such as the melting of the ice in the Arctic, and the melting of the ice sheets in Antarctica. The warmer climate also causes extreme weather events, such as killer typhoons and hurricanes, as well as droughts that compound the problem of dealing with water scarcity. A milestone was reached with the agreement of states to limit the emission of their greenhouse gases at a conference which took place in Paris in December 2015.

From a liberal international point of view, there is a need to effectively implement a set of norms that can serve as a benchmark for the behavior of states in the international system. States have a responsibility not only to protect their own populations, but also vulnerable populations of other states from gross and mass violations of human rights. Human security needs to be given at least as much weight as the traditional concern of national security.

Finally, President Obama left office in 2017, with a legacy of mixed results for his successor. China and Russia continued to challenge the post-World War II international order that had been constructed by the United States. President Obama's Grand Strategy was a combination of realism and idealism, as he reached out to states which had been considered enemies of the United States for decades. For example, in 2016, President Obama reopened diplomatic relations with Cuba, negotiated a nuclear arms agreement with Iran, and was the first sitting US President to visit Laos. The Obama administration also lifted an arms embargo against Vietnam, as ironically, the Vietnamese sought a strategic relationship with America to counter rising Chinese power. On the other hand, President Obama continued to pursue a policy of disengagement from the Middle East which continued to be plagued by civil conflict from Yemen to Libya. On the other hand, the hopes for democratization that had been raised in Egypt with the Arab Spring in 2011 had turned into the Arab Winter, as the Egyptian military crushed the Moslem Brotherhood. Relations with Russia, rather than being reset on a positive course worsened after Russian involvement in the Ukrainian civil war and the annexation of the Crimea. Obama's successor would also have to face the threats posed by a North Korean regime which continued to develop its nuclear arsenal.

Article

Prepared by: Robert Weiner, *University of Massachusetts, Boston*

Our Global Situation and Prospects for the Future

Humanity is making momentous strides forward in health, literacy, and many other critical areas, but also stalling or moving backward on many others, warns The Millennium Project in its latest *State of the Future* report.

JEROME C. GLENN

Learning Outcomes

After reading this article, you will be able to:

- Identify the critical areas in which humanity is not moving forward.
- Discuss the effects of an increasingly interconnected world.

The global situation for humanity continues to improve in general, but at the expense of the environment. Massive transitions from isolated subsistence agriculture and industry to a global, Internet-connected, pluralistic civilization are occurring at unprecedented speed and with never-before-seen levels of uncertainty.

The indicators of progress, from health and education to water and energy, show that we are winning more than we are losing—but where we are losing is very serious. As The Millennium Project has documented over the past 17 years in its annual *State of the Future* reports, humanity clearly has the ideas and resources to address its global challenges, but it has not yet shown the leadership, policies, and management on the scale necessary.

On one hand, people around the world are becoming healthier, wealthier, better educated, more peaceful, and increasingly connected, and they are living longer. The child mortality rate has dropped 47% since 1990, while life expectancy has risen by 10 years to reach 70.5 years today. Extreme poverty in the developing world fell from 50% in 1981 to 21% in 2010, primary-school completion rates grew from 81% in 1990 to 91% in 2011, and only one transborder war occurred in 2013. Furthermore, nearly 40% of humanity is now connected via the Internet.

However, water tables on all continents are falling, glaciers are melting, coral reefs are dying, ocean acidity is increasing, ocean dead zones have doubled every decade since the 1960s, and half the world's topsoil has been destroyed.

Some critical socioeconomic fault lines are worsening, as well: Intrastate conflicts and refugee numbers are rising, income gaps are increasingly obscene, and youth unemployment has reached dangerous proportions. Meanwhile, traffic jams and air pollution are strangling cities. In addition, between $1 trillion and $1.6 trillion is paid in bribes, organized crime takes in twice as much money per year as all military budgets combined, civil liberties are increasingly threatened, and half the world is potentially unstable.

The International Monetary Fund expects the global economy to grow from 3% in 2013 to 3.7% during 2014 and possibly 3.9% in 2015. The world population having grown 1.1% in 2013, global per capita income will be increasing by 2.6% or more a year. Our world is reducing poverty faster than many thought was possible.

Nevertheless, the divide between the rich and poor is growing fast: According to Oxfam, the total wealth of the richest 85 people equals that of 3.6 billion people in the bottom half of the world's economy, and half of the world's wealth is owned by just 1% of the population. We need to continue the successful efforts that are reducing poverty, but we also need to focus far more seriously on reducing income inequality in order to avoid long-term instability.

Instability has already been erupting and expanding in many parts of the world over the last five years, due to a confluence of rising food and energy prices, failing states, falling water tables, climate change, desertification, and increasing migrations resulting from political, environmental, and economic conditions. And, because the world is better educated and increasingly connected, people are becoming less tolerant of the abuse of elite power than in the past. Unless these elites open the conversation about the future with the rest of their populations, unrest and revolutions are likely to continue and increase.

Although wars between states are becoming fewer and fewer, and the numbers of both nuclear weapons and battle-related deaths have been decreasing, conflicts within countries are increasing: A third of Syria's 21 million people are displaced or live as refugees, and the world ignores 6 million war-related deaths in the Congo.

Other fault lines are emerging worldwide in the form of rapidly rising frequency of cyberattacks and espionage, an escalation in territorial tensions among Asian countries, and overlapping jurisdictions for energy access to the melting Arctic. It will be a test of humanity's maturity to resolve all these conflicts peacefully.

Meanwhile, the world is automating jobs far more broadly and quickly than it did in earlier eras. How many truck and taxicab drivers will future self-driving vehicles replace? How many industrial laborers will lose their jobs to robotic manufacturing? How many telephone support personnel will be supplanted by AI telephone systems?

In every industry and sector, the number of employees per business revenue is falling, giving rise to employment-less economic growth. Job seekers will need more opportunities for one-person Internet-based self-employment and for markets for their interests and abilities in other job markets worldwide. Successfully leapfrogging slower linear development processes in lower-income countries is likely to require implementing futuristic possibilities—from 3-D printing to seawater agriculture—and making increasing individual and collective intelligence a national objective of each country.

"Because the world is better educated and increasingly connected, people are becoming less tolerant of the abuse of elite power than in the past."

The explosive, accelerating growth of knowledge in a rapidly changing and increasingly interdependent world gives us so much to know about so many things. Unfortunately, we are also flooded with so much trivial news that serious issues get little attention or interest, and too much time is wasted going through useless information.

At the same time, the world is increasingly engaged in diverse conversations about how to relate to the environment and to our fellow humans, and about what technologies, economics, and laws are right for our common future. These conversations are emerging from countless international negotiations, UN gatherings, and thousands of Internet discussion groups and big-data analyses. Humanity is slowly but surely becoming aware of itself as an integrated system of cultures, economies, technologies, natural and built environments, and governance systems.

Collecting Our Intelligence

These great conversations will be better informed if we realize that the world is improving more than most pessimists know and that future dangers are worse than most optimists indicate. Better ideas, new tools, and creative management approaches are popping up all over the world, but the lack of imagination and courage to make serious change is drowning the innovations needed to make the world work for all.

As a global think tank, The Millennium Project gathers insights from a network of more than 4,500 experts who continuously gather and share data via our online Global Futures System (GFS). GFS can be thought of as a global information utility from which different readers can draw different value for improving their understanding and decisions.

The collective intelligence emerges in GFS from synergies among data/information/knowledge, software/hardware, and experts and others with insight that continually learn from feedback to produce just-in-time knowledge for better decisions than any of these elements acting alone.

In addition to succinct but relatively detailed descriptions of the current situation and forecasts, we also formulate recommendations to address the various global challenges. Some of our recommendations are as follows:

- Establish a U.S.–China 10-year environmental security goal to reduce climate change and improve trust.
- Grow meat without growing animals, to reduce water demand and greenhouse-gas emissions.
- Develop seawater agriculture for biofuels, carbon sink, and food without rain.
- Build global collective intelligence systems for input to long-range strategic plans.
- Create tele-nations connecting brains overseas to the development process back home.
- Establish trans-institutions for more effective implementation of strategies.
- Detail and implement a global strategy to counter organized crime.
- Use the State of the Future Index as an alternative to GDP as a measure of progress for the world and nations with 30 variables that include indicators for social equity and well-being.

The World Report Card

The world is in a race between implementing ever-increasing ways to improve the human condition and the seemingly ever-increasing complexity and scale of global problems. So, how is the world doing in this race? What's the score so far?

The Millennium Project's global State of the Future Index (SOFI), produced annually since 2000, measures the 10-year outlook for the future based on historical data on 30 key variables. In the aggregate, these data depict whether the future promises to be better or worse. The SOFI is intended to show the directions and intensity of change and to identify the responsible factors and the relationships among them.

The current SOFI, shown in Figure 1, indicates a slower progress since 2007, although the overall outlook is promising.

Some Key Trends Affecting the State of the Future

- **Computing.** The EU, United States, Japan, and China have announced programs to understand how the brain works and apply that knowledge to make better computers with better computer–user interfaces. Google also is working to create artificial brains that could serve us as personal artificial-intelligence assistants. Another great race is on to make supercomputer power available to the masses with advances in IBM's Watson and with cloud computing by Amazon and others. About 85% of the world's population is expected to be covered by high-speed mobile Internet in 2017.

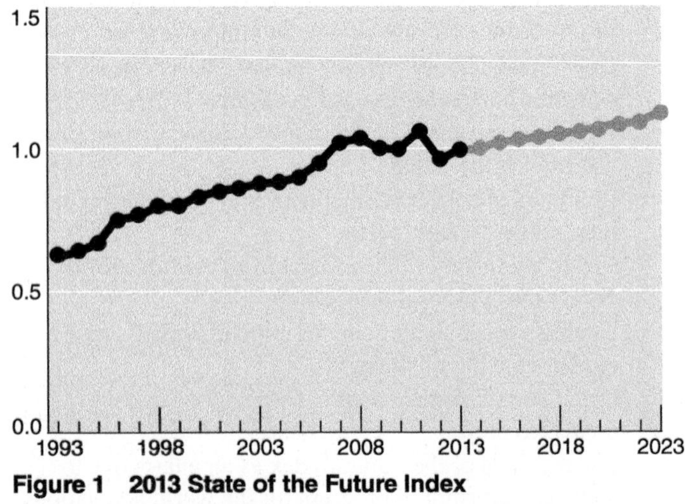

Figure 1 2013 State of the Future Index

Each of the 30 variables making up the index (Box 1) can be examined to show where we are winning, where we are losing, and where there is unclear or little progress.

- **A Web-connected world.** More than 8 billion devices are connected to the Internet of Things, which is expected to grow to 40 billion–80 billion devices by 2020. According to the UN's International Telecommunications Union, nearly 40% of humanity now uses the Internet. This global network is close to becoming the de facto global brain of humanity.

  So what happens when the entire world has access to nearly all the world's knowledge, along with instantaneous access to artificial brains that can solve problems and create new conditions like geniuses, while blurring previous distinctions between virtual realities and physical reality? We have already seen brilliant financial experts—augmented with data and software—making the short-term, selfish, economic decisions that led to the 2008 global financial crisis, continued environmental degradation, and widening income disparities. It is not yet clear that humanity will grow from short-term, me-first thinking to longer-term, we-first, planet-oriented decision making.

"It is not yet clear that humanity will grow from short-term, me-first thinking to longer-term, we-first, planet-oriented decision making."

Humanity may become more responsible and compassionate as the Internet of people and things grows across the planet, making us more aware of humanity as a whole and of our natural and built environments. Yet multi-way interactive media also attracts individuals with common interests into isolated ideological groups, reinforcing social polarization and conflict and forcing some political systems into gridlock.

And although the Internet's growth may make it increasingly difficult for conventional crimes to go undetected, cyberspace has become the medium for new kinds of crimes: According to the cloud-services provider Akamai, there were 628 cyber-attacks over 24 hours on July 24, 2013, the majority of which attacked targets in the United States. Cyber-attacks can be thought of as a new kind of guerrilla warfare. Prevention may involve an endless intellectual arms race of hacking and counter-hacking software, setting cyber traps, exposing sources, and initiating trade sanctions.

- **Civil strife.** The long-range trend toward democracy is strong, but Freedom House reports that world political and civil liberties deteriorated for the eighth consecutive year in 2013, with declines noted in 54 countries and

Box 1

Variables Used in the 2013–14 State of the Future Index

1. GNI per capita, PPP (constant 2005 international $)
2. Economic income inequality (share of top 10%)
3. Unemployment, total (% of world labor force)
4. Poverty headcount ratio at $1.25 a day (PPP) (% of population)
5. Levels of corruption (0 = highly corrupt; 6 = very clean)
6. Foreign direct investment, net inflows (balance of payments, current $, billions)
7. R&D expenditures (% of GDP)
8. Population growth (annual %)
9. Life expectancy at birth (years)
10. Mortality rate, infant (per 1,000 live births)
11. Prevalence of undernourishment
12. Health expenditure per capita (current $)
13. Physicians (per 1,000 people)
14. Improved water source (% of population with access)
15. Renewable internal freshwater resources per capita (thousand cubic meters)
16. Ecological Footprint/Biocapacity ratio
17. Forest area (% of land area)
18. CO_2 emissions from fossil-fuel and cement production (billion tonnes)
19. Energy efficiency [GDP per unit of energy use (constant 2005 PPP $ per kg of oil equivalent)]
20. Electricity production from renewable sources, excluding hydroelectric (% of total)
21. Literacy rate, adult total (% of people ages 15 and above)
22. School enrollment, secondary (% gross)
23. Number of wars (conflicts with more than 1,000 fatalities)
24. Terrorism incidents
25. Number of countries and groups that had or still have intentions to build nuclear weapons
26. Freedom rights (number of countries rated free)
27. Voter turnout (% voting population)
28. Proportion of seats held by women in national parliaments (% of members)
29. Internet users (per 100 people)
30. Prevalence of HIV (% of population age 15–49)

improvements in only 40 countries. At the same time, increasing numbers of educated and mobile-phone/Internet-savvy people are no longer tolerating the abuse of power and may be setting the stage for a long and difficult transition to more global democracy.

- **Climate change.** The Intergovernmental Panel on Climate Change's *Fifth Assessment Report* found that world greenhouse gas emissions grew by an annual average of 2.2% between 2000 and 2010, up from 1.3% per year between 1970 and 2000. Each decade of the past three was warmer than the previous decade. The past 30 years was likely the warmest period in the Northern Hemisphere in the last 1,400 years.

  Furthermore, even if all CO_2 emissions are stopped today, the IPCC report notes that "most aspects of climate change will persist for many centuries." Hence, the world has to take adaptation far more seriously, in addition to reducing emissions, and creating new methods to reduce the greenhouse gases that are already in the atmosphere.

  Without dramatic changes, UN Environment Program projects a 2°C (3.6°F) rise above preindustrial levels in 20–30 years, accelerating changing climate, ocean acidity, changes in disease patterns, and saltwater intrusions into freshwater areas worldwide. The UN Food and Agriculture Organization reports that 87% of global fish stocks are either fully exploited or overexploited. Oceans absorb about 33% of human-generated CO_2, but their ability to continue doing this is being reduced by changing acidity and the die-offs of coral reefs and other living systems.
- **Energy needs.** The world also needs to create enough electrical production capacity for an additional 3.7 billion people by 2050. There are 1.2 billion people without electricity today (17% of the world), and an additional 2.4 billion people will be added to the world's population between now and 2050.

  Compounding this is the requirement to decommission aging nuclear power plants and to replace or retrofit fossil fuel plants. The cost of nuclear power is increasing, while the cost of renewables is falling—wind power passed nuclear as Spain's leading source of electricity. However, fossil fuels (coal, oil, and natural gas) will continue to supply the vast majority of the world's electricity past 2050 unless there are major social and technological changes. If the long-term trends toward a wealthier and more sophisticated world continue, our energy demands by 2050 could be more than expected. However, the convergences of technologies are accelerating rapidly to make energy efficiencies far greater by 2050 than forecast today.
- **Water stress.** Major progress was made over the past 25 years that provided enough clean water for an additional 2 billion people. But as a result of water pollution, accelerating climate change, falling water tables around the world, and an additional 2.4 billion people in just 36 years, some of the people with safe water today may not have it in the future unless significant changes occur. According to the Organisation for Economic Co-operation

and Development (OECD), half the world could be living in areas with severe water stress by 2030.

- **Population growth.** The UN's mid-range forecast is that the world's population, which now totals 7.2 billion people, will number 9.6 billion by 2050. By that date, the number of people over age 65 will equal or surpass the number under 15.

  Average life expectancy at birth has increased from 48 years in 1955 to 70.5 years today. Future scientific and medical breakthroughs could give people longer and more productive lives than most would believe possible today. For example, uses of genetic data, software, and nanotechnology will help detect and treat disease at the genetic or molecular level.

"It is unreasonable to expect the world to cooperatively create and implement strategies to build a better future without some general agreement about what that desirable future is.

- **Accelerating technologies.** Science and technology's continued acceleration is fundamentally changing what is possible, and access to this knowledge is becoming universally available. For example, China's Tianhe-2 supercomputer is the world's fastest computer, at 33.86 petaflops (quadrillion floating point operations per second)—passing the computational speed of a human brain. Individual gene sequencing is now available for $1,000—and the price could go down much further in coming years—a development that will enable individualized genetic medicine for every patient.

  Although advances in synthetic biology, quantum entanglement, Higgs-like particles, and computational science seem remote from improving the human condition, such basic scientific endeavors are necessary to increase the knowledge that scientists can use to develop and improve technologies to benefit humanity. But with little news coverage and educational curricula, the general public seem unaware of the extraordinary changes and consequences that need to be discussed: Is it ethical to clone ourselves, to bring dinosaurs back to life, or to invent thousands of new life forms through synthetic biology? Should basic scientific research be pursued without direct regard for social issues? On the other hand, might social considerations impair progress toward a truthful understanding of reality?
- **Gender equity.** Violence against women is the largest war today, as measured by death and casualties per year. Globally, 35% of women have experienced physical and/or sexual violence. While the gender gaps for health and educational attainment were closed by 96% and 93% respectively, according to the 2013 Global Gender Gap report by the World Economic Forum, the gap in economic participation has been closed by only 60%, and the gap in political outcomes by only 21%: Women account for only 21.3% of the membership of national legislative bodies worldwide, up from 11.3% in 1997.
- **"Hidden" hunger.** Food markets in much of the developing world exhibit an increasing problem of hidden hunger—that is, the intake of calories is sufficient, but those calories contain little in nutritious value, vitamins, and minerals. Although the share of people in the world who are hungry has fallen from over 30% in 1970 to 15% today, concerns are increasing over the variety and nutritional quality of food. The FAO estimates that some 30% of the world population (2 billion people) suffers from hidden hunger.
- **Vulnerable urban coastal zones.** Human construction is diminishing the land structures that the world's coastal zones rely on to blunt the impacts of hurricanes, tsunamis, and pollution. This is a harmful outcome, not only for flora and fauna, but for us, as well, since more than half the world's people live within 120 miles of a coastline. Without appropriate mitigation, prevention, and management of the natural infrastructure within urban coastal zones, billions of people will be increasingly vulnerable to a range of disasters.
- **"Lone wolf" terrorism.** Individuals acting alone can wield increasing amounts of damage. The number of terrorism incidents increased over the past 20 years, reaching 8,441 in 2012 and more than 5,000 in the first half of 2013.

  Of all terrorism, the lone-wolf type is the most insidious, because it is exceedingly difficult to anticipate, given the actions and intent of individuals acting alone. The average opinion of our international panel is that nearly a quarter of terrorist attacks carried out in 2015 might be by a lone wolf, and that the situation might escalate: About half of the participants that we surveyed thought that lone-wolf terrorists might attempt to use weapons of mass destruction by around 2030.

"Global Collective Intelligence Systems" Bring It All Together

It is unreasonable to expect the world to cooperatively create and implement strategies to build a better future without some general agreement about what that desirable future is. Such a future can only be built with awareness of the global situation and of the extraordinary possibilities.

What we need is a global collective intelligence system to track science and technology advances, forecast consequences, and document a range of views on them. The accelerating rates of changes that the world now experiences call for new kinds of decision making with global real-time feedback. The Global Futures System is an early expression of that future direction.

Critical Thinking

1. What critical socioeconomic fault lines are worsening?
2. Pick 10 of the most important variables used in the State of the Future 2013–2014 Index.
3. What are the most important trends in planet-oriented decision-making?

Internet References

The Futures Forum

futuresforum.org

The Millennium Project

Millennium-project.org

The World Futures Society

www.wfs/org/

Jerome C. Glenn is CEO of The Millennium Project and The Global Futures System, www.themp.org. This article is adapted from *2013–14 State of the Future,* co-authored by Glenn with Theodore J. Gordon and Elizabeth Florescu (published by The Millennium Project, millennium-project.org/millennium/201314SOF.html).

Article

Prepared by: Robert Weiner, *University of Massachusetts, Boston*

The Geopolitics of Cyberspace after Snowden

RON DEIBERT

Learning Outcomes

After reading this article, you will be able to:

- Discuss the effects of Snowden's revelations on the freedom of the Internet.
- Explain the major changes that have occurred recently in the use of the Internet.

For several years now, it seems that not a day has gone by without a new revelation about the perils of cyberspace: the networks of Fortune 500 companies breached; cyberespionage campaigns uncovered; shadowy hacker groups infiltrating prominent websites and posting extremist propaganda. But the biggest shock came in June 2013 with the first of an apparently endless stream of riveting disclosures from former US National Security Agency (NSA) contractor Edward Snowden. These alarming revelations have served to refocus the world's attention, aiming the spotlight not at cunning cyber activists or sinister data thieves, but rather at the world's most powerful signals intelligence agencies: the NSA, Britain's Government Communications Headquarters (GCHQ), and their allies.

The public is captivated by these disclosures, partly because of the way in which they have been released, but mostly because cyberspace is so essential to all of us. We are in the midst of what might be the most profound communications evolution in all of human history. Within the span of a few decades, society has become completely dependent on the digital information and communication technologies (ICTs) that infuse our lives. Our homes, our jobs, our social networks—the fundamental pillars of our existence—now demand immediate access to these technologies.

With so much at stake, it should not be surprising that cyberspace has become heavily contested. What was originally designed as a small-scale but robust information-sharing network for advanced university research has exploded into the information infrastructure for the entire planet. Its emergence has unsettled institutions and upset the traditional order of things, while simultaneously contributing to a revolution in economics, a path to extraordinary wealth for Internet entrepreneurs, and new forms of social mobilization. These contrasting outcomes have set off a desperate scramble, as stakeholders with competing interests attempt to shape cyberspace to their advantage. There is a geopolitical battle taking place over the future of cyberspace, similar to those previously fought over land, sea, air, and space.

Three major trends have been increasingly shaping cyberspace: the big data explosion, the growing power and influence of the state, and the demographic shift to the global South. While these trends preceded the Snowden disclosures, his leaks have served to alter them somewhat, by intensifying and in some cases redirecting the focus of the conflicts over the Internet. This essay will identify several focal points where the outcomes of these contests are likely [to] be most critical to the future of cyberspace.

Big Data

Before discussing the implications of cyberspace, we need to first understand its characteristics: What is unique about the ICT environment that surrounds us? There have been many extraordinary inventions that revolutionized communications throughout human history: the alphabet, the printing press, the telegraph, radio, and television all come to mind. But arguably the most far-reaching in its effects is the creation and development of social media, mobile connectivity, and cloud computing—referred to in shorthand as "big data." Although these three technological systems are different in many ways, they share one very important characteristic: a vast and rapidly growing volume of personal information, shared (usually voluntarily) with entities separate from the individuals to whom the information applies. Most of those entities are privately owned companies, often headquartered in political jurisdictions other than the one in which the individual providing the information lives (a critical point that will be further examined below).

We are, in essence, turning our lives inside out. Data that used to be stored in our filing cabinets, on our desktop computers, or even in our minds, are now routinely stored on equipment maintained by private companies spread across the globe. This data we entrust to them includes that which we are conscious of and deliberate about—websites visited, emails sent, texts received, images posted—but a lot of which we are unaware.

For example, a typical mobile phone, even when not in use, emits a pulse every few seconds as a beacon to the nearest WiFi router or cellphone tower. Within that beacon is an extraordinary amount of information about the phone and its owner (known as "metadata"), including make and model, the user's name, and geographic location. And that is just the mobile device itself. Most users have within their devices several dozen applications (more than 50 billion apps have been downloaded from Apple's iTunes store for social networking, fitness, health, games, music, shopping, banking, travel, even tracking sleep patterns), each of which typically gives itself permission to extract data about the user and the device. Some applications take the practice of data extraction several bold steps further, by requesting access to geolocation information, photo albums, contacts, or even the ability to turn on the device's camera and microphone.

We leave behind a trail of digital "exhaust" wherever we go. Data related to our personal lives are compounded by the numerous and growing Internet-connected sensors that permeate our technological environment. The term "Internet of Things" refers to the approximately 15 billion devices (phones, computers, cars, refrigerators, dishwashers, watches, even eyeglasses) that now connect to the Internet and to each other, producing trillions of ever-expanding data points. These data points create an ethereal layer of digital exhaust that circles the globe, forming, in essence, a digital stratosphere.

Given the virtual characteristics of the digital experience, it may be easy to overlook the material properties of communication technologies. But physical geography is an essential component of cyberspace: *Where* technology is located is as important as *what* it is. While our Internet activities may seem a kind of ephemeral and private adventure, they are in fact embedded in a complex infrastructure (material, logistical, and regulatory) that in many cases crosses several borders. We assume that the data we create, manipulate, and distribute are in our possession. But in actuality, they are transported to us via signals and waves, through cables and wires, from distant servers that may or may not be housed in our own political jurisdiction. It is actual matter we are dealing with when we go online, and that matters—a lot. The data that follow us around, that track our lives and habits, do not disappear; they live in the servers of the companies that own and operate the infrastructure. What is done with this information is a decision for those companies to make. The details are buried in their rarely read terms of service, or, increasingly, in special laws, requirements, or policies laid down by the governments in whose jurisdictions they operate.

The vast majority of Internet users now live in the global South.

Big State

The Internet started out as an isolated experiment largely separate from government. In the early days, most governments had no Internet policy, and those that did took a deliberately laissez-faire approach. Early Internet enthusiasts mistakenly understood this lack of policy engagement as a property unique to the technology. Some even went so far as to predict that the Internet would bring about the end of organized government altogether. Over time, however, state involvement has expanded, resulting in an increasing number of Internet-related laws, regulations, standards, and practices. In hindsight, this was inevitable. Anything that permeates our lives so thoroughly naturally introduces externalities—side effects of industrial or commercial activity—that then require the establishment of government policy. But as history demonstrates, linear progress is always punctuated by specific events—and for cyberspace, that event was 9/11.

We continue to live in the wake of 9/11. The events of that day in 2001 profoundly shaped many aspects of society. But no greater impact can be found than the changes it brought to cyberspace governance and security, specifically with respect to the role and influence of governments. One immediate impact was the acceleration of a change in threat perception that had been building for years.

During the Cold War, and largely throughout the modern period (roughly the eighteenth century onward), the primary threat for most governments was "interstate" based. In this paradigm, the state's foremost concern is a cross-border invasion or attack—the idea that another country's military could use force and violence in order to gain control. After the Cold War, and especially since 9/11, the concern has shifted to a different threat paradigm: that a violent attack could be executed by a small extremist group, or even a single human being who could blow himself or herself up in a crowded mall, hijack an airliner, or hack into critical infrastructure. Threats are now dispersed across all of society, regardless of national borders. As a result, the focus of the state's security gaze has become omnidirectional.

Accompanying this altered threat perception are legal and cultural changes, particularly in reaction to what was widely perceived as the reason for the 9/11 catastrophe in the first place: a "failure to connect the dots." The imperative shifted from the micro to the macro. Now, it is not enough to simply look for a needle in the haystack. As General Keith Alexander (former head of the NSA and the US Cyber Command) said, it is now necessary to collect "the entire haystack." Rapidly, new laws have been introduced that substantially broaden the reach of law enforcement and intelligence agencies, the most notable of them being the Patriot Act in the United States—although many other countries have followed suit.

This imperative to "collect it all" has focused government attention squarely on the private sector, which owns and operates most of cyberspace. States began to apply pressure on companies to act as a proxy for government controls—policing their own networks for content deemed illegal, suspicious, or a threat to national security. Thanks to the Snowden disclosures, we now have a much clearer picture of how this pressure manifests itself. Some companies have been paid fees to collude, such as Cable and Wireless (now owned by Vodafone), which was paid tens of millions of pounds by the GCHQ to install surveillance equipment on its networks. Other companies have been subjected to formal or informal pressures, such as court orders, national security letters, the with-holding of operating licenses, or even appeals to patriotism. Still others became the targets of computer exploitation, such as US-based Google, whose back-end data infrastructure was secretly hacked into by the NSA.

This manner of government pressure on the private sector illustrates the importance of the physical geography of cyberspace. Of course, many of the corporations that own and operate the infrastructure—companies like Facebook, Microsoft, Twitter, Apple, and Google—are headquartered in the United States. They are subject to US national security law and, as a consequence, allow the government to benefit from a distinct home-field advantage in its attempt to "collect it all." And that it does—a staggering volume, as it turns out. One top-secret NSA slide from the Snowden disclosures reveals that by 2011, the United States (with the cooperation of the private sector) was collecting and archiving about 15 billion Internet metadata records *every single day.* Contrary to the expectations of early Internet enthusiasts, the US government's approach to cyberspace—and by extension that of many other governments as well—has been anything but laissez-faire in the post-9/11 era. While cyberspace may have been born largely in the absence of states, as it has matured states have become an inescapable and dominant presence.

Domain Domination

After 9/11, there was also a shift in US military thinking that profoundly affected cyberspace. The definition of cyberspace as a single "domain"— equal to land, sea, air, and space—was formalized in the early 2000s, leading to the imperative to dominate and rule this domain; to develop offensive capabilities to fight and win wars within cyberspace. A Rubicon was crossed with the Stuxnet virus, which sabotaged Iranian nuclear enrichment facilities. Reportedly engineered jointly by the United States and Israel, the Stuxnet attack was the first de facto act of war carried out entirely through cyberspace. As is often the case in international security dynamics, as one country reframes its objectives and builds up its capabilities, other countries follow suit. Dozens of governments now have within their armed forces dedicated "cyber commands" or their equivalents.

The race to build capabilities also has a ripple effect on industry, as the private sector positions itself to reap the rewards of major cyber-related defense contracts. The imperatives of mass surveillance and preparations for cyberwarfare across the globe have reoriented the defense industrial base. It is noteworthy in this regard how the big data explosion and the growing power and influence of the state are together generating a political-economic dynamic. The aims of the Internet economy and those of state security converge around the same functional needs: collecting, monitoring, and analyzing as much data as possible. Not surprisingly, many of the same firms service both segments. For example, companies that market facial recognition systems find their products being employed by Facebook on the one hand and the Central Intelligence Agency on the other.

As private individuals who live, work, and play in the cyber realm, we provide the seeds that are then cultivated, harvested, and delivered to market by a massive machine, fueled by the twin engines of corporate and national security needs. The confluence of these two major trends is creating extraordinary tensions in state-society relations, particularly around privacy. But perhaps the most important implications relate to the fact that the market for the cybersecurity industrial complex knows no boundaries—an ominous reality in light of the shifting demographics of cyberspace.

Southern Shift

While the "what" of cyberspace is critical, the "who" is equally important. There is a major demographic shift happening today that is easily overlooked, especially by users in the West, where the technology originates. The vast majority of Internet users now live in the global South. Of the 6 billion mobile devices in circulation, over 4 billion are located in the developing world. In 2001, 8 of every 100 citizens in developing nations owned a mobile subscription. That number has now jumped to 80. In Indonesia, the number of Internet users increases each month by a stunning 800,000. Nigeria had 200,000 Internet users in 2000; today, it has 68 million.

Remarkably, some of the fastest growing online populations are emerging in countries with weak governmental structures or corrupt, autocratic, or authoritarian regimes. Others are developing in zones of conflict, or in countries that have only recently gone through difficult transitions to democracy. Some of the fastest growth rates are in "failed" states, or in countries riven by ethnic rivalries or challenged by religious differences and sensitivities, such as Nigeria, India, Pakistan, Indonesia, and Thailand. Many of these countries do not have long-standing democratic traditions, and therefore lack proper systems of accountability to guard against abuses of power. In some, corruption is rampant, or the military has disproportionate influence.

Consider the relationship between cyberspace and authoritarian rule. We used to mock authoritarian regimes as slow-footed, technologically challenged dinosaurs that would be inevitably weeded out by the information age. The reality has proved more nuanced and complex. These regimes are proving much more adaptable than expected. National-level Internet controls on content and access to information in these countries are now a growing norm. Indeed, some are beginning to affect the very technology itself, rather than vice versa.

In China (the country with the world's most Internet users), "foreign" social media like Facebook, Google, and Twitter are banned in favor of nationally based, more easily controlled alternatives. For example, We Chat—owned by China-based parent company Tencent—is presently the fifth-largest Internet company in the world after Google, Amazon, Alibaba, and eBay, and as of August 2014 it had 438 million active users (70 million outside China) and a public valuation of over $400 billion. China's popular chat applications and social media are required to police the country's networks with regard to politically sensitive content, and some even have hidden censorship and surveillance functionality "baked" into their software. Interestingly, some of We Chat's users outside China began experiencing the same type of content filtering as users inside China, an issue that Tencent claimed was due to a software bug (which it promptly fixed). But the implication of such extraterritorial applications of national-level controls is certainly worth further scrutiny, particularly as China-based companies begin to expand their service offerings in other countries and regions.

It is important to understand the historical context in which this rapid growth is occurring. Unlike the early adopters of the Internet in the West, citizens in the developing world are plugging in and connecting after the Snowden disclosures, and with the model of the NSA in the public domain. They are coming online with cybersecurity at the top of the international agenda, and fierce international competition emerging throughout cyberspace, from the submarine cables to social media. Political leaders in these countries have at their disposal a vast arsenal of products, services, and tools that provide their regimes with highly sophisticated forms of information control. At the same time, their populations are becoming more savvy about using digital media for political mobilization and protest.

While the digital innovations that we take advantage of daily have their origins in high-tech libertarian and free-market hubs like Silicon Valley, the future of cyberspace innovation will be in the global South. Inevitably, the assumptions, preferences, cultures, and controls that characterize that part of the world will come to define cyberspace as much as those of the early entrepreneurs of the information age did in its first two decades.

Who Rules?

Cyberspace is a complex technological environment that spans numerous industries, governments, and regions. As a consequence, there is no one single forum or international organization for cyberspace. Instead, governance is spread throughout numerous small regimes, standard-setting forums, and technical organizations from the regional to the global. In the early days, Internet governance was largely informal and led by non-state actors, especially engineers. But over time, governments have become heavily involved, leading to more politicized struggles at international meetings.

Although there is no simple division of camps, observers tend to group countries into those that prefer a more open Internet and a tightly restricted role for governments versus those that prefer a more centralized and state-led form of governance, preferably through the auspices of the United Nations. The United States, the United Kingdom, other European nations, and Asian democracies are typically grouped in the former, with China, Russia, Iran, Saudi Arabia, and other nondemocratic countries grouped in the latter. A large number of emerging market economies, led by Brazil, India, and Indonesia, are seen as "swing states" that could go either way.

Prior to the Snowden disclosures, the battle lines between these opposing views were becoming quite acute—especially around the December 2012 World Congress on Information Technology (WCIT), where many feared Internet governance would fall into UN (and thus more state-controlled) hands. But the WCIT process stalled, and those fears never materialized, in part because of successful lobbying by the United States and its allies, and by Internet companies like Google. After the

Snowden disclosures, however, the legitimacy and credibility of the "Internet freedom" camp have been considerably weakened, and there are renewed concerns about the future of cyberspace governance.

The original promise of the Internet as a forum for free exchange of information is at risk.

Meanwhile, less noticed but arguably more effective have been lower-level forms of Internet governance, particularly in regional security forums and standards-setting organizations. For example, Russia, China, and numerous Central Asian states, as well as observer countries like Iran, have been coordinating their Internet security policies through the Shanghai Cooperation Organization (SCO). Recently, the SCO held military exercises designed to counter Internet-enabled opposition of the sort that participated in the "color revolutions" in former Soviet states. Governments that prefer a tightly controlled Internet are engaging in partnerships, sharing best practices, and jointly developing information control platforms through forums like the SCO. While many casual Internet observers ruminate over the prospect of a UN takeover of the Internet that may never materialize, the most important norms around cyberspace controls could be taking hold beneath the spotlight and at the regional level.

Technological Sovereignty

Closely related to the questions surrounding cyberspace governance at the international level are issues of domestic-level Internet controls, and concerns over "technological sovereignty." This area is one where the reactions to the Snowden disclosures have been most palpably felt in the short term, as countries react to what they see as the US "home-field advantage" (though not always in ways that are straightforward). Included among the leaked details of US- and GCHQ-led operations to exploit the global communications infrastructure are numerous accounts of specific actions to compromise state networks, or even the handheld devices of government officials—most notoriously, the hacking of German Chancellor Angela Merkel's personal cellphone and the targeting of Brazilian government officials' classified communications. But the vast scope of US-led exploitation of global cyberspace, from the code to the undersea cables and everything in between, has set off shockwaves of indignation and loud calls to take immediate responses to restore "technological sovereignty."

For example, Brazil has spearheaded a project to lay a new submarine cable linking South America directly to Europe, thus bypassing the United States. Meanwhile, many European politicians have argued that contracts with US-based companies that may be secretly colluding with the NSA should be cancelled and replaced with contracts for domestic industry to implement regional and/or nationally autonomous data-routing policies—arguments that European industry has excitedly supported. It is sometimes difficult to unravel whether such measures are genuinely designed to protect citizens, or are really just another form of national industrial protectionism, or both. Largely obscured beneath the heated rhetoric and underlying self-interest, however, are serious questions about whether any of the measures proposed would have any more than a negligible impact when it comes to actually protecting the confidentiality and integrity of communications. As the Snowden disclosures reveal, the NSA and GCHQ have proved to be remarkably adept at exploiting traffic, no matter where it is based, by a variety of means.

A more troubling concern is that such measures may end up unintentionally legitimizing national cyberspace controls, particularly for developing countries, "swing states," and emerging markets. Pointing to the Snowden disclosures and the fear of NSA-led surveillance can be useful for regimes looking to subject companies and citizens to a variety of information controls, from censorship to surveillance. Whereas policy makers previously might have had concerns about being cast as pariahs or infringers on human rights, they now have a convenient excuse supported by European and other governments' reactions.

Spyware Bazaar

One by-product of the huge growth in military and intelligence spending on cybersecurity has been the fueling of a global market for sophisticated surveillance and other security tools. States that do not have an in-house operation on the level of the NSA can now buy advanced capabilities directly from private contractors. These tools are proving particularly attractive to many regimes that face ongoing insurgencies and other security challenges, as well as persistent popular protests. Since the advertised end uses of these products and services include many legitimate needs, such as network traffic management or the lawful interception of data, it is difficult to prevent abuses, and hard even for the companies themselves to know to what ends their products and services might ultimately be directed. Many therefore employ the term "dual-use" to describe such tools.

We leave behind a trail of digital "exhaust" wherever we go.

Research by the University of Toronto's Citizen Lab from 2012 to 2014 has uncovered numerous cases of human rights activists targeted by advanced digital spyware manufactured by Western companies. Once implanted on a target's device, this spyware can extract files and contacts, send emails and text messages, turn on the microphone and camera, and track the location of the user. If these were isolated incidences, perhaps we could write them off as anomalies. But the Citizen Lab's international scan of the command and control servers of these products—the computers used to send instructions to infected devices—has produced disturbing evidence of a global market that knows no boundaries. Citizen Lab researchers found one product, Finspy, marketed by a UK company, Gamma Group, in a total of 25 countries— some with dubious human rights records, such as Bahrain, Bangladesh, Ethiopia, Qatar, and Turkmenistan. A subsequent Citizen Lab report found that 21 governments are current or former users of a spyware product sold by an Italian company called Hacking Team, including 9 that received the lowest ranking, "authoritarian," in the *Economist's* 2012 Democracy Index.

Meanwhile, a 2014 Privacy International report on surveillance in Central Asia says many of the countries in the region have implemented far-reaching surveillance systems at the base of their telecommunications networks, using advanced US and Israeli equipment, and supported by Russian intelligence training. Products that provide advanced deep packet inspection (the capability to inspect data packets in detail as they flow through networks), content filtering, social network mining, cellphone tracking, and even computer attack targeting are being developed by Western firms and marketed worldwide to regimes seeking to limit democratic participation, isolate and identify opposition, and infiltrate meddlesome adversaries abroad.

Pushing Back

The picture of the cyberspace landscape painted above is admittedly quite bleak, and therefore one-sided. The contests over cyberspace are multidimensional and include many groups and individuals pushing for technologies, laws, and norms that support free speech, privacy, and access to information. Here, too, the Snowden disclosures have had an animating effect, raising awareness of risks and spurring on change. Whereas vague concerns about widespread digital spying were voiced by a minority and sometimes trivialized before Snowden's disclosures, now those fears have been given real substance and credibility, and surveillance is increasingly seen as a practical risk that requires some kind of remediation.

The Snowden disclosures have had a particularly salient impact on the private sector, the Internet engineering community, and civil society. The revelations have left many US companies in a public relations nightmare, with their trust weakened and lucrative contracts in jeopardy. In response, companies are pushing back. It is now standard for many telecommunications and social media companies to issue transparency reports about government requests to remove information from websites or share user data with authorities. US-based Internet companies even sued the government over gag orders that bar them from disclosing information on the nature and number of requests for user information. Others, including Google, Microsoft, Apple, Facebook, and WhatsApp, have implemented end-to-end encryption.

Internet engineers have reacted strongly to revelations showing that the NSA and its allies have subverted their security standards-setting processes. They are redoubling efforts to secure communications networks wholesale as a way to shield all users from mass surveillance, regardless of who is doing the spying. Among civil society groups that depend on an open cyberspace, the Snowden disclosures have helped trigger a burgeoning social movement around digital-security tool development and training, as well as more advanced research on the nature and impacts of information controls.

Wild Card

The cyberspace environment in which we live and on which we depend has never been more in flux. Tensions are mounting in several key areas, including Internet governance, mass and targeted surveillance, and military rivalry. The original promise of the Internet as a forum for free exchange of information is at risk. We are at a historical fork in the road: Decisions could take us down one path where cyberspace continues to evolve into a global commons, empowering individuals through access to information and freedom of speech and association, or down another path where this ideal meets its eventual demise. Securing cyberspace in ways that encourage freedom, while limiting controls and surveillance, is going to be a serious challenge.

Trends toward militarization and greater state control were already accelerating before the Snowden disclosures, and seem unlikely to abate in the near future. However, the leaks have thrown a wild card into the mix, creating opportunities for alternative approaches emphasizing human rights, corporate social responsibility, norms of mutual restraint, cyberspace arms control, and the rule of law. Whether such measures will be enough to stem the tide of territorialized controls remains to be seen. What is certain, however, is that a debate over the future of cyberspace will be a prominent feature of world politics for many years to come.

Critical Thinking

1. What is the role of the UN in the governance of the Internet?
2. What countries support a free Internet and what countries support state control of the Internet?
3. What countries are considered "swing" states in the use of the Internet?

Internet References

Berkman Center for Internet and Society
https://cyber.law.harvard.edu

Internet Governance Forum
intgovforum.org

Internet Society
internetsociety.org

National Security Agency
https://www.nsa.gov

Ron Deibert is a professor of political science and director of the Canada Center for Global Security Studies and the Citizen Lab at the University of Toronto. His latest book is *Black Code: Inside the Battle for Cyberspace* (Signal, 2013).

Article

Prepared by: Robert Weiner, *University of Massachusetts, Boston*

The Return of Geopolitics: The Revenge of the Revisionist Powers

WALTER RUSSELL MEAD

Learning Outcomes

After reading this article, you will be able to:

- Understand what the post-Cold War settlement meant.
- Explain what is meant by status quo and revisionism in the international system.

So far, the year 2014 has been a tumultuous one, as geopolitical rivalries have stormed back to center stage. Whether it is Russian forces seizing Crimea, China making aggressive claims in its coastal waters, Japan responding with an increasingly assertive strategy of its own, or Iran trying to use its alliances with Syria and Hezbollah to dominate the Middle East, old-fashioned power plays are back in international relations.

The United States and the EU, at least, find such trends disturbing. Both would rather move past geopolitical questions of territory and military power and focus instead on ones of world order and global governance: trade liberalization, nuclear nonproliferation, human rights, the rule of law, climate change, and so on. Indeed, since the end of the Cold War, the most important objective of U.S. and EU foreign policy has been to shift international relations away from zero-sum issues toward win-win ones. To be dragged back into old-school contests such as that in Ukraine doesn't just divert time and energy away from those important questions; it also changes the character of international politics. As the atmosphere turns dark, the task of promoting and maintaining world order grows more daunting.

But Westerners should never have expected old-fashioned geopolitics to go away. They did so only because they fundamentally misread what the collapse of the Soviet Union meant: the ideological triumph of liberal capitalist democracy over communism, not the obsolescence of hard power. China, Iran, and Russia never bought into the geopolitical settlement that followed the Cold War, and they are making increasingly forceful attempts to overturn it. That process will not be peaceful, and whether or not the revisionists succeed, their efforts have already shaken the balance of power and changed the dynamics of international politics.

A False Sense of Security

When the Cold War ended, many Americans and Europeans seemed to think that the most vexing geopolitical questions had largely been settled. With the exception of a handful of relatively minor problems, such as the woes of the former Yugoslavia and the Israeli-Palestinian dispute, the biggest issues in world politics, they assumed, would no longer concern boundaries, military bases, national self-determination, or spheres of influence.

One can't blame people for hoping. The West's approach to the realities of the post-Cold War world has made a great deal of sense, and it is hard to see how world peace can ever be achieved without replacing geopolitical competition with the construction of a liberal world order. Still, Westerners often forget that this project rests on the particular geopolitical foundations laid in the early 1990s.

In Europe, the post-Cold War settlement involved the unification of Germany, the dismemberment of the Soviet Union, and the integration of the former Warsaw Pact states and the Baltic republics into NATO and the EU. In the Middle East, it entailed the dominance of Sunni powers that were allied with the United States (Saudi Arabia, its Gulf allies, Egypt, and Turkey) and the double containment of Iran and Iraq. In Asia, it meant the uncontested dominance of the United States, embedded in a series of security relationships with Japan, South Korea, Australia, Indonesia, and other allies.

This settlement reflected the power realities of the day, and it was only as stable as the relationships that held it up. Unfortunately, many observers conflated the temporary geopolitical conditions of the post-Cold War world with the presumably more final outcome of the ideological struggle between liberal democracy and Soviet communism. The political scientist Francis Fukuyama's famous formulation that the end of the Cold War meant "the end of history" was a statement about ideology. But for many people, the collapse of the Soviet Union didn't just mean that humanity's ideological struggle was over for good; they thought geopolitics itself had also come to a permanent end.

At first glance, this conclusion looks like an extrapolation of Fukuyama's argument rather than a distortion of it. After all, the idea of the end of history has rested on the geopolitical consequences of ideological struggles ever since the German philosopher Georg Wilhelm Friedrich Hegel first expressed it at the beginning of the nineteenth century. For Hegel, it was the Battle of Jena, in 1806, that rang the curtain down on the war of ideas. In Hegel's eyes, Napoleon Bonaparte's utter destruction of the Prussian army in that brief campaign represented the triumph of the French Revolution over the best army that prerevolutionary Europe could produce. This spelled an end to history, Hegel argued, because in the future, only states that adopted the principles and techniques of revolutionary France would be able to compete and survive.

Adapted to the post-Cold War world, this argument was taken to mean that in the future, states would have to adopt the principles of liberal capitalism to keep up. Closed, communist societies, such as the Soviet Union, had shown themselves to be too uncreative and unproductive to compete economically and militarily with liberal states. Their political regimes were also shaky, since no social form other than liberal democracy provided enough freedom and dignity for a contemporary society to remain stable.

To fight the West successfully, you would have to become like the West, and if that happened, you would become the kind of wishy-washy, pacifistic milquetoast society that didn't want to fight about anything at all. The only remaining dangers to world peace would come from rogue states such as North Korea, and although such countries might have the will to challenge the West, they would be too crippled by their obsolete political and social structures to rise above the nuisance level (unless they developed nuclear weapons, of course). And thus former communist states, such as Russia, faced a choice. They could jump on the modernization bandwagon and become liberal, open, and pacifistic, or they could cling bitterly to their guns and their culture as the world passed them by.

At first, it all seemed to work. With history over, the focus shifted from geopolitics to development economics and nonproliferation, and the bulk of foreign policy came to center on questions such as climate change and trade. The conflation of the end of geopolitics and the end of history offered an especially enticing prospect to the United States: the idea that the country could start putting less into the international system and taking out more. It could shrink its defense spending, cut the State Department's appropriations, lower its profile in foreign hotspots—and the world would just go on becoming more prosperous and more free.

This vision appealed to both liberals and conservatives in the United States. The administration of President Bill Clinton, for example, cut both the Defense Department's and the State Department's budgets and was barely able to persuade Congress to keep paying U.S. dues to the UN. At the same time, policymakers assumed that the international system would become stronger and wider-reaching while continuing to be conducive to U.S. interests. Republican neo-isolationists, such as former Representative Ron Paul of Texas, argued that given the absence of serious geopolitical challenges, the United States could dramatically cut both military spending and foreign aid while continuing to benefit from the global economic system.

After 9/11, President George W. Bush based his foreign policy on the belief that Middle Eastern terrorists constituted a uniquely dangerous opponent, and he launched what he said would be a long war against them. In some respects, it appeared that the world was back in the realm of history. But the Bush administration's belief that democracy could be implanted quickly in the Arab Middle East, starting with Iraq, testified to a deep conviction that the overall tide of events was running in America's favor.

President Barack Obama built his foreign policy on the conviction that the "war on terror" was overblown, that history really was over, and that, as in the Clinton years, the United States' most important priorities involved promoting the liberal world order, not playing classical geopolitics. The administration articulated an extremely ambitious agenda in support of that order: blocking Iran's drive for nuclear weapons, solving the Israeli-Palestinian conflict, negotiating a global climate change treaty, striking Pacific and Atlantic trade deals, signing arms control treaties with Russia, repairing U.S. relations with the Muslim world, promoting gay rights, restoring trust with European allies, and ending the war in Afghanistan. At the same time, however, Obama planned to cut defense spending dramatically and reduced U.S. engagement in key world theaters, such as Europe and the Middle East.

An Axis of Weevils?

All these happy convictions are about to be tested. Twenty-five years after the fall of the Berlin Wall, whether one focuses on the rivalry between the EU and Russia over Ukraine, which led Moscow to seize Crimea; the intensifying competition between China and Japan in East Asia; or the subsuming of sectarian conflict into international rivalries and civil wars in the Middle East, the world is looking less post-historical by the day. In very different ways, with very different objectives, China, Iran, and Russia are all pushing back against the political settlement of the Cold War.

The relationships among those three revisionist powers are complex. In the long run, Russia fears the rise of China. Tehran's worldview has little in common with that of either Beijing or Moscow. Iran and Russia are oil-exporting countries and like the price of oil to be high; China is a net consumer and wants prices low. Political instability in the Middle East can work to Iran's and Russia's advantage but poses large risks for China. One should not speak of a strategic alliance among them, and over time, particularly if they succeed in undermining U.S. influence in Eurasia, the tensions among them are more likely to grow than shrink.

What binds these powers together, however, is their agreement that the status quo must be revised. Russia wants to reassemble as much of the Soviet Union as it can. China has no intention of contenting itself with a secondary role in global affairs, nor will it accept the current degree of U.S. influence in Asia and the territorial status quo there. Iran wishes to replace the current order in the Middle East—led by Saudi Arabia and dominated by Sunni Arab states—with one centered on Tehran.

Leaders in all three countries also agree that U.S. power is the chief obstacle to achieving their revisionist goals. Their hostility toward Washington and its order is both offensive and defensive: not only do they hope that the decline of U.S. power will make it easier to reorder their regions, but they also worry that Washington might try to overthrow them should discord within their countries grow. Yet the revisionists want to avoid direct confrontations with the United States, except in rare circumstances when the odds are strongly in their favor (as in Russia's 2008 invasion of Georgia and its occupation and annexation of Crimea this year). Rather than challenge the status quo head on, they seek to chip away at the norms and relationships that sustain it.

Since Obama has been president, each of these powers has pursued a distinct strategy in light of its own strengths and weaknesses. China, which has the greatest capabilities of the three, has paradoxically been the most frustrated. Its efforts to assert itself in its region have only tightened the links between the United States and its Asian allies and intensified nationalism in Japan. As Beijing's capabilities grow, so will its sense of frustration. China's surge in power will be matched by a surge in Japan's resolve, and tensions in Asia will be more likely to spill over into global economics and politics.

Iran, by many measures the weakest of the three states, has had the most successful record. The combination of the United States' invasion of Iraq and then its premature withdrawal has enabled Tehran to cement deep and enduring ties with significant power centers across the Iraqi border, a development that has changed both the sectarian and the political balance of power in the region. In Syria, Iran, with the help of its longtime ally Hezbollah, has been able to reverse the military tide and prop up the government of Bashar al-Assad in the face of strong opposition from the U.S. government. This triumph of realpolitik has added considerably to Iran's power and prestige. Across the region, the Arab Spring has weakened Sunni regimes, further tilting the balance in Iran's favor. So has the growing split among Sunni governments over what to do about the Muslim Brotherhood and its offshoots and adherents.

Russia, meanwhile, has emerged as the middling revisionist: more powerful than Iran but weaker than China, more successful than China at geopolitics but less successful than Iran. Russia has been moderately effective at driving wedges between Germany and the United States, but Russian President Vladimir Putin's preoccupation with rebuilding the Soviet Union has been hobbled by the sharp limits of his country's economic power. To build a real Eurasian bloc, as Putin dreams of doing, Russia would have to underwrite the bills of the former Soviet republics—something it cannot afford to do.

Nevertheless, Putin, despite his weak hand, has been remarkably successful at frustrating Western projects on former Soviet territory. He has stopped NATO expansion dead in its tracks. He has dismembered Georgia, brought Armenia into his orbit, tightened his hold on Crimea, and, with his Ukrainian adventure, dealt the West an unpleasant and humiliating surprise. From the Western point of view, Putin appears to be condemning his country to an ever-darker future of poverty and marginalization. But Putin doesn't believe that history has ended, and from his perspective, he has solidified his power at home and reminded hostile foreign powers that the Russian bear still has sharp claws.

The Powers That Be

The revisionist powers have such varied agendas and capabilities that none can provide the kind of systematic and global opposition that the Soviet Union did. As a result, Americans have been slow to realize that these states have undermined the Eurasian geopolitical order in ways that complicate U.S. and European efforts to construct a post-historical, win-win world.

Still, one can see the effects of this revisionist activity in many places. In East Asia, China's increasingly assertive stance has yet to yield much concrete geopolitical progress, but it has fundamentally altered the political dynamic in the region with the fastest-growing economies on earth. Asian politics today revolve around national rivalries, conflicting territorial claims, naval buildups, and similar historical issues. The nationalist revival in Japan, a direct response to China's agenda, has set up a process in which rising nationalism in one country feeds off the same in the other. China and Japan are escalating their rhetoric, increasing their military budgets, starting bilateral crises with greater frequency, and fixating more and more on zero-sum competition.

Although the EU remains in a post-historical moment, the non-EU republics of the former Soviet Union are living in a very different age. In the last few years, hopes of transforming the former Soviet Union into a post-historical region have faded. The Russian occupation of Ukraine is only the latest in a series of steps that have turned eastern Europe into a zone of sharp geopolitical conflict and made stable and effective democratic governance impossible outside the Baltic states and Poland.

In the Middle East, the situation is even more acute. Dreams that the Arab world was approaching a democratic tipping point—dreams that informed U.S. policy under both the Bush and the Obama administrations—have faded. Rather than building a liberal order in the region, U.S. policymakers are grappling with the unraveling of the state system that dates back to the 1916 Sykes-Picot agreement, which divided up the Middle Eastern provinces of the Ottoman Empire, as governance erodes in Iraq, Lebanon, and Syria. Obama has done his best to separate the geopolitical issue of Iran's surging power across the region from the question of its compliance with the Nuclear Nonproliferation Treaty, but Israeli and Saudi fears about Iran's regional ambitions are making that harder to do. Another obstacle to striking agreements with Iran is Russia, which has used its seat on the UN Security Council and support for Assad to set back U.S. goals in Syria.

Russia sees its influence in the Middle East as an important asset in its competition with the United States. This does not mean that Moscow will reflexively oppose U.S. goals on every occasion, but it does mean that the win-win outcomes that Americans so eagerly seek will sometimes be held hostage to Russian geopolitical interests. In deciding how hard to press Russia over Ukraine, for example, the White House cannot avoid calculating the impact on Russia's stance on the Syrian war or Iran's nuclear program. Russia cannot make itself a richer country or a much larger one, but it has made itself a more important factor in U.S. strategic thinking, and it can use that leverage to extract concessions that matter to it.

If these revisionist powers have gained ground, the status quo powers have been undermined. The deterioration is sharpest in Europe, where the unmitigated disaster of the common currency has divided public opinion and turned the EU's attention in on itself. The EU may have avoided the worst possible consequences of the euro crisis, but both its will and its capacity for effective action beyond its frontiers have been significantly impaired.

The United States has not suffered anything like the economic pain much of Europe has gone through, but with the country facing the foreign policy hangover induced by the Bush-era wars, an increasingly intrusive surveillance state, a slow economic recovery, and an unpopular health-care law, the public mood has soured. On both the left and the right, Americans are questioning the benefits of the current world order and the competence of its architects. Additionally, the public shares the elite consensus that in a post-Cold War world, the United States ought to be able to pay less into the system and get more out. When that doesn't happen, people blame their leaders. In any case, there is little public appetite for large new initiatives at home or abroad, and a cynical public is turning away from a polarized Washington with a mix of boredom and disdain.

Obama came into office planning to cut military spending and reduce the importance of foreign policy in American politics while strengthening the liberal world order. A little more than halfway through his presidency, he finds himself increasingly bogged down in exactly the kinds of geopolitical rivalries he had hoped to transcend. Chinese, Iranian, and Russian revanchism haven't overturned the post-Cold War settlement in Eurasia yet, and may never do so, but they have converted an uncontested status quo into a contested one. U.S. presidents no longer have a free hand as they seek to deepen the liberal system; they are increasingly concerned with shoring up its geopolitical foundations.

The Twilight of History

It was 22 years ago that Fukuyama published *The End of History and the Last Man*, and it is tempting to see the return of geopolitics as a definitive refutation of his thesis. The reality is more complicated. The end of history, as Fukuyama reminded readers, was Hegel's idea, and even though the revolutionary state had triumphed over the old type of regimes for good, Hegel argued, competition and conflict would continue. He predicted that there would be disturbances in the provinces, even as the heartlands of European civilization moved into a post-historical time. Given that Hegel's provinces included China, India, Japan, and Russia, it should hardly be surprising that more than two centuries later, the disturbances haven't ceased. We are living in the twilight of history rather than at its actual end.

A Hegelian view of the historical process today would hold that substantively little has changed since the beginning of the nineteenth century. To be powerful, states must develop the ideas and institutions that allow them to harness the titanic forces of industrial and informational capitalism. There is no alternative; societies unable or unwilling to embrace this route will end up the subjects of history rather than the makers of it. But the road to postmodernity remains rocky. In order to increase its power, China, for example, will clearly have to go through a process of economic and political development that will require the country to master the problems that modern Western societies have confronted. There is no assurance, however, that China's path to stable liberal modernity will be any less tumultuous than, say, the one that Germany trod. The twilight of history is not a quiet time.

The second part of Fukuyama's book has received less attention, perhaps because it is less flattering to the West. As Fukuyama investigated what a post-historical society would look like, he made a disturbing discovery. In a world where the great questions have been solved and geopolitics has been subordinated to economics, humanity will look a lot like the nihilistic "last man" described by the philosopher Friedrich Nietzsche: a narcissistic consumer with no greater aspirations beyond the next trip to the mall.

In other words, these people would closely resemble today's European bureaucrats and Washington lobbyists. They are competent enough at managing their affairs among post-historical people, but understanding the motives and countering the strategies of old-fashioned power politicians is hard for them. Unlike their less productive and less stable rivals, post-historical people are unwilling to make sacrifices, focused on the short term, easily distracted, and lacking in courage.

The realities of personal and political life in post-historical societies are very different from those in such countries as China, Iran, and Russia, where the sun of history still shines. It is not just that those different societies bring different personalities and values to the fore; it is also that their institutions work differently and their publics are shaped by different ideas.

Societies filled with Nietzsche's last men (and women) characteristically misunderstand and underestimate their supposedly primitive opponents in supposedly backward societies—a blind spot that could, at least temporarily, offset their countries' other advantages. The tide of history may be flowing inexorably in the direction of liberal capitalist democracy, and the sun of history may indeed be sinking behind the hills. But even as the shadows lengthen and the first of the stars appears, such figures as Putin still stride the world stage. They will not go gentle into that good night, and they will rage, rage against the dying of the light.

Critical Thinking

1. Why is Francis Fukuyama's idea of the end of history relevant in the contemporary international system?
2. Why should China, Russia, and Iran be considered revisionist powers?
3. What have been the major foreign policy goals of the Obama administration, and why have they shifted?

Internet References

Ministry of Foreign Affairs of the People's Republic of China
www.fmprc.gov.cn/eng/

National Security Strategy 2015
whitehouse.gov

The Ministry of Foreign Affairs of the Russian Federation
en.mid.ru

Walter Russell Mead is James Clarke Chace Professor of Foreign Affairs and Humanities at Bard College and Editor-at-Large of The American Interest. Follow him on Twitter @wrmead.

Article

Prepared by: Robert Weiner, *University of Massachusetts, Boston*

Drifting to 2016

NIKOLAS K. GVOSDEV

Learning Outcomes

After reading this article, you will be able to:

- Explain how the limitations of U.S. resources affects its foreign policy.
- Discuss the foreign policy legacy of Obama.

The 2016 presidential campaign is turning into a mirror of the 2008 race. Democratic front-runner Hillary Clinton seeks to distinguish herself from the policies of the sitting chief executive, while sundry Republican candidates maintain that Obama's incompetence has made America less safe and diminished its position in the world. No one seeking to become Barack Obama's successor is promising to continue his approach in foreign policy, just as, in 2008, no one ran on a platform of adopting the policies of the George W. Bush administration.

As much as Democratic partisans may resent the comparison, President Obama, entering the final stage of his second term, seems to be presiding over a foreign affairs trajectory similar to the final years of the Bush administration. Obama's tenure has been defined by a deterioration of the U.S. position in the international order; growing anti-American sentiment as reflected in public opinion surveys around the globe; an increased willingness of rising and resurgent powers to challenge American presence abroad; and difficulty in assuring friends that Congress will honor the key agreements the president conducts with foreign leaders. These difficulties are fomenting profound unease among the American electorate about its future safety and prosperity. According to Republican foreign policy practitioners, the Obama administration will leave office in a year's time stymied by the same obstacles that bedeviled his predecessor's administration: the inability to understand Russia's position in a post-Cold War world without alienating American allies and the struggle to set Afghanistan and Iraq on sustainable paths to peace and stability. The challenges in those areas endanger in turn a third U.S. goal—pivoting toward East Asia. Democrats today no longer enjoy any advantage over Republicans in terms of competence in foreign policy or national security. The final quarter of the Obama presidency has eroded any of the gains made by Democrats over the past 10 years in that area. This is quite an unexpected reversal of fortune.

Barack Obama was inaugurated in January 2009 amid general optimism that his administration would repair the damage done to U.S. foreign policy during the tenure of George W. Bush in the White House. The expectation was that Obama would remove the United States from entanglements in the Middle East and heal tensions with Russia. Under Obama's leadership, Democrats were eager to demonstrate their superior skill in handling the nation's security. Unlike their Republican predecessors, they knew how to use military force more effectively, build more lasting and comprehensive international coalitions (especially with the trans-Atlantic allies), and produce tangible results. By Obama's second term, the European Union would become a major contributor to global security, relations with Russia would stabilize, the first seedlings of democracy would take root in Middle East, and the grand rebalance to the Asia-Pacific would be within reach.

The roots of Obama era dysfunction precede his election by at least a decade, arising from the still-unhealed 2002 Democratic schism over the impending war in Iraq.

This ambitious plan was codified in the 2011 Defense Strategic Guidance, which was supposed to steer U.S. defense spending for the remainder of Obama's tenure in office. Five years later, however, these expectations are not aligning with reality. Hopes that Washington could usher the Arab Spring into a glorious summer of democracy have been replaced by the pessimism of an Arab Winter, with states collapsing and extremism on the rise. The Obama administration is preparing to leave office with the Iran nuclear issue essentially frozen for a decade—they were potentially successful in preventing a short-term Iranian dash to weapons capability, but they have left larger concerns about Iran's intentions unresolved. Russia's resurgence and its unwillingness to accept the post-Cold War settlement in Europe, together with the European Union's own ongoing internal travails, have dashed hopes that Europe could become a security provider to augment U.S. efforts elsewhere. A rising China seems prepared to test American commitments in Asia as it seeks to redefine a regional order that the United States has underwritten for many decades. The fate of landmark trade deals that would put the United States at the center of both trans-Atlantic and trans-Pacific economic areas remains in doubt. Polling data collected at the end of 2015 suggests that Americans feel more unsafe and believe that the Obama administration's policies are not sufficient to ensure the safety and security of the United States in what they see as an increasingly dangerous and chaotic world.

Why has this happened? Why has the Obama administration, like its predecessor, been unable to set America on the road to a durable, enduring, bipartisan foreign policy consensus that can guide the United States towards success in the world of the twenty-first century? Why is every candidate hoping to succeed Obama in the White House finding fault with how he has conducted foreign affairs?

The president's supporters maintain that an intransigent international and congressional leadership is to blame. Abroad, Vladimir Putin, Bashar al-Assad, Xi Jinping, and others are responsible for the failure of U.S. policies; at home, Republican leaders like Mitch McConnell and John Boehner are the ones who should be held responsible for deficiencies in U.S. policy, having stood in the way of the president's agenda. Detractors—not only Republicans in opposition but also some Democrats looking to create some distance between Obama's legacy and the Democratic Party's "brand" in national security—focus the blame squarely on the president. According to an unnamed former White House official in a Michael Crowley essay in *Politico* last October, U.S. national security policy "is driven by one man, and one man only, and it is Barack Obama."

It is true that any president does much to define his administration's foreign policy priorities and processes, and former secretary of defense Robert Gates' December 2015 op-ed in the *Washington Post* lays out the personal characteristics that a chief executive must possess to be successful in governing effectively. While personal characteristics are important, it feeds into an American tendency to see mistakes in U.S. national security policy not as a sum of errors endemic to a complex system, but as the fault of a particular presidential administration or even a particular president. This popular but mistaken view holds that the setbacks experienced in American foreign policy are the results of errors in programming and execution made by a national security team. This avoids the more rigorous question: are those errors instead attributable to fundamental flaws in America's perception of the world and its own place in it?

Indeed, the roots of Obama era dysfunction precede his election by at least a decade, before Obama was even a national figure; they arise from the still unhealed 2002 Democratic schism over the impending war in Iraq. Beyond that, the problems experienced during the Obama years reflect an increasing sclerosis in the U.S. policy process itself that has made it far more difficult for Washington to implement effective policy. If nothing is done, this dysfunction is likely to continue into the next administration, regardless of party.

Like Senator Obama, Governor George W. Bush played down any Wilsonian beliefs he may have held during his 2000 presidential campaign in favor of running as a pragmatist with a more realistic and restrained approach to American foreign policy. Bush contrasted his rhetoric with the triumphalist Clinton administration's assertion that the United States was the world's "indispensable nation," but the events of 9/11 changed all of that. In responding to the September 11 attacks, Bush was faced with a choice: to frame the forthcoming campaign as a limited, antiterrorist operation, or to fall for the siren song that an opportunity had arisen to reshape the Middle Eastern and global orders by harnessing American power in the service of American ideals. By choosing this more expansive option, the Bush administration—along with many Democrats—fell into the trap that this vast endeavor could be executed on the cheap, without much sacrifice or expenditure. The source of the hubris that defined the initial plans for the war in Iraq was an oversimplified belief that an operation could both oust Saddam Hussein and put the country on the path to a sustainable democracy by Christmas 2003 with minimal U.S. forces, with Iraq footing the bill from its oil revenue.

A "not Bush" approach could not provide a template for an alternative set of workable foreign policies.

While only a few House Republicans (and some prominent members of the Republican corps of grey eminences) spoke out against the proposed invasion as a costly and reckless gamble, Democrats were much more divided. The traditional anti-interventionist left opposed attacking Iraq, but the centrist core had no particular ideological objections to intervention per se. When the tally of the vote in Congress was taken in 2002 on whether to give President Bush the authority to use force in Iraq, the schism among Democrats was in full display: 39 percent of House Democrats and 58 percent of Democrats in the Senate—among them the party's presumptive future presidential nominees John Kerry and Hillary Clinton—voted in favor of invading Iraq.

This initial bipartisanship, solidified by Baghdad's rapid fall, crumbled as the war dragged on and public support waned. Yet the Democratic Party could not reconcile whether the idea of intervention itself was a bad thing—or whether Hussein's removal could have been better orchestrated by Democrats (who might have been able to get a United Nations resolution, secure meaningful support from allies or otherwise run the occupation more effectively). The seesawing between these two positions led to the memorable moment in the 2004 campaign when Senator John Kerry tried to explain how he could simultaneously be for and against the Iraqi operation.

Kerry's flip-flopping message hurt his bid for the Oval Office and cost Democrats the House. As the Democrats geared up for the 2006 midterms, grassroots antiwar activists highlighted the fact that highly visible mainstream Democrats had been prowar. It was clear that the party's chances for future electoral success would be greatly diminished if an internecine conflict between interventionists and noninterventionists continued. Rather than settle the question of what Democrats were for, it was easier to form a coalition of voters and candidates who agreed that President Bush was the problem. Focusing on Bush's blunders worked for both the anti-interventionist and pacifist wings, but it also covered those Democrats who argued that if they had been in charge, they could have made intervention work.

This strategy allowed the Democrats to have a caucus that encompassed the two Independent outliers—Senators Bernie Sanders and Joseph Lieberman. Democratic opponents of missile defense could join with supporters who disagreed with specifics of Bush's plans for implementation. Democrats who felt that John Mearsheimer and Stephen F. Cohen's reading—that the West provoked Russia by supporting NATO expansion—was right could embrace the Democrats who argued that Bush was too trusting of Putin. On a whole set of issues—from how to cope with Iran's nuclear program, whether to pursue free-trade agreements, whether to rapidly withdraw from Iraq, the best way to deal with a rising China—Democrats who could not find a common unifying approach could still unify to oppose the Bush administration's policies. As an electoral strategy, it worked, delivering control of both houses of Congress to the Democrats. But a "not Bush" approach could not provide a template for an alternative set of workable foreign policies.

Obama, on the other hand, opposed the Iraq war from the beginning, while supporting the military actions taken to degrade and destroy Al-Qaeda and its affiliates. Unlike other candidates who were still attempting to "Kerry straddle" when it came to the Iraq war, Obama voiced his opposition clearly and with conviction. Similar to Governor Bush in 2000, Senator Obama laid out a restrained, focused foreign policy vision, most notably in his 2007 address to the Chicago Council on Global Affairs, but it was not yet clear how committed the senator was when it came to implementing this vision were he to be elected, just as Governor Bush's humble restraint disappeared shortly after he became president. It was also uncertain whether Obama's campaign staff wanted him to develop a binding foreign policy approach beyond citing differences with Hillary Clinton and President Bush to mobilize his base of supporters. Indeed, Obama himself acknowledged that he was a "political Rorschach test"—with different groups of voters projecting their policy expectations onto him. He could, based on his audience, reflect the ideas of a noninterventionist, a pragmatic realist or a Wilsonian idealist. This impression was reinforced by his staffing choices. Obama selected a foreign policy team that was drawn from a variety of different foreign policy perspectives, yet the palpable relief that the new president was "not Bush"—and the anticipation of renewal created by the announcement of policy reviews and fresh starts in relationships with America's European allies, the Middle East, Russia and the states of the Asia-Pacific—allowed the Obama administration to enjoy a domestic and international honeymoon that culminated in a Nobel Peace Prize.

If Obama was ever committed to any sort of progressive-realist vision he did not use those principles to guide the selection of his national security team. His "team of rivals" was only united in their determination to walk the United States back from what they agreed had been the principal mistakes of the Bush administration (at least until 2006); they, in turn, had to filter their policy recommendations through a group of campaign operatives brought into the White House whose guiding star was preserving the president's poll numbers, reelection chances and overall legacy. There would be, under their watch, no new Iraq-level commitments to distract from a domestic agenda and destroy the president's credibility.

The campaign trail—where Obama's rhetorical gifts and his skills as an orator were crucial in winning both the primary and general elections—also colored the approach to foreign policy. Obama's close advisers from the campaign had an inordinate faith that the power of Obama's speeches would bring about major changes. Inspired by the president's soaring rhetoric, other governments would voluntarily comply with American preferences. Because Obama was "not Bush," NATO allies would be prepared to increase their contributions to the war in Afghanistan and to defense spending to secure the European continent; Iran would come to the table to settle the nuclear dispute, the Kremlin would reset relations with the United States and the Chinese would embrace an American vision for regional and global order. If the right words were said, policy would fall into place.

Throughout Obama's first term, the administration attempted to balance its interventionist, idealistic voices with a more pragmatic approach. The Arab Spring upset this balancing act.

When this strategy failed, however, problems arose, and the unresolved schism in Democratic foreign policy thinking further complicated matters. Aside from the main speech, there was a disunity of voices. For instance, Chinese leaders wondered whether the secretary of state's criticisms of Chinese policy or the national security adviser's quiet reassurances reflected the real position of the United States. The vaunted "reset" with Russia disappointed the Kremlin when the substance of many U.S. policies did not change but it also aggravated America's Central European partners' concerns about the strength of the U.S. commitment to their security and well-being.

The debate on whether to increase military presence in Afghanistan exemplified this problem. The president approved what appeared to be a satisfying compromise between two coalitions. Eschewing the limited mission-set focus on counterterrorism advised by Vice President Joseph Biden, Obama embarked instead on a broader set of transformative objectives in Afghanistan and increased U.S. troop commitments and expenditures to satisfy those in his administration who maintained that under Democratic stewardship a major exercise of U.S. power could produce effective results. At the same time, the compromise incorporated the concerns of the political team, which did not want the Obama administration involved in a long, drawn-out escalation in the Middle East, so strict limits on the number of troops and time allotted for the mission were established. The result was a compromise that satisfied no one. For the advocates of intervention, the surge was halfhearted—not enough forces, resources or time to undertake a major transformation of Afghanistan. For those who wanted to refocus the mission on more limited, achievable aims, the surge would end up being an 18-month-long waste of resources and energy with no lasting results.

Similarly, the announcement that relations with Russia would be reset did not produce an internal consensus within the administration over what the United States was to do to make it happen. The main reason that U.S.–Russia relations soured in the second term of the Bush administration was because of the United States' inability to reconcile the expansion of Euro-Atlantic institutions into the former Soviet space—viewed, rightly or wrongly, by the Kremlin as a threat to Russian interests—with the previously stated objective to maintain a partnership with Russia. One of the Obama administration's early actions—the cancellation of the land deployment of components of a theater ballistic-missile-defense system vehemently opposed by Russia in September 2009—was denounced by both Republicans and Central European leaders as Kremlin appeasement. But the subsequent announcement that a sea-based system remained on track infuriated Moscow, who saw this as the old system in a new configuration. The U.S.–Russia reset could only gain momentum once elections in Ukraine and Georgia—the key flash points in the U.S.–Russia relations during the Bush administration—brought new, more pragmatic governments to power. Indeed, the lack of any substantial resolution of these festering issues was made all too clear when the Maidan movement swept the Ukrainian regime of Viktor Yanukovych from power in spring 2014—empowering a new pro-Western government and immediately returning Ukraine to the geopolitical chessboard.

Throughout Obama's first term, the administration attempted to balance its interventionist, idealistic voices with a more pragmatic approach rooted both in the limits of American power and a deep concern on the part of the political advisors to avoid any foreign policy ventures that might drag the Obama administration into new quagmires and distract from its domestic agenda. The Arab Spring, however, upset this balancing act. As authoritarian governments began to fall all across the Middle East, it seemed that a Democratic administration was being presented with its own version of the *annus mirabilis* of 1989 and the rapid collapse of the Soviet bloc. Obama would do for the Middle East what George Bush the elder had done 20 years earlier in Eastern Europe: preside over the collapse of authoritarianism and the triumph of democracy across a critical region of the world. The voices of those within the Obama administration warning about the need for measured transitions could not resist the same siren song that had so captivated the younger Bush administration—the prospect of U.S. power aiding and abetting a sweeping, massive transformation in the Middle East, with the hopes that this time, it would be under Democratic administration. President Obama would succeed where Bush and his Republicans had failed. Nowhere was this more apparent than the decision to intervene in Libya. In places like Bahrain and Azerbaijan, security concerns had led the United States to side with existing authoritarian governments even when faced with popular pressure for democratization. But once Muammar el-Qaddafi, the last remaining poster child for bloody tyrants in the Arab world, issued his bombastic threats to wipe out the civilians in Benghazi who had facilitated rebellion against his rule, it seemed an appropriate time to unleash American power against him.

The form which the Libyan intervention took was guided by the lessons which the interventionists and the political advisors had drawn from the Iraq War. For the liberal hawks, Libya was the "right" way for the United States to intervene: a Democratic administration had received the proper authorizations from the Arab League and the United Nations Security Council; had assembled *a real* coalition of NATO allies and Arab partners (although the United States ended up absorbing more than 75 percent of the costs); and deposed an Arab tyrant with the aid of a provisional government that espoused democratic ideals. For the political advisers, the near-absolute ban on the deployment of any U.S. ground forces (other than a handful of special operatives) avoided the possibility of the United States being sucked into a new land war in the Middle East. U.S. air power was deemed sufficient to achieve U.S. objectives. Libya seemed to herald the emergence of a new form of low-cost, no (U.S.) casualty intervention which would avoid the mistakes of the Iraq war.

The desire to vindicate Democratic interventionism combined with a fear of Iraq-style quagmires produced the worst kind of compromise: a mismatch between grandiose goals and limited resources.

Qaddafi's fall seemed occur to at a fraction of the cost of Hussein's, but the former's removal produced shock waves that proved damaging to U.S. interests. Unlike Saddam Hussein, Qaddafi had renounced terrorism and his weapons of mass destruction program and was generally cooperating with the West on security matters. At the same time, his claims that Islamist extremists were behind the rebellion were not entirely self-serving—there was a genuine radical presence in the eastern part of the country that was empowered as a result of Qaddafi's removal. While a moderate opposition was the face of the new Libya on the surface, the real power rested in the hands of Islamist fighters and clan militias that were uninterested in any commitment to liberal values or democratic governance. Yet the principal lesson from Iraq—that the fall of an authoritarian regime creates a vacuum defined by disorder and instability unless "boots on the ground" were present in sufficient numbers to guarantee order—was downplayed in Libya, in part to avoid pressures to send U.S. and NATO forces to secure the peace. Thus, the desire to vindicate Democratic interventionism combined with a fear of Iraq-style quagmires produced the worst kind of compromise: a mismatch between grandiose goals and limited resources. The Libyan intervention was neither reduced to a more small-scale operation nor expanded when it became clear that a strategy of "leading from behind" would not produce a successful outcome. Moreover, as Libya began to unravel, it became apparent that the operation did not inspire fear in the hearts of U.S. enemies, testify to American power, revitalize NATO or encourage our partners to spend more on defense. Despite the rhetoric about an Asian pivot, both China and U.S. allies in the region concluded that the United States could still be easily distracted in the Middle East. Washington's handling of the fate of Hosni Mubarak in Egypt in addition to the Libyan operation sent mixed signals to our allies about whether the United States truly "had their back" and accelerated efforts to hedge against U.S. unreliability. Partners and competitors alike concluded that the United States was trying to find a way to take bold action in the service of ambitious goals but was not willing to pay the costs or take the risks to do so. Additionally, the manner in which Qaddafi was deposed ended once and for all any hope that a Libya-style denuclearization deal would ever be embraced by Iran or North Korea: the only agreement that Iran would eventually contemplate signing would be one that legitimized its continued possession of the building blocks of a nuclear program, not the complete dismantling that the Obama administration had claimed in 2009 was its nonnegotiable goal.

The Libyan intervention also soured relations with both Beijing and Moscow. UN Security Council resolution 1973 was ratified as a way to create safe havens for civilians to find refuge from the fighting—a condition Russia and China were prepared to accept in order not to veto its passage in the Security Council. Almost immediately, the Western-led intervention focused not on ending a threat to civilians but entering the Libyan conflict as cobelligerents on the side of the opposition. Eventually, U.S. and NATO airpower overwhelmed Qaddafi's military—and the Libyan despot ended up being captured and executed by rebel forces. Russia and China, concluding that they were fooled by the Obama administration, have subsequently resisted U.S. efforts to push for humanitarian action in other conflicts, notably in Syria, where the opposition has concluded that if the United States had intervened in Libya to stop a planned massacre in Benghazi, it would take action against the much more tangible crimes of the Bashar al-Assad regime. The Syrian crisis has thus festered for more than four years. Combined with the effective collapse of Libya as a state, spurring a migration crisis which has seriously destabilized the European

Union and allowed for militants to find a base from which to spread their influence through Africa, the Libyan and Syrian wars have facilitated the rise of the Islamic State as a new and more potent replacement for Al Qaeda, one that is also developing a reach capable of striking targets in the West, including the U.S. homeland. The same warnings that Brent Scowcroft sounded in 2002 prior to the start of the Iraq war were also voiced in the run-up to the Libya intervention, and dismissed by a Democratic administration almost as quickly. Libya today is no more a model of successful intervention in 2016 than Iraq was in 2007, with the one saving grace that the United States is not expending large amounts of blood and treasure.

The Obama administration has settled into a pattern of delaying tough choices for as long as possible, and then being not fully on board with owning the results. For instance, a national security goal is for America to enjoy energy independence by decreasing dependence on the Middle East, and offering an alternative to Russian sources of energy supply to our European allies. An ambitious and expensive program of alternative options (such as biofuels) would be unable to achieve this on its own—but tapping further sources of hydrocarbons in North America might.

The paradigm of low-cost, no-casualty intervention is a bipartisan construction. It will endure with only minor modifications after 2017.

Yet there has been sustained opposition on environmental grounds to expanding production and development of unconventional sources, as well as constructing the necessary infrastructure to bring them on the market. The Obama administration delayed making a decision on the Keystone XL pipeline for years for fear of alienating different domestic constituencies, but was also unwilling to make the case that prioritizing environment and climate concerns trumped national security concerns about global energy markets. The result has been that other countries that choose to be proactive can drive results. A Saudi Arabian decision to increase its oil supply to the international market, followed by a Russian one not to cut supplies, lowered energy prices to levels that made the cost of North American projects like the Keystone XL pipeline prohibitively high. Yet the long-term strategic questions remain unresolved: the United States and its allies are once again becoming addicted to cheap foreign oil, which can kill off alternative-energy programs—and makes it much harder to achieve another announced strategic goal of being able to rebalance from the Middle East at any point in the near future.

The current crop of candidates vying to succeed Obama have all offered variants of the same message: that a change in the occupant of the Oval Office will produce vastly different (and more preferable) outcomes for U.S. foreign policy. If Hillary Clinton succeeds in her quest, she is unlikely to embrace the team-of-rivals approach to governance and seems much more committed to following a more hawkish, liberal-interventionist line. All of the Republican challengers also signal that they would be "different" than Obama. Yet Clinton or any of the Republicans will find it extremely hard to break out of the morass Obama finds himself in. Here's why.

The candidates have criticized the Obama administration's responses to the crises in Ukraine and Syria and to the growing threat of the Islamic State. Yet a closer examination of the accusations does not reveal fundamentally different approaches. Instead, they indicate that a different president would somehow be more effective in carrying out existing policies. For instance, across the spectrum, different politicians continue to express the opinion that the solution to the chaos in Libya and Syria is to find that illusive species of local moderates prepared to fight against extremist forces and establish liberal-leaning, pro-American regimes to obviate the need for a large deployment of U.S. ground forces. The red lines that political advisers that surround the current president and also his potential successors insist cannot be crossed for fear of triggering another Iraq are the same.

Obama has also received tremendous criticism—some of which is justified—for how he has handled Putin and the relationship with Russia. Yet a good deal of the Obama administration's Russia policy has been shaped by self-imposed U.S. constraints. Beyond the standard trope of Russia as a nuclear superpower that cannot be subjected to much direct pressure, Americans want a policy of confronting Russia that limits the risks they will be asked to bear—for example, the requirement that economic sanctions imposed on Russia to punish it for the seizure of Crimea and its operations in Eastern Ukraine have minimal fallout for U.S. economic and business interests. Additionally, for the Iran nuclear deal to work, Russian cooperation is needed, and there is a growing desire for Moscow to better align its operations in Syria with American preferences—shifting most of its military strikes against Islamic State targets while using its influence to persuade al-Assad to step down. At the same time, while proposals to settle the Ukraine crisis by formally designating Ukraine as a neutral state is a nonviable option because the U.S. does not want to be seen as appeasing Moscow, it is also reluctant to spend the necessary resources to pull Ukraine into the Euro-Atlantic orbit. One principal policy divergence between Obama and his potential successors is the question of providing aid to the Ukrainian military. Yet those who criticize Obama's refusal to take a more aggressive stance on this issue still search for a way for the United States to supply weapons to Kiev to pressure Russia to reverse its course while being able to disavow U.S. responsibility for how those arms might be used by Kiev. Most of the 2016 candidates' stance on Russia policy involves some variant of being "tougher" on Putin and showing "resolve," but not showing much willingness to own the only two plausible policy options: a commitment to renewed and robust containment of Russia—requiring a much higher expenditure than anyone in the U.S. seems willing to pay, plus the risks of losing Russian cooperation on other issues—or a search for accommodation.

Such a loosely defined Ukraine policy represents in a microcosm what an October 2015 RAND report indicates is the prevailing problem for U.S. national security policy today: the pronounced imbalance between resources and requirements. The United States must either be prepared to increase what it is willing to spend in terms of funds, personnel and attention, or it must be willing to scale back its ambitions and to redefine what it considers to be acceptable outcomes.

Currently, there is no sign that the American foreign policy establishment is any readier to contemplate hard choices and entertain unpleasant tradeoffs in the coming years. The paradigm of low-cost, no-casualty intervention is a bipartisan construction that will endure with only minor modifications after 2017 in the absence of a truly existential threat to U.S. security. A new approach in 2016? Don't believe a word of it.

Critical Thinking

1. Should the U.S. provide military aid to Ukraine?
2. Why does the author think that there will be no new U.S. foreign policy initiatives in 2017?
3. Why do Americans see their world as increasingly dangerous and chaotic?

Internet References

Council on Foreign Relations
cfr.org

Foreign Policy Initiative
www.foreign policyi.org

National Security Council
https.www.whitehouse.gov/administration/eop/nsc

Nikolas K. Gvosdev is a contributing editor at the *National Interest* and coauthor of *U.S. Foreign Policy and Defense Strategy* (Georgetown, 2015). The views expressed here are his own.

Article

Prepared by: Robert Weiner, *University of Massachusetts, Boston*

The Once and Future Hegemon

SALVATORE BABONES

Learning Outcomes

After reading this article, you will be able to:

- Identify the major elements of the Declinist argument.
- Better understand what is meant by hegemony in global affairs.

Is retreat from global hegemony in America's national interest? No idea has percolated more widely over the past decade—and none is more bogus. The United States is not headed for the skids and there is no reason it should be. The truth is that America can and should seek to remain the world's top dog.

The idea of American hegemony is as old as Benjamin Franklin, but has its practical roots in World War II. The United States emerged from that war as the dominant economic, political and technological power. The only major combatant to avoid serious damage to its infrastructure, its housing stock or its demographic profile, the United States ended the war with the greatest naval order of battle ever seen in the history of the world. It became the postwar home of the United Nations, the International Monetary Fund and the World Bank. And, of course, the United States had the bomb. America was, in every sense of the word, a hegemon.

"Hegemony" is a word used by social scientists to describe leadership within a system of competing states. The Greek historian Thucydides used the term to characterize the position of Athens in the Greek world in the middle of the fifth century BC. Athens had the greatest fleet in the Mediterranean; it was the home of Socrates and Plato, Sophocles and Aeschylus; it crowned its central Acropolis with the solid-marble temple to Athena known to history as the Parthenon. Athens had a powerful rival in Sparta, but no one doubted that Athens was the hegemon of the time until Sparta defeated it in a bitter 27-year war.

America's only global rival in the twentieth century was the Soviet Union. The Soviet Union never produced more than about half of America's total national output. Its nominal allies in Eastern Europe were in fact restive occupied countries, as were many of its constituent republics. Its client states overseas were at best partners of convenience, and at worst expensive drains on its limited resources. The Soviet Union had the power to resist American hegemony, but not to displace it. It had the bomb and an impressive space program, but little else.

When the Soviet Union finally disintegrated in 1991, American hegemony was complete. The United States sat at the top of the international system, facing no serious rivals for global leadership. This "unipolar moment" lasted a mere decade. September 11, 2001, signaled the emergence of a new kind of threat to global stability, and the ensuing rise of China and reemergence of Russia put paid to the era of unchallenged American leadership. Now, America's internal politics have deadlocked and the U.S. government shrinks from playing the role of global policeman. In the second decade of the twenty-first century, American hegemony is widely perceived to be in terminal decline.

American hegemony is now as firm as or firmer than it has ever been, and will remain so for a long time to come.

Or so the story goes. In fact, reports of the passing of U.S. hegemony are greatly exaggerated. America's costly wars in Iraq and Afghanistan were relatively minor affairs considered in long-term perspective. The strategic challenge posed by China has also been exaggerated. Together with its inner circle of unshakable English-speaking allies, the United States possesses near-total control of the world's seas, skies, airwaves and cyberspace, while American universities, think tanks and journals dominate the world of ideas. Put aside all the alarmist punditry. American hegemony is now as firm as or firmer than it has ever been, and will remain so for a long time to come.

The massive federal deficit, negative credit-agency reports, repeated debt-ceiling crises and the 2013 government shutdown all created the impression that the U.S. government is bankrupt, or close to it. The U.S. economy imports half a trillion dollars a year more than it exports. Among the American population, poverty rates are high and ordinary workers' wages have been stagnant (in real terms) for decades. Washington seems to be paralyzed by perpetual gridlock. On top of all this, strategic exhaustion after two costly wars in Afghanistan and Iraq has substantially degraded U.S. military capabilities. Then, at the very moment the military needed to regroup, rebuild and rearm, its budget was hit by sequestration.

If economic power forms the long-term foundation for political and military power, it would seem that America is in terminal decline. But policy analysts tend to have short memories. Cycles of hegemony run in centuries, not decades (or seasons). When the United Kingdom finally defeated Napoleon at Waterloo in 1815, its national resources were completely exhausted. Britain's public-debt-to-GDP ratio was over 250 percent, and early nineteenth-century governments lacked access to the full range of fiscal and financial tools that are available today. Yet the British Century was only just beginning. The *Pax Britannica* and the elevation of Queen Victoria to become empress of India were just around the corner.

By comparison, America's current public-debt-to-GDP ratio of less than 80 percent is relatively benign. Those with even a limited historical memory may remember the day in January 2001 when the then chairman of the Federal Reserve, Alan Greenspan, testified to the Senate Budget Committee that "if current policies remain in place, the total unified surplus will reach $800 billion in fiscal year 2011. . . . The emerging key fiscal policy need is to address the implications of maintaining surpluses." As the poet said, bliss was it in that dawn to be alive! Two tax cuts, two wars and one financial crisis later, America's budget deficit was roughly the size of the projected surplus that so worried Greenspan.

This is not to argue that the U.S. government should ramp up taxes and spending, but it does illustrate the fact that it has enormous potential fiscal resources available to it, should it choose to use them. Deficits come and go. America's fiscal capacity in 2015 is stupendously greater than Great Britain's was in 1815. Financially, there is every reason to think that America's century lies in the future, not in the past.

The same is true of the supposed exhaustion of the U.S. military. On the one hand, thirteen years of continuous warfare have reduced the readiness of many U.S. combat units, particularly in the army. On the other hand, U.S. troops are now far more experienced in actual combat than the forces of any other major military in the world. In any future conflict, the advantage given by this experience would likely outweigh any decline in effectiveness due to deferred maintenance and training. Constant deployment may place an unpleasant and unfair burden on U.S. service personnel and their families, but it does not necessarily diminish the capability of the U.S. military. On the contrary, it may enhance it.

America's limited wars in Afghanistan and Iraq were hardly the final throes of a passing hegemon. They are more akin to Britain's bloody but relatively inconsequential conflicts in Afghanistan and Crimea in the middle of the nineteenth century. Brutal wars like these repeatedly punctured, but never burst, British hegemony. In fact, Britain engaged in costly and sometimes disastrous conflicts throughout the century-long *Pax Britannica.* British hegemony did not come to an end until the country faced Germany head-on in World War I. Even then, Britain ultimately prevailed (with American help). Its empire reached its maximum extent not before World War I but immediately after, in 1922.

Ultimately, it is inevitable that in the long run American power will weaken and American hegemony over the rest of the world will fade. But how long is the long run? There are few factual indications that American decline has begun—or that it will begin anytime soon. Short-term fluctuations should not be extrapolated into long-term trends. Without a doubt, 1991 was a moment of supreme U.S. superiority. But so was 1946, after which came the Soviet bomb, Korea and Vietnam. American hegemony has waxed and waned over the last seventy years, but it has never been eclipsed. And it is unlikely that the eclipse is nigh.

When pundits scope out the imminent threats to U.S. hegemony, the one country on their radar screens is China. While the former Soviet Union never reached above 45 percent of U.S. total national income, the Chinese economy may already have overtaken the American economy, and if not it certainly will soon. If sheer economic size is the foundation of political and military power, China is positioned for future global hegemony. Will it build on this foundation? Can it?

Much depends on the future of China's relationships with its neighbors. China lives in a tough neighborhood. It faces major middle-tier powers on three sides: Russia to the north, South Korea and Japan to the east and Vietnam and India to the south. To the west it faces a series of weak and failing states, but that may be more of a burden than a blessing: China's own western regions are also sites of persistent instability.

It is perhaps realistic to imagine China seeking to expand to the north at the expense of Russia and Mongolia. Ethnic Russians are abandoning Siberia and the Pacific coast in droves, and strategic areas along Russia's border with China have been demographically and economically overwhelmed by Chinese immigration. Twenty-second-century Russia may find it difficult to hold the Far East against China. But that is not a serious threat to U.S. hegemony. If anything, increasing Sino-Russian tensions may reinforce U.S. global hegemony, much as Sino-Soviet tensions did in the 1970s.

There will be no Chinese-sponsored Asian equivalent of NATO or the Warsaw Pact.

To the southeast, China clearly seeks to dominate the South China Sea and beyond. The main barrier to its doing so is the autonomy of Taiwan. Were Taiwan ever to be reintegrated with China, it would be difficult for other regional powers to successfully challenge a united China for control of the basin. In the future, it is entirely possible that China will come to dominate these, its own coastal waters. This would be a minor setback to an America accustomed to dominating all of the world's seas, but it would not constitute a serious strategic threat to the United States.

Across the East China Sea, China faces Japan and South Korea—two of the most prosperous, technologically advanced and militarily best-equipped countries in the world. Historical enmities ensure that China will never expand in that direction. Worse for China, it is quite likely that any increase in China's ability to project power beyond its borders will be matched with similar steps by a wary, remilitarizing Japan.

The countries on China's southern border are so large, populous and poor that it is difficult to imagine China taking much interest in the region beyond simple resource exploitation. Chinese companies may seek profit opportunities in Cambodia, Myanmar and Pakistan, but there is little for China to gain from strategic domination of the region. There will be no Chinese-sponsored Asian equivalent of NATO or the Warsaw Pact.

Farther abroad, much has been made of China's strategic engagement in Africa and Latin America. Investment-starved countries in these regions have been eager to access Chinese capital and in many cases have welcomed Chinese investment, expertise and even immigration. But it is hard to imagine them welcoming Chinese military bases, and equally hard to imagine China asking them for bases. The American presence in Africa is in large part the legacy of centuries of European colonialism. China has no such legacy to build on.

Above all, however, the prospects for future Chinese hegemony depend on the prospects for future Chinese economic growth. Measured in per capita terms, China is still poorer than Mexico. That China will catch up to Mexico seems certain. That China will continue its extraordinary growth trajectory once it has caught up to Mexico is less obvious. In 2011, when the Chinese economy was growing by more than 10 percent a year, I predicted that China was headed for much slower growth. At the time, the IMF was projecting a long-term growth rate of 9.5 percent. Today, the same IMF projections assume 7 percent growth.

Even at 7 percent annual growth, the Chinese economy would account for more than half of total global output by 2050. The United States in its post-World War II heyday never achieved that level of dominance. But exponential extrapolations are inherently tricky. If China continues to grow at 7 percent while the world economy as a whole grows by 3 percent per year, China will account for 90 percent of global economic output by 2100 and 100 percent by 2110. After that, China's economy will be even larger than the world's economy, which of course is impossible unless China moves a large portion of its production off-planet.

A more reasonable assumption is that China's economic growth will eventually settle down to global average rates. The only question is when. Existing demographic trends make it almost certain that the answer is: soon. The U.S. Census Bureau has projected that China's working-age population would reach its peak in 2014 and then go into long-term decline. In the twenty years from 2014 to 2034, China's working-age population will fall by 87 million, while its elderly population will rise by 149 million. In the language of economic punditry, China will "grow old before it grows rich."

The U.S. population, by contrast, is young and growing. In 2034, the U.S. population is projected to be growing at a rate of 0.6 percent per year (compared to −0.2 percent in China), with substantial immigration of talented, productive people (compared to net emigration from China). The U.S. median age of 39.2 will be significantly younger than the Chinese median age of 44.8. Over the long term these trends may change, but the twenty-year scenario is almost certain, because for the most part it has already happened. Economic trends can turn on a dime, but demographic trends are mostly immutable: tomorrow's child-bearers have already been born.

In the ancient Mediterranean world, Rome rose to regional hegemony a century or two after the passing of the Athenian empire. The hegemonic Roman Republic was a hybrid political entity. It consisted of Rome itself, Roman colonies, Roman protectorates, cities conquered by Rome and cities allied to Rome. For four hundred years before 91 BC, the Italian cities allied to Rome were effectively part of the Roman state despite their formal political independence. They participated in Rome's wars under Roman command. They did not pay taxes or tribute to Rome, but they were fully incorporated into a political system centered on Rome. When Hannibal crossed the Alps in 218 BC, most of the Italian cities did not rise up against Rome as he expected. They stood with Rome because they were effectively part of Rome.

In a similar way, the effective borders of the American polity extend well beyond the Atlantic and Pacific coasts. If the Edward Snowden leaks have revealed nothing else, they have shown the depth of intelligence cooperation between the United States and its English-speaking allies Australia, Canada, New Zealand and the United Kingdom. These are the so-called Five

Eyes countries. These English-speaking allies work so closely with the United States on security issues that they resemble ancient Rome's Italian allies. Despite their formal political independence, they do not make major strategic decisions without considering America's interests as well as their own.

Curiously, America's English-speaking allies resemble the United States in their demographic structures as well. While East Asia's birthrates have fallen well below replacement levels and parts of continental Europe face outright depopulation, the English-speaking countries have stable birthrates and substantial immigration. The most talented people in the world don't always move to the United States, but more often than not they move to English-speaking countries. It doesn't hurt that English is the global lingua franca as well as the language of the Internet.

One surprising result of these trends is that the once-unfathomable demographic gap between China and the English-speaking world is narrowing. According to U.S. Census Bureau projections, in 2050 the U.S. population will be 399 million and rising by 0.5 percent per year while the Chinese population will be 1.304 billion and falling by 0.5 percent per year. Throw in America's four English-speaking allies, and the combined five-country population will be 546 million—nearly 42 percent of China's population—with a growth rate of 0.4 percent per year. No longer will China have the overwhelming demographic advantage that has historically let it punch above its economic weight.

Is it reasonable to treat America's English-speaking allies as integrated components of the U.S. power structure? Of course, they are not formally integrated into the U.S. state. But the real, effective borders of countries are much fuzzier than the legal lines drawn on maps. The United States exercises different levels of influence over its sovereign territory, extraterritorial possessions, the English-speaking allies, NATO allies, other treaty allies, nontreaty allies, client states, spheres of influence, exclusionary zones and even enemy territories. All of these categories are fluid in their memberships and meanings, but taken together they constitute more than just a network of relationships. They constitute a cooperative system of shared sovereignty, something akin to the power structure of the Roman Republic.

America's allies constitute more than just a network of relationships. They are a cooperative system of shared sovereignty, something akin to the power structure of the Roman Republic.

No other country in the world possesses, has ever possessed, or is likely to possess in this century such a world-straddling vehicle for the enforcement of its will. More to the point, the U.S.-dominated system shows no signs of falling apart. Even the revelation that America and its English-speaking allies have been spying on the leaders of their NATO peers has not led to calls for the dissolution of NATO. The American system may not last forever, but its remaining life may be measured in centuries rather than decades. Cycles of hegemony turn very slowly because systems of hegemony are very robust. The American power network is much bigger, much stronger and much more resilient than the formal American state as such.

A recurring meme is the idea that the whole world should be able to vote in U.S. presidential elections because the whole world has a stake in the outcome. This argument is not meant to be taken seriously. It is made to prove a point: that the United States is uniquely and pervasively important in the world. At least since the Suez crisis of 1956, it has been clear to everyone that the other countries of the world, whether alone or in concert, are unable to project power beyond their shores without American support. Mere American acquiescence is not enough. In global statecraft, the United States is the indispensable state.

One widely held definition of a state is that a state is a body that successfully claims a monopoly on the legitimate use of force within a territory. The German sociologist Max Weber first proposed this definition in 1919, in the chaotic aftermath of World War I. Interestingly, he included the qualifier "successfully" in his definition. To constitute a real state, a government cannot merely claim the sole right to use force; it must make this claim stick. It must be successful in convincing its people, civil-society groups and, most importantly, other states to accept its claim.

In the twenty-first century, the United States effectively claims a monopoly on the legitimate use of force worldwide. Whether or not it makes this claim in so many words, it makes it through its policies and actions, and America's monopoly on the legitimate use of force is generally accepted by most of the governments (if not the peoples) of the world. That is not to say that all American uses of force are accepted as legitimate, but that all uses of force that are accepted as legitimate are either American or actively supported by the United States. The world condemns Russian intervention in Ukraine but accepts Saudi intervention in Yemen, and of course it looks to the United States to solve conflicts in places like Libya, Syria and Iraq. The United States has not conquered the world, but most of the world's governments (with the exceptions of countries such as Russia, Iran and China) and major intergovernmental organizations accept America's lead. Very often they ask for it.

This American domination of global affairs extends well beyond hegemony. In the nineteenth century, the United Kingdom was a global hegemon. Britannia ruled the waves, and from its domination of the oceans it derived extraordinary influence over global affairs. But China, France, Germany, Russia and later Japan continually challenged the legitimacy of British domination and tested it at every turn. Major powers certainly believed that they could engage independently in global statecraft and acted on that belief. France did not seek British permission to conquer its colonies; Germany did not seek British permission to conquer France.

No one ever likes an empire, but despite Ronald Reagan's memorable phrase, the word "empire" is not inseparably linked to the word "evil."

Twenty-first-century America dominates the world to an extent completely unmatched by nineteenth-century Britain. There is no conflict anywhere in the world in which the United States is not in some way involved. More to the point, participants in conflicts everywhere in the world, no matter how remote, expect the United States to be involved. Revisionists ranging from pro-Russian separatists in eastern Ukraine to Bolivian peasant farmers who want to chew coca leaves see the United States as the power against which they are rebelling. The United States is much more than the world's policeman. It is the world's lawgiver.

The world state of so many fictional utopias and dystopias is here, and it is not a nameless postmodern entity called global governance. It is America. Another word for a world state that dominates all others is an "empire," a word that Americans of all political persuasions abhor. For FDR liberals it challenges cherished principles of internationalism and fair play. For Jeffersonian conservatives it reeks of foreign adventurism. For today's neoliberals it undermines faith in the primacy of market competition over political manipulation. And for neoconservatives it implies an unwelcome responsibility for the welfare of the world beyond America's shores.

In fact, it is difficult to avoid the conclusion that the United States has become an imperial world state—a world-empire—that sets the ground rules for smooth running of the global economy, imposes its will largely without constraint and without consideration of the reasonable desires of other countries, and severely punishes those few states and nonstate actors that resist its dictates.

No one ever likes an empire, but despite Ronald Reagan's memorable phrase, the word "empire" is not inseparably linked to the word "evil." When it comes to understanding empire, history is probably a better guide than science fiction. Consider the Roman Empire. For several centuries after the ascension of Augustus, life under Rome was generally freer, safer and more prosperous than it had been under the previously independent states. Perhaps it was not better for the enslaved or for the Druids, and certainly not for the Jews, but for most people of the ancient Mediterranean, imperial Rome brought vast improvements most of the time.

Ancient analogies notwithstanding, no one would seriously suggest that the United States should attempt to directly rule the rest of the world, and there is no indication that the rest of the world would let it. But the United States could manage its empire more effectively, which is something that the rest of the world would welcome. A winning strategy for low-cost, effective management of empire would be for America to work with and through the system of global governance that America itself has set up, rather than systematically seeking to blunt its own instruments of power.

For example, the United States was instrumental in setting up the International Criminal Court, yet Washington will not place itself under the jurisdiction of the ICC and will not allow its citizens to be subject to the jurisdiction of the ICC. Similarly, though the United States is willing to use UN Security Council resolutions to censure its enemies, it is not willing to accept negotiated limits on its own freedom of action. From a purely military-political standpoint, the United States is sufficiently powerful to go it alone. But from a broader realist standpoint that takes account of the full costs and unintended consequences of military action, that is a suboptimal strategy. Had the United States listened to dissenting opinions on the Security Council before the invasion of Iraq, it would have saved hundreds of billions of dollars and hundreds of thousands of lives. The United States might similarly have done well to have heeded Russian reservations over Libya, as it ultimately did in responding to the use of chemical weapons in Syria.

A more responsible (and consequently more effective) United States would subject itself to the international laws and agreements that it expects others to follow. It would genuinely seek to reduce its nuclear arsenal in line with its commitments under the Nuclear Non-Proliferation Treaty. It would use slow but sure police procedures to catch terrorists, instead of quick but messy drone strikes. It would disavow all forms of torture. All of these policies would save American treasure while increasing American power. They would also increase America's ability to say "no" to its allies when they demand expensive U.S. commitments to protect their interests abroad.

Such measures would not ensure global peace, nor would they necessarily endear the United States to everyone across the world. But they would reduce global tensions and make it easier for America to act in its national interests where those interests are truly at stake. Both the United States and the world as a whole would be better off if Washington did not waste time, money and diplomatic capital on asserting every petty sovereign right it is capable of enforcing. A more strategic United States would preside over a more peaceful and prosperous world.

In pondering its future course, Washington might consider this tale from the ancient world: When Cyrus the Great conquered the neighboring kingdom of Lydia, he allowed his army to loot and pillage Lydia's capital city, Sardis. The deposed Lydian king Croesus became his captive and slave. After Cyrus taunted Croesus by asking him how it felt to see his capital city being plundered, Croesus responded: "It's not my city that your troops are plundering; it's your city." Cyrus ordered an immediate end to the destruction.

Critical Thinking

1. Compare the differences and similarities between British and U.S. hegemony.
2. Does the United States have an empire?
3. Why isn't China a strategic threat to the U.S.?

Internet References

Center for a New American Security
www.cnas.org/

The Atlantic Council
www.atlanticcouncil.org

The Hudson Institute
www.hudson.org/

Salvatore Babones is an associate professor of sociology and social policy at the University of Sydney and an associate fellow at the Institute for Policy Studies.

Article

Prepared by: Robert Weiner, *University of Massachusetts, Boston*

The Once and Future Superpower

Why China Won't Overtake the United States

STEPHEN G. BROOKS and WILLIAM C. WOHLFORTH

Learning Outcomes

After reading this article, you will be able to:

- Explain the relationship between military technology and power.
- Discuss what the grand strategy of the U.S. consists of.

After two and a half decades, is the United States' run as the world's sole superpower coming to an end? Many say yes, seeing a rising China ready to catch up to or even surpass the United States in the near future. By many measures, after all, China's economy is on track to become the world's biggest, and even if its growth slows, it will still outpace that of the United States for many years. Its coffers overflowing, Beijing has used its new wealth to attract friends, deter enemies, modernize its military, and aggressively assert sovereignty claims in its periphery. For many, therefore, the question is not whether China will become a superpower but just how soon.

But this is wishful, or fearful, thinking. Economic growth no longer translates as directly into military power as it did in the past, which means that it is now harder than ever for rising powers to rise and established ones to fall. And China—the only country with the raw potential to become a true global peer of the United States—also faces a more daunting challenge than previous rising states because of how far it lags behind technologically. Even though the United States' economic dominance has eroded from its peak, the country's military superiority is not going anywhere, nor is the globe-spanning alliance structure that constitutes the core of the existing liberal international order (unless Washington unwisely decides to throw it away). Rather than expecting a power transition in international politics, everyone should start getting used to a world in which the United States remains the sole superpower for decades to come.

Lasting preeminence will help the United States ward off the greatest traditional international danger, war between the world's major powers. And it will give Washington options for dealing with nonstate threats such as terrorism and transnational challenges such as climate change. But it will also impose burdens of leadership and force choices among competing priorities, particularly as finances grow more straitened. With great power comes great responsibility, as the saying goes, and playing its leading role successfully will require Washington to display a maturity that U.S. foreign policy has all too often lacked.

The Wealth of Nations

In forecasts of China's future power position, much has been made of the country's pressing domestic challenges: its slowing economy, polluted environment, widespread corruption, perilous financial markets, non-existent social safety net, rapidly aging population, and restive middle class. But as harmful as these problems are, China's true Achilles' heel on the world stage is something else: its low level of technological expertise compared with the United States. Relative to past rising powers, China has a much wider technological gap to close with the leading power. China may export container after container of high-tech goods, but in a world of globalized production, that doesn't reveal much. Half of all Chinese exports consist of what economists call "processing trade," meaning that parts are imported into China for assembly and then exported afterward. And the vast majority of these Chinese exports are directed not by Chinese firms but by corporations from more developed countries.

When looking at measures of technological prowess that better reflect the national origin of the expertise, China's true position becomes clear. World Bank data on payments for the use of intellectual property, for example, indicate that the United States is far and away the leading source of innovative technologies, boasting $128 billion in receipts in 2013—more than four times as much as the country in second place, Japan. China, by contrast, imports technologies on a massive scale yet received less than $1 billion in receipts in 2013 for the use of its intellectual property. Another good indicator of the technological gap is the number of so-called triadic patents, those registered in the United States, Europe, and Japan. In 2012, nearly 14,000 such patents originated in the United States, compared with just under 2,000 in China. The distribution of highly influential articles in science and engineering—those in the top one percent of citations, as measured by the National Science Foundation—tells the same story, with the United States accounting for almost half of these articles, more than eight times China's share. So does the breakdown of Nobel Prizes in Physics, Chemistry, and Physiology or Medicine. Since 1990, 114 have gone to U.S.-based researchers. China-based researchers have received two.

Precisely, because the Chinese economy is so unlike the U.S. economy, the measure fueling expectations of a power shift, GDP, greatly underestimates the true economic gap between the two countries. For one thing, the immense destruction that China is now wreaking on its environment counts favorably toward its GDP, even though it will reduce economic capacity over time by shortening life spans and raising cleanup and health-care costs. For another thing, GDP was originally designed to measure mid-20th century manufacturing economies, and so the more knowledge-based and globalized a country's production is, the more its GDP underestimates its economy's true size.

A new statistic developed by the UN suggests the degree to which GDP inflates China's relative power. Called "inclusive wealth," this measure represents economists' most systematic effort to date to calculate a state's wealth. As a UN report explained, it counts a country's stock of assets in three areas: "(i) manufactured capital (roads, buildings, machines, and equipment), (ii) human capital (skills, education, health), and (iii) natural capital (subsoil resources, ecosystems, the atmosphere)." Added up, the United States' inclusive wealth comes to almost $144 trillion—4.5 times China's $32 trillion.

The true size of China's economy relative to the United States' may lie somewhere in between the numbers provided by GDP and inclusive wealth, and admittedly, the latter measure has yet to receive the same level of scrutiny as GDP. The problem with GDP, however, is that it measures a flow (typically, the value of goods and services produced in a year), whereas inclusive wealth measures a stock. As *The Economist* put it, "Gauging an economy by its GDP is like judging a company by its quarterly profits, without ever peeking at its balance-sheet." Because inclusive wealth measures the pool of resources a government can conceivably draw on to achieve its strategic objectives, it is the more useful metric when thinking about geopolitical competition.

But no matter how one compares the size of the U.S. and Chinese economies, it is clear that the United States is far more capable of converting its resources into military might. In the past, rising states had levels of technological prowess similar to those of leading ones. During the late nineteenth and early twentieth centuries, for example, the United States didn't lag far behind the United Kingdom in terms of technology, nor did Germany lag far behind the erstwhile Allies during the interwar years, nor was the Soviet Union backward technologically compared with the United States during the early Cold War. This meant that when these challengers rose economically, they could soon mount a serious military challenge to the dominant power. China's relative technological backwardness today, however, means that even if its economy continues to gain ground, it will not be easy for it to catch up militarily and become a true global strategic peer, as opposed to a merely a major player in its own neighborhood.

Barriers to Entry

The technological and economic differences between China and the United States wouldn't matter much if all it took to gain superpower status were the ability to use force locally. But what makes the United States a superpower is its ability to operate globally, and the bar for that capability is high. It means having what the political scientist Barry Posen has called "command of the commons"—that is, control over the air, space, and the open sea, along with the necessary infrastructure for managing these domains. When one measures the 14 categories of systems that create this capability (everything from nuclear attack submarines to satellites to transport aircraft), what emerges is an overwhelming U.S. advantage in each area, the result of decades of advances on multiple fronts. It would take a very long time for China to approach U.S. power on any of these fronts, let alone all of them.

For one thing, the United States has built up a massive scientific and industrial base. China is rapidly enhancing its technological inputs, increasing its R&D spending and its numbers of graduates with degrees in science and engineering. But there are limits to how fast any country can leap forward in such matters, and there are various obstacles in China's way—such as a lack of effective intellectual property protections and inefficient methods of allocating capital—that will be extremely hard to change given its rigid political system. Adding to the difficulty, China is chasing a moving target. In 2012, the United States spent $79 billion on military R&D, more than 13 times as much as China's estimated amount, so even rapid Chinese advances might be insufficient to close the gap.

A giant economy alone won't make China the world's second superpower.

Then, there are the decades the United States has spent procuring advanced weapons systems, which have grown only more complex over time. In the 1960s, aircraft took about five years to develop, but by the 1990s, as the number of parts and lines of code ballooned, the figure reached ten years. Today, it takes 15–20 years to design and build the most advanced fighter aircraft, and military satellites can take even longer. So even if another country managed to build the scientific and industrial base to develop the many types of weapons that give the United States command of the commons, there would be a lengthy lag before it could actually possess them. Even Chinese defense planners recognize the scale of the challenge.

Command of the commons also requires the ability to supervise a wide range of giant defense projects. For all the hullabaloo over the evils of the military-industrial complex and the "waste, fraud, and abuse" in the Pentagon, in the United States, research labs, contractors, and bureaucrats have painstakingly acquired this expertise over many decades, and their Chinese counterparts do not yet have it. This kind of "learning by doing" experience resides in organizations, not in individuals. It can be transferred only through demonstration and instruction, so cybertheft or other forms of espionage are not an effective shortcut for acquiring it.

China's defense industry is still in its infancy, and as the scholar Richard Bitzinger and his colleagues have concluded, "Aside from a few pockets of excellence such as ballistic missiles, the Chinese military industrial complex has appeared to demonstrate few capacities for designing and producing relatively advanced conventional weapon systems." For example, China still cannot mass-produce high-performance aircraft engines, despite the immense resources it has thrown at the effort, and relies instead on second-rate Russian models. In other areas, Beijing has not even bothered competing. Take undersea warfare. China is poorly equipped for antisubmarine warfare and is doing very little to improve. And only now is the country capable of producing nuclear-powered attack submarines that are comparable in quietness to the kinds that the U.S. Navy commissioned in the 1950s. Since then, however, the U.S. government has invested hundreds of billions of dollars and six decades of effort in its current generation of Virginia-class submarines, which have achieved absolute levels of silencing.

Finally, it takes a very particular set of skills and infrastructure to actually use all these weapons. Employing them is difficult not just because the weapons themselves tend to be so complex but also because they typically need to be used in a coordinated manner. It is an incredibly complicated endeavor, for example, to deploy a carrier battle group; the many associated ships and aircraft must work together in real time. Even systems that may seem simple require a complex surrounding architecture in order to be truly effective. Drones, for example, work best when a military has the highly trained personnel to operate them and the technological and organizational capacity to rapidly gather, process, and act on information collected from them. Developing the necessary infrastructure to seek command of the commons would take any military a very long time. And since the task places a high premium on flexibility and delegation, China's centralized and hierarchical forces are particularly ill suited for it.

This Time Is Different

In the 1930s alone, Japan escaped the depths of depression and morphed into a rampaging military machine, Germany transformed from the disarmed loser of World War I into a juggernaut capable of conquering Europe, and the Soviet Union recovered from war and revolution to become a formidable land power. The next decade saw the United States' own sprint from military also ran to global superpower, with a nuclear Soviet Union close on its heels. Today, few seriously anticipate another world war, or even another cold war, but many observers argue that these past experiences reveal just how quickly countries can become dangerous once they try to extract military capabilities from their economies.

But what is taking place now is not your grandfather's power transition. One can debate whether China will soon reach the first major milestone on the journey from great power to superpower: having the requisite economic resources. But a giant economy alone won't make China the world's second superpower, nor would overcoming the next big hurdle, attaining the requisite technological capacity. After that lies, the challenge of transforming all this latent power into the full range of systems needed for global power projection and learning how to use them. Each of these steps is time consuming and fraught with difficulty. As a result, China will, for a long time, continue to hover somewhere between a great power and a superpower. You might call it "an emerging potential superpower": thanks to its economic growth, China has broken free from the great-power pack, but it still has a long way to go before it might gain the economic and technological capacity to become a superpower.

This is not your grandfather's power transition.

China's quest for superpower status is undermined by something else, too: weak incentives to make the sacrifices required. The United States owes its far-reaching military capabilities to the existential imperatives of the Cold War. The country would never have borne the burden it did had policymakers not faced the challenge of balancing the Soviet Union, a superpower with the potential to dominate Eurasia. (Indeed, it is no surprise that two and a half decades after the Soviet Union collapsed, it is Russia that possesses the second-greatest military capability in the world.) Today, China faces nothing like the Cold War pressures that led the United States to invest so much in its military. The United States is a far less threatening superpower than the Soviet Union was: however aggravating Chinese policymakers find U.S. foreign policy, it is unlikely to engender the level of fear that motivated Washington during the Cold War.

Stacking the odds against China even more, the United States has few incentives to give up power, thanks to the web of alliances it has long boasted. A list of U.S. allies reads as a who's who of the world's most advanced economies, and these partners have lowered the price of maintaining the United States' superpower status. U.S. defense spending stood at around three percent of GDP at the end of the 1990s, rose to around five percent in the next decade on account of the wars in Afghanistan and Iraq, and has now fallen back to close to three percent. Washington has been able to sustain a global military capacity with relatively little effort thanks in part to the bases its allies host and the top-end weapons they help develop. China's only steadfast ally is North Korea, which is often more trouble than it is worth.

Given the barriers thwarting China's path to superpower status, as well as the low incentives for trying to overcome them, the future of the international system hinges most on whether the United States continues to bear the much lower burden of sustaining what we and others have called "deep engagement," the globe-girdling grand strategy it has followed for some 70 years. And barring some odd change of heart that results in a true abnegation of its global role (as opposed to overwrought, politicized charges sometimes made about its already having done so), Washington will be well positioned for decades to maintain the core military capabilities, alliances, and commitments that secure key regions, backstop the global economy, and foster cooperation on transnational problems.

A world of lasting U.S. military preeminence and declining U.S. economic dominance will test the United States' capacity for restraint.

The benefits of this grand strategy can be difficult to discern, especially in light of the United States' foreign misadventures in recent years. Fiascos such as the invasion of Iraq stand as stark reminders of the difficulty of using force to alter domestic politics abroad. But power is as much about preventing unfavorable outcomes as it is about causing favorable ones, and here Washington has done a much better job than most Americans appreciate.

For a largely satisfied power leading the international system, having enough strength to deter or block challengers is in fact more valuable than having the ability to improve one's position further on the margins. A crucial objective of U.S. grand strategy over the decades has been to prevent a much more dangerous world from emerging, and its success in this endeavor can be measured largely by the absence of outcomes common to history: important regions destabilized by severe security dilemmas, tattered alliances unable to contain breakout challengers, rapid weapons proliferation, great-power arms races, and a descent into competitive economic or military blocs.

Were Washington to truly pull back from the world, more of these challenges would emerge, and transnational threats would likely loom even larger than they do today. Even if such threats did not grow, the task of addressing them would become immeasurably harder if the United States had to grapple with a much less stable global order at the same time. And as difficult as it sometimes is today for the United States to pull together coalitions to address transnational challenges, it would be even harder to do so if the country abdicated its leadership role and retreated to tend its garden, as a growing number of analysts and policymakers—and a large swath of the public—are now calling for.

Lead Us Not into Temptation

Ever since the Soviet Union's demise, the United States' dramatic power advantage over other states has been accompanied by the risk of self-inflicted wounds, as occurred in Iraq. But the slippage in the United States' economic position may have the beneficial effect of forcing U.S. leaders to focus more on the core mission of the country's grand strategy rather than being sucked into messy peripheral conflicts. Indeed, that has been the guiding logic behind President Barack Obama's foreign policy. Nonetheless, a world of lasting U.S. military preeminence and declining U.S. economic dominance will continue to test the United States' capacity for restraint, in four main ways.

First is the temptation to bully or exploit American allies in the pursuit of self-interested gain. U.S. allies are dependent on Washington in many ways, and leaning on them to provide favors in return—whether approving of controversial U.S. policies, refraining from activities the United States opposes, or agreeing to lopsided terms in mutually beneficial deals—seems like something only a chump would forgo. (Think of the Republican presidential candidate Donald Trump's frequent claims that the United States always loses in its dealings with foreigners, including crucial allies, and that he would restore the country's ability to win.) But the basic contract at the heart of the contemporary international order is that if its members put aside the quest for relative military advantage, join a dense web of institutional networks, and agree to play by common rules, then the United States will not take advantage of its dominance to extract undue returns from its allies. It would be asking too much to expect Washington to never use its leverage to seek better deals, and a wide range of presidents—including John F. Kennedy, Ronald Reagan, George W. Bush, and Obama—have done so at various times. But if Washington too often uses its power to achieve narrowly self-interested gains, rather than to protect and advance the system as a whole, it will run a real risk of eroding the legitimacy of both its leadership and the existing order.

Second, the United States will be increasingly tempted to overreact when other states—namely, China—use their growing economic clout on the world stage. Most of the recent rising powers of note, including Germany, Japan, and the Soviet Union, were stronger militarily than economically. China, by contrast, will for decades be stronger economically than militarily. This is a good thing, since military challenges to global order can turn ugly quickly. But it means that China will mount economic challenges instead, and these will need to be handled wisely. Most of China's efforts along these lines will likely involve only minor or cosmetic alterations to the existing order, important for burnishing Beijing's prestige but not threatening to the order's basic arrangements or principles. Washington should respond to these gracefully and with forbearance, recognizing that paying a modest price for including Beijing within the order is preferable to risking provoking a more fundamental challenge to the structure in general.

The recent fracas over the Asian Infrastructure Investment Bank is a good example of how not to behave. China proposed the AIIB in 2013 as a means to bolster its status and provide investment in infrastructure in Asia. Although its criteria for loans might turn out to be less constructive than desired, it is not likely to do major harm to the region or undermine the structure of the global economy. And yet the United States responded by launching a public diplomatic campaign to dissuade its allies from joining. They balked at U.S. opposition and signed up eagerly. By its reflexive opposition both to a relatively constructive Chinese initiative and to its allies' participation in it, Washington created an unnecessary zero-sum battle that ended in a humiliating diplomatic defeat. (A failure by the U.S. Congress to pass the Trans-Pacific Partnership as negotiated, meanwhile, would be an even greater fiasco, leading to serious questions abroad about U.S. global leadership.)

Third, the United States will still face the temptation that always accompanies power, to intervene in places where its core national interests are not in play (or to expand the definition of its core national interests so much as to hollow out the concept). That temptation can exist in the midst of a superpower struggle—the United States got bogged down in Vietnam during the Cold War, as did the Soviet Union in Afghanistan—and it clearly exists today, at a time when the United States has no peer rivals. Obama has carefully guarded against this temptation. He attracted much criticism for elevating "Don't do stupid stuff" to a grand-strategic maxim. But if doing stupid stuff threatens the United States' ability to sustain its grand strategy and associated global presence, then he had a point. Missing, though, was a corollary: "Keep your eye on the ball." And for nearly seven decades, that has meant continuing Washington's core mission of fostering stability in key regions and keeping the global economy and wider order humming.

Finally, Washington will need to avoid adopting overly aggressive military postures even when core interests are at stake, such as with China's increasingly assertive stance in its periphery. It is true that Beijing's "anti-access/area-denial" capabilities have greatly raised the costs and risks of operating U.S. aircraft and surface ships (but not submarines) near China. How Washington should respond to Beijing's newfound local military capability, however, depends on what Washington's strategic goals are. To regain all the military freedom of action the United States enjoyed during its extraordinary dominance throughout the 1990s would indeed be difficult, and the actions necessary would increase the risk of future confrontations. Yet if Washington's goals are more limited—securing regional allies and sustaining a favorable institutional and economic order—then the challenge should be manageable.

By adopting its own area-denial strategy, for example, the United States could still deter Chinese aggression and protect U.S. allies despite China's rising military power. Unlike the much-discussed Air-Sea Battle doctrine for a Pacific conflict, this approach would not envision hostilities rapidly escalating to strikes on the Chinese mainland. Rather, it would be designed to curtail China's ability during a conflict to operate within what is commonly known as "the first island chain," encompassing parts of Japan, the Philippines, and Taiwan. Under this strategy, the United States and its allies would employ the same mix of capabilities—such as mines and mobile anti-ship missiles—that China itself has used to push U.S. surface ships and aircraft away from its coast. And it could turn the tables and force China to compete in areas where it remains very weak, most notably, undersea warfare.

By adopting its own area-denial strategy, the United States could still deter Chinese aggression and protect U.S. allies.

The premise of such a strategy is that even if China were able to deny U.S. surface forces and aircraft access to the area near its coast, it would not be able to use that space as a launching pad for projecting military power farther during a conflict. China's coastal waters, in this scenario, would turn into a sort of no man's sea, in which neither state could make much use of surface ships or aircraft. This would be a far cry from the situation that prevailed during the 1990s, when China could not stop the world's leading military power from enjoying unfettered access to its airspace and ocean right up to its territorial border. But the change needs to be put in perspective: it is only natural that after spending tens of billions of dollars over decades, China has begun to reverse this unusual vulnerability, one the United States would never accept for itself.

While this area-denial strategy would help solve a long-term problem, it would do little to address the most immediate challenge from China: the military facilities it is steadily building on artificial islands in the South China Sea. There is no easy answer, but Washington should avoid too aggressive a reaction, which could spark a conflict. After all, these small, exposed islands arguably leave the overall military balance unchanged, since they would be all but impossible to defend in a conflict. China's assertiveness may even be backfiring. Last year, the Philippines—real islands with extremely valuable basing facilities—welcomed U.S. forces back onto its shores after a 24-year absence. And the United States is now in talks to base long-range bombers in Australia.

To date, the Obama administration has chosen to conduct so-called freedom-of-navigation operations in order to contest China's maritime claims. But as the leader of the order it largely shaped, the United States has many other arrows in its quiver. To place the burden of escalation on China, the United States—or, even better, its allies—could take a page from China's playbook and ramp up quasi-official research voyages in the area. Another asset Washington has is international law. Pressure is mounting on China to submit its territorial disputes to arbitration in international courts, and if Beijing continues to resist doing so, it will lose legitimacy and could find itself a target of sanctions and other diplomatic punishments. And if Beijing tried to extract economic gains from contested regions, Washington could facilitate a process along the lines of the proportional punishment strategy it helped make part of the World Trade Organization: let the Permanent Court of Arbitration, in The Hague, determine the gains of China's illegal actions, place a temporary tariff on Chinese exports to collect exactly that much revenue while the sovereignty claims are being adjudicated, and then distribute them once the matter is settled before the International Court of Justice. Whatever approach is adopted, what matters for U.S. global interests is not the islands themselves or the nature of the claims per se but what these provocations do to the wider order.

Although China can "pose problems without catching up," in the words of the political scientist Thomas Christensen, the bottom line is that the United States' global position gives it room to maneuver. The key is to exploit the advantages of standing on the defensive: as a raft of strategic thinkers have pointed out, challenging a settled status quo is very hard to do.

Know Thyself

Despite China's ascent, the United States' superpower position is more secure than recent commentary would have one believe—so secure, in fact, that the chief threat to the world's preeminent power arguably lies within. As U.S. dominance ebbs slightly from its peak two decades ago, Washington may be tempted to overreact to the setbacks inherent in an admittedly frustrating and hard-to-manage world by either lashing out or coming home—either way abandoning the patient and constructive approach that has been the core of its grand strategy for many decades. This would be a grave mistake. That grand strategy has been far more successful and beneficial than most people realize, since they take for granted its chief accomplishment—preventing the emergence of a much less congenial world.

One sure way to generate a wrong-headed push for retrenchment would be to undertake another misadventure like the war in Iraq. That America has so far weathered that disaster with its global position intact is a testament to just how robust its superpower status is. But that does not mean that policymakers can make perpetual blunders with impunity. In a world in which the United States retains its overwhelming military preeminence as its economic dominance slips, the temptation to overreact to perceived threats will grow—even as the margin of error for absorbing the costs of the resulting mistakes will shrink. Despite what is being said on the campaign trail these days, the United States is hardly in an unusually perilous global situation. But nor is its standing so secure that irresponsible policies by the next president won't take their toll.

Critical Thinking

1. Why did China become a peer competitor of the United States?
2. What are the goals of U.S. grand strategy in the international system?
3. Do the authors believe that China is not a threat to U.S. core national interests? Why or why not?

Internet References

Asian Infrastructure Investment Bank
www.aib.org/

One Belt, One Road

https://en.wikipedia.org/wiki/One_Belt_One_Road

The Regional Comprehensive Economic Partnership

https://en.wikipedia.org/wiki/Regional_Comprehensive_Economic-Partnership

Stephen G. Brooks is an associate professor of government at Dartmouth College.

William C. Wohlforth is Daniel Webster Professor of Government at Dartmouth College.

Article

Prepared by: Robert Weiner, *University of Massachusetts, Boston*

American Imperium

Untangling Truth and Fiction in an Age of Perpetual War

ANDREW J. BACEVICH

Learning Outcomes

After reading this article, you will be able to:

- Discuss why the author believes that the U.S. is not a peaceful country.
- Understand the author's narrative of U.S. military history.

Republicans and Democrats disagree today on many issues, but they are united in their resolve that the United States must remain the world's greatest military power. This bipartisan commitment to maintaining American supremacy has become a political signature of our times. In its most benign form, the consensus finds expression in extravagant and unremitting displays of affection for those who wear the uniform. Considerably less benign is a pronounced enthusiasm for putting our soldiers to work "keeping America safe." This tendency finds the United States more or less permanently engaged in hostilities abroad, even as presidents from both parties take turns reiterating the nation's enduring commitment to peace.

To be sure, this penchant for military activism attracts its share of critics. Yet dissent does not imply influence. The trivializing din of what passes for news drowns out the antiwar critique. One consequence of remaining perpetually at war is that the political landscape in America does not include a peace party. Nor, during presidential election cycles, does that landscape accommodate a peace candidate of voter consequence. The campaign now in progress has proved no exception. Candidates calculate that tough talk wins votes. They are no more likely to question the fundamentals of U.S. military policy than to express skepticism about the existence of a deity. Principled opposition to war ranks as a disqualifying condition, akin to having once belonged to the Communist Party or the KKK. The American political scene allows no room for the intellectual progeny of Jane Addams, Eugene V. Debs, Dorothy Day, or Martin Luther King Jr.

So, this November, voters will choose between rival species of hawks. Each of the finalists will insist that freedom's survival hinges on having in the Oval Office a president ready and willing to employ force, even as each will dodge any substantive assessment of what acting on that impulse has produced of late. In this sense, the outcome of the general election has already been decided. As regards so-called national security, victory is ensured. The status quo will prevail, largely unexamined and almost entirely intact.

Citizens convinced that U.S. national security policies are generally working well can therefore rest easy. Those not sharing that view, meanwhile, might wonder how it is that military policies that are manifestly defective—the ongoing accumulation of unwon wars providing but one measure—avoid serious scrutiny, with critics of those policies consigned to the political margins.

History provides at least a partial answer to this puzzle. The constructed image of the past to which most Americans habitually subscribe prevents them from seeing other possibilities, a condition for which historians themselves bear some responsibility. Far from encouraging Americans to think otherwise, these historians have effectively collaborated with those interests that are intent on suppressing any popular inclination toward critical reflection. This tunnel vision affirms certain propositions that are dear to American hearts, preeminently the conviction that history itself has summoned the United States to create a global order based on its own self-image. The resulting metanarrative unfolds as a drama in four acts: in the first, Americans respond to but then back away from history's charge; in the second, they indulge in an interval of adolescent folly, with dire consequences; in the third, they reach maturity and shoulder their providentially assigned responsibilities; in the fourth, after briefly straying off course, they stage an extraordinary recovery. When the final curtain in this drama falls, somewhere around 1989, the United States is the last superpower standing.

A Conviction that Is Dear to American Hearts Is the Proposition that History Itself Has Summoned the United States to Create A Global Order Based on Its Own Self-Image

For Americans, the events that established the twentieth-century as their century occurred in the military realm: two misleadingly named "world wars" separated by an "interwar period" during which the United States ostensibly took a time-out, followed by a so-called Cold War that culminated in decisive victory despite being inexplicably marred by Vietnam. To believe in the lessons of this melodrama—which warn above all against the dangers of isolationism and appeasement—is to accept that the American Century should last in perpetuity. Among Washington insiders, this view enjoys a standing comparable to belief in the Second Coming among devout Christians.

Unfortunately, in the United States these lessons retain little relevance. Whatever the defects of current U.S. policy, isolationism and appeasement do not number among them. With its military active in more than 150 countries, the United States today finds itself, if anything, overextended. Our principal security challenges—the risks to the planet posed by climate change, the turmoil enveloping much of the Islamic world and now spilling into the West, China's emergence as a potential rival to which Americans have mortgaged their prosperity—will not yield to any solution found in the standard Pentagon repertoire. Yet when it comes to conjuring up alternatives, the militarized history to which Americans look for instruction has little to offer.

Prospects for thinking otherwise require an altogether different historical frame. Shuffling the deck—reimagining our military past—just might produce lessons that speak more directly to our present predicament.

Consider an alternative take on the twentieth-century U.S. military experience, with a post-9/11 codicil included for good measure. Like the established narrative, this one also consists of four episodes: a Hundred Years' War for the Hemisphere, launched in 1898; a War for Pacific Dominion, also initiated in 1898, petering out in the 1970s but today showing signs of reviving; a War for the West, already under way when the United States entered it in 1917 and destined to continue for seven more decades; and a War for the Greater Middle East, dating from 1980 and ongoing still with no end in sight.

In contrast to the more familiar four-part narrative, these several military endeavors bear no more than an incidental relationship to one another. Even so, they resemble one another in this important sense: each found expression as an expansive yet geographically specific military enterprise destined to extend across several decades. Each involved the use (or threatened use) of violence against an identifiable adversary or set of adversaries.

Yet for historians inclined to think otherwise, the analytically pertinent question is not against whom U.S. forces fought but why. It's what the United States was seeking to accomplish that matters most. Here, briefly, is a revised account of the wars defining the (extended) American Century, placing purpose or motive at the forefront.

In February 1898, the battleship U.S.S. Maine, at anchor in Havana Harbor, blew up and sank, killing 266 American sailors. Widely viewed at the time as an act of state-sponsored terrorism, this incident initiated what soon became a War for the Hemisphere.

Two months later, vowing to deliver Cubans from oppressive colonial rule, the United States Congress declared war on Spain. Within weeks, however, the enterprise evolved into something quite different. After ousting Cuba's Spanish overseers, the United States disregarded the claims of nationalists calling for independence, subjected the island to several years of military rule, and then converted it into a protectorate that was allowed limited autonomy. Under the banner of anti-imperialism, a project aimed at creating an informal empire had commenced.

America's intervention in Cuba triggered a bout of unprecedented expansionism. By the end of 1898, U.S. forces had also seized Puerto Rico, along with various properties in the Pacific. These actions lacked a coherent rationale until Theodore Roosevelt, elevated to the presidency in 1901, took it on himself to fill that void. An American-instigated faux revolution that culminated with a newly founded Republic of Panama signing over to the United States its patrimony—the route for a trans-isthmian canal—clarified the hierarchy of U.S. interests. Much as concern about Persian Gulf oil later induced the United States to assume responsibility for policing that region, so concern for securing the as yet unopened canal induced it to police the Caribbean.

In 1904, Roosevelt's famous "corollary" to the Monroe Doctrine, claiming for the United States authority to exercise "international police power" in the face of "flagrant . . . wrongdoing or impotence," provided a template for further action. Soon thereafter, U.S. forces began to intervene at will throughout the Caribbean and Central America, typically under the guise of protecting American lives and property but in fact to position the United States as regional suzerain. Within a decade, Haiti, the Dominican Republic, and Nicaragua joined Cuba and Panama on the roster of American protectorates. Only in Mexico, too large to occupy and too much in the grip of revolutionary upheaval to tame, did U.S. military efforts to impose order come up short.

"Yankee imperialism" incurred costs, however, not least of all by undermining America's preferred self-image as benevolent and peace-loving, and therefore unlike any other great power in history. To reduce those costs, beginning in the 1920s successive administrations sought to lower the American military profile in the Caribbean basin. The United States was now content to allow local elites to govern so long as they respected parameters established in Washington. Here was a workable formula for exercising indirect authority, one that prioritized order over democracy, social justice, and the rule of law.

By 1933, when Franklin Roosevelt inaugurated his Good Neighbor policy with the announcement that "the definite policy of the United States from now on is one opposed to armed intervention," the War for the Hemisphere seemed largely won. Yet neighborliness did not mean that U.S. military forces were leaving the scene. As insurance against backsliding, Roosevelt left intact the U.S. bases in Cuba and Puerto Rico, and continued to garrison Panama.

So rather than ending, the Hundred Years' War for the Hemisphere had merely gone on hiatus. In the 1950s, the conflict resumed and even intensified, with Washington now defining threats to its authority in ideological terms. Leftist radicals rather than feckless caudillos posed the problem. During President Dwight D. Eisenhower's first term, a CIA-engineered coup in Guatemala tacitly revoked FDR's nonintervention pledge and appeared to offer a novel way to enforce regional discipline without actually committing U.S. troops. Under President John F. Kennedy, the CIA tried again, in Cuba. That was just for starters.

Between 1964 and 1994, U.S. forces intervened in the Dominican Republic, Grenada, Panama, and Haiti, in most cases for the second or third time. Nicaragua and El Salvador also received sustained American attention. In the former, Washington employed methods that were indistinguishable from terrorism to undermine a regime it viewed as illegitimate. In the latter, it supported an ugly counterinsurgency campaign to prevent leftist guerrillas from overthrowing right-wing oligarchs. Only in the mid-1990s did the Hundred Years' War for the Hemisphere once more begin to subside. With the United States having forfeited its claim to the Panama Canal and with U.S.–Cuban relations now normalized, it may have ended for good.

Today, the United States enjoys unquestioned regional primacy, gained at a total cost of fewer than a thousand U.S. combat fatalities, even counting the luckless sailors who went down with the Maine. More difficult to say with certainty is whether a century of interventionism facilitated or complicated U.S. efforts to assert primacy in its "own back yard." Was coercion necessary? Or might patience have produced a similar outcome? Still, in the end, Washington got what it wanted. Given the gaping imbalance of power between the Colossus of the North and its neighbors, we may wonder whether the final outcome was ever in doubt.

During its outward thrust of 1898, the United States seized the entire Philippine archipelago, along with smaller bits of territory such as Guam, Wake, and the Hawaiian Islands. By annexing the Philippines, U.S. authorities enlisted in a high-stakes competition to determine the fate of the Western Pacific, with all parties involved viewing China as the ultimate prize. Along with traditional heavyweights such as France, Great Britain, and Russia, the ranks of the competitors included two emerging powers. One was the United States, the other imperial Japan. Within two decades, thanks in large part to the preliminary round of the War for the West, the roster had thinned considerably, putting the two recent arrivals on the path for a showdown.

The War for Pacific Dominion confronted the U.S. military with important preliminary tasks. Obliging Filipinos to submit to a new set of colonial masters entailed years of bitter fighting. More American soldiers died pacifying the Philippines between 1899 and 1902 than were to lose their lives during the entire Hundred Years' War for the Hemisphere. Yet even as U.S. forces were struggling in the Philippines, orders from Washington sent them venturing more deeply into Asia. In 1900, several thousand American troops deployed to China to join a broad coalition (including Japan) assembled to put down the so-called Boxer Rebellion. Although the expedition had a nominally humanitarian purpose—Boxers were murdering Chinese Christians while laying siege to legations in Peking's diplomatic quarter—its real aim was to preserve the privileged status accorded foreigners in China. In that regard, it succeeded, thereby giving a victory to imperialism.

Through its participation in this brief campaign, the United States signaled its own interest in China. A pair of diplomatic communiqués known as the Open Door Notes codified Washington's position by specifying two non-negotiable demands: first, to preserve China's territorial integrity; and second, to guarantee equal opportunity for all the foreign powers engaged in exploiting that country. Both of these demands would eventually put the United States and Japan at cross-purposes. To substantiate its claims, the United States established a modest military presence in China. At Tientsin, two days march from Peking, the U.S. Army stationed an infantry regiment. The U.S. Navy ramped up its patrols on the Yangtze River between Shanghai and Chungking—more or less the equivalent of Chinese gunboats today traversing the Mississippi River between New Orleans and Minneapolis.

U.S. and Japanese interests in China proved to be irreconcilable. In hindsight, a violent collision between these two rising powers appears almost unavoidable. As wide as the Pacific might be, it was not wide enough to accommodate the ambitions of both countries. Although a set of arms-limiting treaties negotiated at the Washington Naval Conference of 1921–1922 put a momentary brake on the rush toward war, that pause could not withstand the crisis of the Great Depression. Once Japanese forces invaded Manchuria in 1931 and established the puppet state of Manchukuo, the options available to the United States had reduced to two: either allow the Japanese a free hand in China or muster sufficient power to prevent them from having their way. By the 1930s, the War for Pacific Dominion had become a zero-sum game.

To recurring acts of Japanese aggression in China, Washington responded with condemnation and, eventually, punishing economic sanctions. What the United States did not do, however, was reinforce its Pacific outposts to the point where they could withstand serious assault. Indeed, the Navy and War departments all but conceded that the Philippines, impulsively absorbed back in the heady days of 1898, were essentially indefensible.

At odds with Washington over China, Japanese leaders concluded that the survival of their empire hinged on defeating the United States in a direct military confrontation. They could see no alternative to the sword. Nor, barring an unexpected Japanese capitulation to its demands, could the United States. So the December 7, 1941, attack on Pearl Harbor came as a surprise only in the narrow sense that U.S. commanders underestimated the prowess of Japan's aviators.

The December 7, 1941, Attack on Pearl Harbor Came as a Surprise Only in the Narrow Sense that U.S. Commanders Underestimated the Prowess of Japan's Aviators

That said, the ensuing conflict was from the outset a huge mismatch. Only in willingness to die for their country did the Japanese prove equal to the Americans. By every other measure—military-age population, raw materials, industrial capacity, access to technology—they trailed badly. Allies exacerbated the disparity, since Japan fought virtually alone. Once FDR persuaded his countrymen to go all out to win—after Pearl Harbor, not a difficult sell—the war's eventual outcome was not in doubt. When the incineration of Hiroshima and Nagasaki ended the fighting, the issue of Pacific dominion appeared settled. Having brought their principal foe to its knees, the Americans were now in a position to reap the rewards.

In the event, things were to prove more complicated. Although the United States had thwarted Japan's efforts to control China, developments within China itself soon dashed American expectations of enjoying an advantageous position there. The United States "lost" it to communist revolutionaries who ousted the regime that Washington had supported against the Japanese. In an instant, China went from ally to antagonist.

So U.S. forces remained in Japan, first as occupiers and then as guarantors of Japanese security (and as a check on any Japanese temptation to rearm). That possible threats to Japan were more than theoretical became evident in the summer of 1950, when war erupted on the nearby Korean peninsula. A mere five years after the War for Pacific Dominion had seemingly ended, G.I.s embarked on a new round of fighting.

The experience proved an unhappy one. Egregious errors of judgment by the Americans drew China into the hostilities, making the war longer and more costly than it might otherwise have been. When the end finally came, it did so in the form of a painfully unsatisfactory draw. Yet with the defense of South Korea now added to Washington's list of obligations, U.S. forces stayed on there as well.

In the eyes of U.S. policymakers, Red China now stood as America's principal antagonist in the Asia-Pacific region. Viewing the region through rose-tinted glasses, Washington saw communism everywhere on the march. So in American eyes, a doomed campaign by France to retain its colonies in Indochina became part of a much larger crusade against communism on behalf of freedom. When France pulled the plug in Vietnam, in 1954, the United States effectively stepped into its role. An effort extending across several administrations to erect in Southeast Asia a bulwark of anticommunism aligned with the United States exacted a terrible toll on all parties involved and produced only one thing of value: machinations undertaken by President Richard Nixon to extricate the United States from a mess of its own making persuaded him to reclassify China not as an ideological antagonist but as a geopolitical collaborator.

As a consequence, the rationale for waging war in Vietnam in the first place—resisting the onslaught of the Red hordes—also faded. With it, so too did any further impetus for U.S. military action in the region. The War for Pacific Dominion quieted down appreciably, though it didn't quite end. With China now pouring its energies into internal development, Americans found plentiful opportunities to invest and indulge their insatiable appetite for consumption. True, a possible renewal of fighting in Korea remained a perpetual concern. But when your biggest worry is a small, impoverished nation-state that is unable even to feed itself, you're doing pretty well.

As far as the Pacific is concerned, Americans may end up viewing the last two decades of the twentieth century and the first decade of the twenty-first century as a sort of golden interlude. The end of that period may now be approaching. Uncertainty about China's intentions as a bona fide superpower is spooking other nearby nations, not least of all Japan. That another round of competition for the Pacific now looms qualifies at the very least as a real possibility.

For the United States, the War for the West began in 1917, when President Woodrow Wilson persuaded Congress to enter a stalemated European conflict that had been under way since 1914. The proximate cause of the U.S. decision to intervene was the resumption of German U-boat attacks on American shipping. To that point, U.S. policy had been one of formal neutrality, a posture that had not prevented the United States from providing Germany's enemies, principally Great Britain and France, with substantial assistance, both material and financial. The Germans had reason to be miffed.

For the war's European participants, the issue at hand was as stark as it was straightforward. Through force of arms, Germany was bidding for continental primacy; through force of arms, Great Britain, France, and Russia were intent on thwarting that bid. To the extent that ideals figured among the stated war aims, they served as mere window dressing. Calculations related to Machtpolitik overrode all other considerations.

Among the Bequests that Europeans Handed Off to the United States as they Wearied of Exercising Power, None Can Surpass the Greater Middle East in Its Problematic Consequences

President Wilson purported to believe that America's entry into the war, ensuring Germany's defeat, would vanquish war itself, with the world made safe for democracy—an argument that he advanced with greater passion and eloquence than logic. Here was the cause for which Americans sent their young men to fight in Europe: the New World was going to redeem the Old.

It didn't work out that way. The doughboys made it to the fight, but belatedly. Even with 116,000 dead, their contribution to the final outcome fell short of being decisive. When the Germans eventually quit, they appealed for a Wilsonian "peace without victory." The Allies had other ideas. Their conception of peace was to render Germany too weak to pose any further danger. Meanwhile, Great Britain and France wasted little time claiming the spoils, most notably by carving up the Ottoman Empire and thereby laying the groundwork for what would eventually become the War for the Greater Middle East.

When Wilson's grandiose expectations of a world transformed came to naught, Americans concluded—not without cause—that throwing in with the Allies had been a huge mistake. What observers today mischaracterize as "isolationism" was a conviction, firmly held by many Americans during the 1920s and 1930s, that the United States should never again repeat that mistake.

According to myth, that conviction itself produced an even more terrible conflagration, the European conflict of 1939–1945, which occurred (at least in part) because Americans had second thoughts about their participation in the war of 1914–1918 and thereby shirked their duty to intervene. Yet this is the equivalent of blaming a drunken brawl between rival street gangs on members of Alcoholics Anonymous meeting in a nearby church basement.

Although the second European war of the twentieth century differed from its predecessor in many ways, it remained at root a contest to decide the balance of power. Once again, Germany, now governed by nihilistic criminals, was making a bid for primacy. This time around, the Allies had a weaker hand, and during the war's opening stages they played it poorly. Fortunately, Adolf Hitler came to their rescue by committing two unforced errors. Even though Joseph Stalin was earnestly seeking to avoid a military confrontation with Germany, Hitler removed that option by invading the Soviet Union in June 1941. Franklin Roosevelt had by then come to view the elimination of the Nazi menace as a necessity, but only when Hitler obligingly declared war on the United States, days after Pearl Harbor, did the American public rally behind that proposition.

In terms of the war's actual conduct, only the United States was in a position to exercise any meaningful choice, whereas Great Britain and the Soviet Union responded to the dictates of circumstance. Exercising that choice, the Americans left the Red Army to bear the burden of fighting. In a decision that qualifies as shrewd or perfidious depending on your point of view, the United States waited until the German army was already on the ropes in the east before opening up a real second front.

The upshot was that the Americans (with Anglo-Canadian and French assistance) liberated the western half of Europe while conceding the eastern half to Soviet control. In effect, the prerogative of determining Europe's fate thereby passed into non-European hands. Although out of courtesy U.S. officials continued to indulge the pretense that London and Paris remained centers of global power, this was no longer actually the case. By 1945, the decisions that mattered were made in Washington and Moscow.

So rather than ending with Germany's second defeat, the War for the West simply entered a new phase. Within months, the Grand Alliance collapsed and the prospect of renewed hostilities loomed, with the United States and the Soviet Union each determined to exclude the other from Europe. During the decades-long armed standoff that ensued, both sides engaged in bluff and bluster, accumulated vast arsenals that included tens of thousands of nuclear weapons, and mounted impressive displays of military might, all for the professed purpose of preventing a "cold" war from turning "hot."

Germany remained a source of potential instability, because that divided country represented such a coveted (or feared) prize. Only after 1961 did a semblance of stability emerge, as the erection of the Berlin Wall reduced the urgency of the crisis by emphasizing that it was not going to end anytime soon. All parties concerned concluded that a Germany split in two was something they could live with.

By the 1960s, armed conflict (other than through gross miscalculation) appeared increasingly improbable. Each side devoted itself to consolidating its holdings while attempting to undermine the other side's hold on its allies, puppets, satellites, and fraternal partners. For national security elites, managing this competition held the promise of a bountiful source of permanent employment. When Mikhail Gorbachev decided, in the late 1980s, to call the whole thing off, President Ronald Reagan numbered among the few people in Washington willing to take the offer seriously. Still, in 1989 the Soviet-American rivalry ended. So, too, if less remarked on, did the larger struggle dating from 1914 within which the so-called Cold War had formed the final chapter.

In what seemed, misleadingly, to be the defining event of the age, the United States had prevailed. The West was now ours.

Among the bequests that Europeans handed off to the United States as they wearied of exercising power, none can surpass the Greater Middle East in its problematic consequences. After the European war of 1939–1945, the imperial overlords of the Islamic world, above all Great Britain, retreated. In a naive act of monumental folly, the United States filled the vacuum left by their departure.

For Americans, the War for the Greater Middle East kicked off in 1980, when President Jimmy Carter designated the Persian Gulf a vital U.S. national security interest. The Carter Doctrine, as the president's declaration came to be known, initiated the militarizing of America's Middle East policy, with next to no appreciation for what might follow.

During the successive "oil shocks" of the previous decade, Americans had made clear their unwillingness to tolerate any disruption to their oil-dependent lifestyle, and, in an immediate sense, the purpose of the War for the Greater Middle East was to prevent the recurrence of such disagreeable events. Yet in its actual implementation, the ensuing military project became much more than simply a war for oil.

In the decades, since Carter promulgated his eponymous doctrine, the list of countries in the Islamic world that U.S. forces have invaded, occupied, garrisoned, bombed, or raided, or where American soldiers have killed or been killed, has grown very long indeed. Since 1980, that list has included Iraq and Afghanistan, of course, but also Iran, Lebanon, Libya, Turkey, Kuwait, Saudi Arabia, Qatar, Bahrain, the United Arab Emirates, Jordan, Bosnia, Kosovo, Yemen, Sudan, Somalia, Pakistan, and Syria. Of late, several West African nations with very large or predominantly Muslim populations have come in for attention. At times, U.S. objectives in the region have been specific and concrete. At other times, they have been broad and preposterously gauzy. Overall, however, Washington has found reasons aplenty to keep the troops busy. They arrived variously promising to keep the peace, punish evildoers, liberate the oppressed, shield the innocent, feed the starving, avert genocide or ethnic cleansing, spread democracy, and advance the cause of women's rights. Rarely have the results met announced expectations.

In sharp contrast with the Hundred Years' War for the Hemisphere, U.S. military efforts in the Greater Middle East have not contributed to regional stability. If anything, the reverse is true. Hopes of achieving primacy comparable to what the United States gained by 1945 in its War for Pacific Dominion remain unfulfilled and appear increasingly unrealistic. As for "winning," in the sense that the United States ultimately prevailed in the War for the West, the absence of evident progress in the theaters that have received the most U.S. military attention gives little cause for optimism.

To be fair, U.S. troops have labored under handicaps. Among the most severe has been the absence of common agreement regarding the mission. Apart from the brief period of 2002–2006 when George W. Bush fancied that what ailed the Greater Middle East was the absence of liberal democracy (with his Freedom Agenda the needed antidote), policymakers have struggled to define the mission that American troops are expected to fulfill. The recurring inclination to define the core issue as "terrorism," with expectations that killing "terrorists" in sufficient numbers should put things right, exemplifies this difficulty. Reliance on such generic terms amounts to a de facto admission of ignorance.

When contemplating the world beyond their own borders, many Americans—especially those in the midst of campaigning for high office—reflexively adhere to a dichotomous teleology of good versus evil and us versus them. The very "otherness" of the Greater Middle East itself qualifies the region in the eyes of most Americans as historically and culturally alien. U.S. military policy there has been inconsistent, episodic, and almost entirely reactive, with Washington cobbling together a response to whatever happens to be the crisis of the moment. Expediency and opportunism have seldom translated into effectiveness.

Consider America's involvement in four successive Gulf Wars over the past 35 years. In Gulf War I, which began in 1980, when Iraq invaded Iran, and lasted until 1988, the United States provided both covert and overt support to Saddam Hussein, even while secretly supplying arms to Iran. In Gulf War II, which began in 1990, when Iraq invaded Kuwait, the United States turned on Saddam. Although the campaign to oust his forces from Kuwait ended in apparent victory, Washington decided to keep U.S. troops in the region to "contain" Iraq. Without attracting serious public attention, Gulf War II thereby continued through the 1990s. In Gulf War III, the events of 9/11 having rendered Saddam's continued survival intolerable, the United States in 2003 finished him off and set about creating a new political order more to Washington's liking. U.S. forces then spent years vainly trying to curb the anarchy created by the invasion and subsequent occupation of Iraq.

Unfortunately, the eventual withdrawal of U.S. troops at the end of 2011 marked little more than a brief pause. Within three years, Gulf War IV had commenced. To prop up a weak Iraqi state now besieged by a new enemy, one whose very existence was a direct result of previous U.S. intervention, the armed forces of the United States once more returned to the fight. Although the specifics varied, U.S. military actions since 1980 in Islamic countries as far afield as Afghanistan, Lebanon, Libya, and Somalia have produced similar results—at best they have been ambiguous, more commonly disastrous.

As for the current crop of presidential candidates vowing to "smash the would-be caliphate" (Hillary Clinton), "carpet bomb them into oblivion" (Ted Cruz), and "bomb the hell out of the oilfields" (Donald Trump), Americans would do well to view such promises with skepticism. If U.S. military power offers a solution to all that ails the Greater Middle East, then why hasn't the problem long since been solved?

Lessons drawn from this alternative narrative of twentieth-century U.S. military history have no small relevance to the present day. Among other things, the narrative demonstrates that the bugaboos of isolationism and appeasement are pure inventions.

If isolationism defined U.S. foreign policy during the 1920s and 1930s, someone forgot to let the American officer corps in on the secret. In 1924, for example, Brigadier General Douglas MacArthur was commanding U.S. troops in the Philippines. Lieutenant Colonel George C. Marshall was serving in China as the commander of the 15th Infantry. Major George S. Patton was preparing to set sail for Hawaii and begin a stint as a staff officer at Schofield Barracks. Dwight D. Eisenhower's assignment in the Pacific still lay in the future; in 1924, Major Eisenhower's duty station was Panama. The indifference of the American people may have allowed that army to stagnate intellectually and materially. But those who served had by no means turned their backs on the world.

Apart from Taking an Occasional Breather, the United States has Shown a Consistent Preference for Activism over Restraint, and for Projecting Power Abroad rather than Husbanding it for Self-defense

As for appeasement, hang that tag on Neville Chamberlain and Édouard Daladier, if you like. But as a description of U.S. military policy over the past century, it does not apply. Since 1898, apart from taking an occasional breather, the United States has shown a strong and consistent preference for activism over restraint and for projecting power abroad rather than husbanding it for self-defense. Only on rare occasions have American soldiers and sailors had reason to complain of being underemployed. So although the British may have acquired their empire "in a fit of absence of mind," as apologists once claimed, the same cannot be said of Americans in the twentieth century. Not only in the Western Hemisphere but also in the Pacific and Europe, the United States achieved preeminence because it sought preeminence.

In the Greater Middle East, the site of our most recent war, a similar quest for preeminence has now foundered, with the time for acknowledging the improbability of it ever succeeding now at hand. Such an admission just might enable Americans to see how much the global landscape has changed since the United States made its dramatic leap into the ranks of great powers more than a century ago, as well as to extract insights of greater relevance than hoary old warnings about isolationism and appeasement.

The first insight pertains to military hegemony, which turns out to be less than a panacea. In the Western Hemisphere, for example, the undoubted military supremacy enjoyed by the United States is today largely beside the point. The prospect of hostile outside powers intruding in the Americas, which U.S. policymakers once cited as a justification for armed intervention, has all but disappeared.

Yet when it comes to actually existing security concerns, conventional military power possesses limited utility. Whatever the merits of gunboat diplomacy as practiced by Teddy Roosevelt and Wilson or by Eisenhower and JFK, such methods won't stem the flow of drugs, weapons, dirty money, and desperate migrants passing back and forth across porous borders.

Even ordinary Americans have begun to notice that the existing paradigm for managing hemispheric relations isn't working—hence the popular appeal of Donald Trump's promise to "build a wall" that would remove a host of problems with a single stroke. However bizarre and impractical, Trump's proposal implicitly acknowledges that with the Hundred Years' War for the Hemisphere now a thing of the past, fresh thinking is in order. The management of hemispheric relations requires a new paradigm, in which security is defined chiefly in economic rather than in military terms and policing is assigned to the purview of police agencies rather than to conventional armed forces. In short, it requires the radical demilitarization of U.S. policy. In the Western Hemisphere, apart from protecting the United States itself from armed attack, the Pentagon needs to stand down.

The second insight is that before signing up to fight for something, we ought to make sure that something is worth fighting for. When the United States has disregarded this axiom, it has paid dearly. In this regard, the annexation of the Philippines, acquired in a fever of imperial enthusiasm at the very outset of the War for Pacific Dominion, was a blunder of the first order. When the fever broke, the United States found itself saddled with a distant overseas possession for which it had little use and which it could not properly defend. Americans may, if they wish, enshrine the ensuing saga of Bataan and Corregidor as glorious chapters in U.S. military history. But pointless sacrifice comes closer to the truth.

By committing itself to the survival of South Vietnam, the United States replicated the error of its Philippine commitment. The fate of the Vietnamese south of the 17th parallel did not constitute a vital interest of the United States. Yet once we entered the war, a reluctance to admit error convinced successive administrations that there was no choice but to press on. A debacle of epic proportions ensued.

Jingoists keen to insert the United States today into minor territorial disputes between China and its neighbors should take note. Leave it to the likes of John Bolton, a senior official during the George W. Bush Administration, to advocate "risky brinkmanship" as the way to put China in its place. Others will ask how much value the United States should assign to the question of what flag flies over tiny island chains such as the Paracels and Spratlys. The answer, measured in American blood, amounts to milliliters.

During the twentieth century, achieving even transitory dominion in the Pacific came at a very high price. In three big fights, the United States came away with one win, one draw, and one defeat. Seeing that one win as a template for the future would be a serious mistake. Few if any of the advantages that enabled the United States to defeat Japan seventy years ago will pertain to a potential confrontation with China today. So unless Washington is prepared to pay an even higher price to maintain Pacific dominion, it may be time to define U.S. objectives there in more modest terms.

A third insight encourages terminating obligations that have become redundant. Here the War for the West is particularly instructive. When that war abruptly ended in 1989, what had the United States won? As it turned out, less than met the eye. Although the war's conclusion found Europe "whole and free," as U.S. officials incessantly proclaimed, the epicenter of global politics had by then moved elsewhere. The prize for which the United States had paid so dearly had in the interim lost much of its value.

Americans drawn to the allure of European culture, food, and fashion have yet to figure this out. Hence the far greater attention given to the occasional terrorist attack in Paris than to comparably deadly and more frequent incidents in places such as Nigeria or Egypt or Pakistan. Yet events in those countries are likely to have as much bearing, if not more, on the fate of the planet than anything occurring in the tenth or eleventh arrondissement.

Furthermore, "whole and free" has not translated into "reliable and effective." Visions of a United States of Europe partnering with the United States of America to advance common interests and common values have proved illusory. The European Union actually resembles a loose confederation, with little of the cohesion that the word "union" implies. Especially in matters related to security, the EU combines ineptitude with irresolution, a point made abundantly clear during the Balkan crises of the 1990s and reiterated since.

Granted, Americans rightly prefer a pacified Europe to a totalitarian one. Yet rather than an asset, Europe today has become a net liability, with NATO having evolved into a mechanism for indulging European dependency. The Western alliance that was forged to deal with the old Soviet threat has survived and indeed expanded ever eastward, having increased from 16 members in 1990 to 28 today. As the alliance enlarges, however, it sheds capability. Allowing their own armies to waste away, Europeans count on the United States to pick up the slack. In effect, NATO provides European nations an excuse to dodge their most fundamental responsibility: self-defense.

Nearly a century after Americans hailed the kaiser's abdication, more than seventy years after they celebrated Hitler's suicide, and almost 30 years after they cheered the fall of the Berlin Wall, a thoroughly pacified Europe cannot muster the wherewithal to deal even with modest threats such as post-Soviet Russia. For the United States to indulge this European inclination to outsource its own security might make sense if Europe itself still mattered as much as it did when the War for the West began. But it does not. Indeed, having on three occasions over the course of eight decades helped prevent Europe from being dominated by a single hostile power, the United States has more than fulfilled its obligation to defend Western civilization. Europe's problems need no longer be America's.

Finally, there is this old lesson, evident in each of the four wars that make up our alternative narrative but acutely present in the ongoing War for the Greater Middle East. That is the danger of allowing moral self-delusion to compromise political judgment. Americans have a notable penchant for seeing U.S. troops as agents of all that is good and holy pitted against the forces of evil. On rare occasions, and even then only loosely, the depiction has fit. Far more frequently, this inclination has obscured both the moral implications of American actions and the political complexities underlying the conflict to which the United States has made itself a party.

Indulging the notion that we live in a black-and-white world inevitably produces military policies that are both misguided and morally dubious. In the Greater Middle East, the notion has done just that, exacting costs that continue to mount daily as the United States embroils itself more deeply in problems to which our military power cannot provide an antidote. Perseverance is not the answer; it's the definition of insanity. Thinking otherwise would be a first step toward restoring sanity. Reconfiguring the past so as to better decipher its meaning offers a first step toward doing just that.

Critical Thinking

1. Why does the author believe that U.S. militarism is of limited utility for its national interest?
2. Has the U.S. engaged in isolationism and appeasement?
3. Is it in the core interest of the U.S. to challenge China's claim to sovereignty over the disputed islands in the South China and East China Seas?

Internet References

National Security Strategy 2015
Nssarchive.us/national-security-strategy-2015/

Quadrennial Defense Review 2014
archive.defense.gov/pubs/2014_Quadriennial_Defense_Review.pdf

Andrew J. Bacevich is the author of America's War for the Greater Middle East: A Military History, just out from Random House.

Article

Prepared by: Robert Weiner, *University of Massachusetts, Boston*

Our Incoherent China Policy

The proposed Trans-Pacific Partnership Is Bad Economics, and Even Worse as Containment of China

CLYDE PRESTOWITZ

Learning Outcomes

After reading this article, you will be able to:

- Explain the relationship between geopolitics and trade.
- Discuss the connection between globalization and trade.

In the summer of 2009, I was invited with a few other policy analysts to the White House for a briefing on the newly proposed Trans-Pacific Partnership (TPP). At that time, the potential participants included Canada, Mexico, Peru, Chile, New Zealand, Australia, Singapore, Brunei, Malaysia, Vietnam, and, of course, the United States. Whether or not Japan would be invited to join had not yet been decided.

Noting that the United States already had free trade agreements with Canada, Mexico, Peru, Chile, Australia, and Singapore, I asked why we needed an agreement that added only the tiny economies of New Zealand, Brunei, Malaysia, and Vietnam. The reply from a member of the National Security Council staff was that it would reassure our Asian allies that America was back; that this agreement would be the economic complement to the increased military deployments of the recently announced "Pivot to Asia" foreign policy, obviously aimed at counterbalancing the spread of Chinese power and influence. Along with health care and a possible treaty on nuclear weapons with Iran, TPP would be a major part of the president's hoped for legacy.

My first reaction was surprise. How could America come *back* to Asia? As far as I could tell, it had never left. The U.S. Seventh Fleet was in its 66th year of patrolling the western Pacific and keeping the seas safe for the mushrooming trade that was making the region rich. The United States still maintained almost 100,000 troops in Japan, South Korea, and Australia, and on the seas to maintain stability. Trade was burgeoning. The enormous U.S. trade deficit with Asia continued to grow as Americans bought everything Asian, and U.S. corporations transferred much of their production and employment, along with most of their technology, to Asia. Thus a policy aimed at correcting an absence seemed to be based on a false assumption.

Of course, this would not be the first time that false assumptions had guided U.S. policy (Vietnam War, Iraq War, War on Drugs, etc.). But it seemed to be a U.S. habit when it came to proposing and negotiating international trade agreements. That was due largely to the fact that after World War II, the U.S. foreign policy elite tightly embraced the classical free trade catechism. As Britain's nineteenth-century free-trade crusader Richard Cobden had put it, "Free trade is God's diplomacy."

Trade was taken to be mutually beneficial for the countries involved, leading to prosperity, democratization, and ultimately to peace among nations. It was assumed that even unilateral free trade, in which Country A opens its markets while Country B does not, is still a win–win proposition, because the more open country would get cheaper imports and the closed nation was just harming itself. The possibility of strategic use of trade by nations, an insight for which Paul Krugman won the Nobel, was not part of the story.

Thus it was natural to conclude that concessions on trade, such as unilateral market opening, could yield geopolitical objectives with no negative economic consequences. For example, the United States might want Japan to stop manipulating its currency, but it also might want Japan to vote with it on something in the UN. Washington has invariably yielded on the economic issue in order to prevail on the geopolitical issue. Geopolitics trumped economics. The result has been a long series of international trade agreements that tended to disadvantage the United States.

Two recent examples are the deal under which the United States agreed to bring China into the World Trade Organization (WTO) in 2001, and the U.S.–Korea Free Trade Agreement of 2012. In the case of China, U.S. strategic thinking was heavily influenced by the notion that by adopting capitalism (or what the Chinese called socialism with Chinese characteristics), China would more and more become like America. As former Deputy Secretary of State Robert Zoellick noted, we wanted to encourage China to become a "responsible stakeholder in the global system."

It was widely assumed that globalization would make China and other developing countries rich; that by being rich they would become democratic; and that by being democratic, they would be at peace because democracies tended not to fight each other (or so we told ourselves). Thus, admitting China to the WTO was seen primarily as a way of encouraging the nation's democratization, but there was also thought to be an economic bonus. Because China's tariffs were much higher than America's, it was thought by most economists (who ignored the fact that Japan's market had remained closed despite its low tariffs) that U.S. exports would gain proportionately more than China's as a result of trade liberalization through Chinese admission to the WTO. Most of the econometric models projected that America's 2001 trade deficit of $83 billion with China would shrink dramatically as a result of Chinese tariff reductions providing better access for U.S. goods and services to the Chinese market.

In fact, however, the deficit doubled in three years and by now has redoubled. At the same time, China appears to have become less, rather than more, democratic. The same pattern can be found in the case of the U.S.–Korea Free Trade Agreement of 2012. That deal was mainly aimed at strengthening the U.S.–Korean alliance and has been used as a kind of template for the TPP negotiations. In both cases, a trade deal was done for primarily geopolitical reasons and produced neither the desired political nor economic results—largely because both the geopolitical and economic assumptions were mostly wrong.

It's Implausible that the TPP Would Limit or Retard the Expansion of Chinese Influence and Power in Any Way.

The TPP falls very much into this tradition. The proposed agreement has been widely debunked in this magazine and elsewhere as dubious economics, a deal crafted by and for corporate elites [see below]. However, the administration has sidestepped the economic criticisms by insisting that the TPP is essential geopolitics—a necessary counterweight to the rise of China. But if anything, the TPP is even less plausible as a China-containment strategy.

As President Barack Obama has worked to sell the agreement, he has declared that Congress must ratify the TPP, lest China write the trade rules of the future. Yet China already has concluded free trade agreements with Singapore, Australia, Chile, New Zealand, Pakistan, Peru, Costa Rica, Iceland, Switzerland, and Korea. It is currently negotiating free trade agreements with the Gulf Cooperation Council, Norway, and Sri Lanka, and is discussing a trilateral deal with Korea and Japan. Beijing is also leading negotiation of the Regional Comprehensive Economic Partnership (RCEP), which includes the 10 member states of ASEAN (Singapore, Philippines, Laos, Cambodia, Indonesia, Malaysia, Thailand, Myanmar, Brunei, and Vietnam), plus Australia, New Zealand, China, India, Japan, and South Korea.

Most of these are also part of the TTP. Countries are understandably playing China and America against each other and hedging their bets on both sides. Consider Singapore. Even as it works to conclude the TPP and warns America of the danger of withdrawal from Asia, it is negotiating assiduously to complete the China-led RCEP.

So it's implausible that TPP would limit or retard the expansion of Chinese influence and power in any way. Conversely, the notion that America will find itself excluded from Asia if it does not adopt the TPP is ridiculous.

Singapore's Foreign Minister K. Shanmugam, for instance, recently warned that failure to adopt the TPP would result in the United States being "crowded out" from the region. Surely, Shanmugam was not suggesting that Singapore would rescind its defense arrangements with the U.S. Navy, or that Japan would expel the U.S. Seventh Fleet, or that Apple would be forced to withdraw the billions of dollars it has stashed in Singapore if Washington does not ratify the TPP. Nor will adoption of the TPP stop any of the Chinese-led deals from going forward.

Thus, to some extent, China will inevitably be writing some of the future rules of world trade. It won't dictate all the rules, but it will surely have a major role in their writing. But that is only the beginning of China's increasing global influence, and the TTP does nothing to counter the other elements. China is on a massive campaign to barter investment in third-world infrastructure projects for access to the raw materials that its economy needs. This offensive naturally yields diplomatic influence.

In late July, *The New York Times* reported on vast infrastructure and oil-exploration projects in Ecuador, based on loans from Chinese government institutions and state-owned enterprises that are to be repaid from claims on Ecuadoran oil and natural resources. According to the *Times*, China takes about 90 percent of Ecuadoran oil and requires that contracts for construction and even for labor be funneled to Chinese firms—something the United States cannot do under international rules because it is classed as a developed country, while China retains the developing country label. Similar deals are ongoing throughout Africa.

The money for this Chinese geoeconomic offensive comes from China's nearly $4 trillion of foreign exchange reserves—and these come primarily from the enormous trade surpluses that China has racked up with the United States over the past 25 years. This deficit was not foreordained. Consider just one emblematic example—the transfer of General Electric's avionics (aircraft electronics) business to China.

The Economics of TPP

No comprehensive economic analysis preceded the launching of the negotiations to establish the TPP. No economic judgments were made as to which countries were most suited to join a new agreement that seemed to be aimed at something closer to an economic union than a mere free trade agreement. The U.S. government has made no estimates of the gains or losses to which the deal might give rise.

Although there are no official estimates of the likely economic impact of the TPP, a number of private analysts, such as the center-right Peter G. Peterson Institute for International Economics and the center-left Center for Economic and Policy Research (CEPR), have done estimates. They all tend to conclude that the cumulative U.S. gains by 2025 would be on the order of 0.1 percent to 3 percent of GDP. This is, of course, in the range of a rounding error. Even if the forecast gains are achieved, they will accrue mainly to the top end of the income distribution, with the average working person actually taking a loss.

The main reason for this is that the TPP includes many provisions that may be good for footloose global companies but not necessarily for real people. Provisions for strong intellectual-property protection could result in delayed introduction of generic drugs and in higher drug prices worldwide. Evergreening of patents through use of small technology upgrades could lock in large corporate profits for many years while preventing real innovation by small, new companies.

The TPP does make a stab at stronger labor and environmental standards, but these have long proven notoriously difficult to enforce. On the other hand, provisions for the easing of regulations on foreign investment and Internet traffic, protection of corporations from losses resulting from changes in government policies and continued approval of inducement subsidies for investment could well result in further offshoring of American investment, jobs, and technology.

Perhaps, the major reason why the TPP will not produce even the minor gains for the United States that are forecast by the models is the impact of exchange rates. These models do not assume that nations manipulate their exchange rates, but several TPP members do just that. Indeed, as I write, China has just devalued its currency by about 3 percent amid slowing growth and falling stock markets. When currencies can move 3 percent in just a day, or 2 percent over a period of several months, negotiation of tariffs, subsidy rules, and market-opening measures loses much of its meaning.

Because the TPP will not change this situation, it almost automatically cannot be good for the U.S. economy or the average American. Indeed, it is important to underline that no one in the Asia–Pacific or Latin American Regions believes the TPP will in any significant way change the structure of regional trade that has long resulted in huge American trade deficits and lost American jobs, as the United States has played the role of global consumer of last resort.

In view of this, it has been extremely disappointing to hear President Obama demonstrate complete misunderstanding of the realities of global trade and of the TPP. For instance, he has asked rhetorically in his speeches how anyone could be opposed to opening the Japanese auto market to enable Fords and Chevys to roll on Japanese roads. The very question showed his ignorance of the fact that nothing on the TPP agenda alters the Japanese manufacturers' control of Japanese dealerships, which is the main mechanism of closure of the Japanese auto market. For example, even though the Koreans are low-cost auto producers and have been taking market share from the Japanese in the European, Chinese, and U.S. markets, they have completely withdrawn from trying to sell in Japan. Since that is not being discussed, the TPP cannot put American or Korean or European or any other mass-market cars on Japanese roads.

Economically, the TPP is not going to be good for America because many of its fundamental assumptions are simply at odds with reality. —*C.P.*

On January 18, 2011, the same day Chinese President Hu Jintao arrived in Washington for a state visit, GE Chairman and CEO Jeff Immelt announced a new joint venture between GE's avionics business and Avic, a Chinese state-owned aviation products maker, which did not previously produce avionics of any significance. Under the announced deal, Immelt, who also served at the time as chairman of Obama's Council on Jobs and Competitiveness, said GE would transfer its leading-edge avionics technology to Avic and move the manufacture of its products from the United States to China. He emphasized that the arrangement had been cleared by the Departments of Defense and Commerce, presumably in keeping with a statement in a joint press conference with Hu that America would help China develop its own indigenous commercial jetliner.

The important points to keep in mind here are that avionics rely on advanced technology that China did not yet possess. Production of avionics is not labor-intensive, and China was therefore not a low-cost location for manufacturing. Indeed, the low-cost producing country was the United States. So, in effect, GE was moving its avionics production to a higher-cost location. All the laws of economics and free trade seemed to dictate that China should import its avionics from America. Yet GE was actually setting up to transfer production and jobs to China and eventually to export avionics from China to America. And this was being done by the guy who chaired the president's Council on Jobs and Competitiveness.

What was going on here? Well, China has a big, rapidly growing market for aircraft and it has made no bones about pursuing a policy of developing its own, homegrown aircraft and avionics capabilities. This is strategic trade par excellence. China was doing here what it had done in most other industries—making sales to China contingent on transferring technology and manufacturing to China. Immelt may be a major-league CEO in Washington, but here he was in Beijing kowtowing, just like all the other peons.

Most incredible of all was the fact that he was being aided in his kowtow by the U.S. government. The president kept him on as chair of the Council on Jobs and Competitiveness, even as he transferred jobs to China. Amazing as it may seem, this did not cause a ripple in U.S. political or media circles. There were no calls for a congressional investigation and no *New York Times* or *Wall Street Journal* editorials pointing out the contradictions and insisting on the conduct of truly reciprocal free trade with China.

TPP Provisions May Be Good for Footloose Global Companies but not Necessarily for Real People input.

These didn't appear because it all seemed so ordinary. It wasn't just the policy of Obama. A whole succession of presidents starting with Ronald Reagan had accepted and even promoted a nonreciprocal and hugely unbalanced trade relationship with China, urging and even indirectly subsidizing U.S. companies to invest in China and to move their production and jobs there.

While America's leaders seemed to have no concern for the structure of their economy and what it produced, Beijing steadfastly pursued a comprehensive industrial-development policy that aimed to make China self-sufficient in most major industries and technologies. Its Internet whizzes systematically hacked the databases of the U.S. government and of the major American industrial and technology companies. It was no accident that the Chinese stealth fighter unveiled during Defense Secretary Robert Gates's January 2011 visit looked very much like America's F-22 Raptor. In conjunction with this thrust for industrial and technological leadership, China also sought to achieve control of key minerals and natural resource reserves around the globe. This single-minded drive had now brought China to the status of the world's second-largest economy—one that was inevitably a major trade and investment partner with most of the rest of the world.

China's Manipulation of Its currency, in violation of the norm that markets should set currency values, is very much part of this strategy. Washington criticizes the practice, yet refuses to use serious leverage to counter it.

I was present at the Asian-Pacific Economic Cooperation (APEC) heads of state meeting in Honolulu in November 2011, when Obama warned in his speech to a group of APEC-country CEOs that China had to stop following its policy of intervening in the international currency markets to keep the Yuan undervalued versus the U.S. dollar. Of course, China had been following this policy for many of the preceding 20 years. Nor was the policy unique to China. Japan and the Asian Tigers (South Korea, Taiwan, Singapore) had invented it and were still following it. By keeping their exports inexpensive while causing their imports to be overpriced, this approach had given them large structural trade surpluses with the United States and big dollar reserves.

Along with the strategic trade policies and practices noted above, this was the major source of the $3.5 trillion treasure chest backing the loans, and of the investments China was using to gain broad global influence and control of critical resources. Of key importance in this regard was and is the fact that the $3.5 trillion is not just the net sum of China's dollar reserves. It is the strategic investment fund of the Chinese government and of the Chinese Communist Party. By contrast, in countries such as Japan, Germany, and Korea that also have large dollar reserves, the investment of these funds is directed at maximum financial returns. But in China, the treasure chest often stands behind government-directed lending and investment aimed at achieving the strategic objectives of central authorities. China is the major investor in some of the world's riskiest markets. According to *The New York Times*, it accounts for 57 percent of all foreign investment in Ecuador, 70 percent in Sierra Leone, 82 percent in Zimbabwe, 79 percent in Afghanistan, and 38 percent in Iraq. Not surprisingly, these are places with a lot of oil and key minerals.

In short, China has a comprehensive national strategy with clear objectives, a coordinated team aimed at achieving the objectives, and the money to pay for whatever it takes to achieve success.

Obama's Plea that he Must have the TPP in order to contain China is embarrassingly superficial as well as disingenuous.

If Obama were serious, he would challenge China's policy of making market access contingent on technology and manufacturing transfers. He would have blocked Jeff Immelt's kowtow by directing the departments of Commerce and Defense to withhold any authorization for transfer of militarily useful technology to China. He would have removed Immelt from the chairmanship of the Jobs and Competitiveness Council. Imposing a China-like "you have to make it here to sell it here" rule on Chinese sales in the U.S. market would have done far more to contain China and to change the structure of trade in the whole Asia-Pacific region than any TPP.

Or, alternatively, inviting China to join the TPP and thus be subject to its much-touted (by the Obama administration) "high standard" rules would at least have put China on the same playing field as the United States. But making a deal with Japan and four minor economies in the Asia–Pacific region will not affect China or its trading relationships at all.

Let's face it: The TPP partners are not significant enough or geographically well-placed enough for there to be any hope of the deal having an impact on China. While it is being labeled Trans-Pacific, it would be more accurate to call the agreement Trans-American. Of the 750 million people who would be covered by the arrangement, about 500 million are in Canada, the United States, Mexico, Peru, and Chile. Of the remaining 250 million, about 125 million are in Japan, which was not an original TPP country and only came into the deal later for its own domestic political reasons, and for the geopolitical purpose of keeping the U.S. happy in the context of the U.S. unilateral commitment to defend Japan. So the number of people in the Asia-Pacific region truly committed to the purposes of the TPP is about 125 million, of whom about 100 million are in Vietnam. This is just not a consortium of players that is going to have any impact on China.

The one thing that might have been done in the TPP negotiations that, while not affecting China too much directly, would have had the potential to change the whole structure of trade and production in the region would have been to ban currency manipulation among the members and to call upon the International Monetary Fund to enforce its rules against such manipulation globally. Based on Obama's statement at the Honolulu APEC meetings, one might have thought that his administration would strongly support such an approach. But it didn't.

Each year Congress requires the secretary of the Treasury to indicate which countries are engaging in currency manipulation. Although it was common knowledge that China was often manipulating, the Obama Treasury steadfastly refused to report that to Congress. Moreover, when setting the agenda for the TPP, the Obama administration again ducked even putting currency manipulation on the agenda.

It's a shame, really. Obama has done some very important things—the Affordable Care Act, economic recovery, withdrawal from Afghanistan and Iraq, progress on civil rights and the environment. But on trade, globalization, competitiveness, and dealing with China, Obama has continued the perverse policies of his predecessors. To propose the TPP as a remedy, something that will affect the U.S. economy negatively while having no effect on China, is worse than embarrassing. It's delusional.

What Should a Serious China policy be? It depends on our perceptions of America's national interests. One serious approach would be to back off. That sounds surprising, but think about it for a moment. Does China represent a direct threat to the United States in any way? Is China really going to invade the United States? Consider that the U.S. Navy actively patrols the coast of China and sends surveillance aircraft along its borders every day. How would we Americans react if Chinese aircraft carrier task forces were actively patrolling our Pacific coast around Los Angeles and San Francisco?

Take the Senkaku Islands, to which Japan and China both make claims but which the United States has promised to defend on behalf of Japan. Do we Americans really want to go to war with China over islands that have absolutely no significance to us? In the 19th and early 20th centuries, when the United States was a rising power, Great Britain faced a choice. Should it try to maintain supremacy in the Caribbean Sea and western Atlantic Ocean, or should it back off and let the Americans maintain stability in the region? The Brits decided to yield and what might have been a dangerous rivalry was avoided. Maybe we should forget the Pivot to Asia and the whole notion of containing China. The Chinese economy actually doesn't look too healthy these days, and its problems are only likely to become more difficult as its population rapidly ages and its environmental degradation becomes worse. Maybe there's really nothing to contain. Maybe we should pivot to America and direct the resources now devoted to containment to rebuilding America.

If Obama Were Serious, He Would Challenge China's Policy of Conditioning Market Access on Technology and Manufacturing Transfer.

Or, perhaps the analogy to Great Britain and the rising America is false. After all, both countries were democracies under a rule of law and shared a tradition of freedom of thought and speech, something which is not at all the case between China and the United States. So, perhaps there is a real rationale for containing China. But if that is the case, then America needs a real, comprehensive strategy—not a toothless TPP.

A serious American approach would begin with a demand that the IMF enforce its rules on currency manipulation. It would limit or condition U.S. investment in China and Chinese direct investment in the United States while taxing Chinese investment in U.S. financial instruments. It would use U.S. influence in the IMF to prevent any possibility of the Chinese Yuan being made a reserve currency. There is much more that could be included here, but you get the idea.

My main point is that if you're a dove, the TPP does nothing for you because it will simply increase the trade deficit while worsening the circumstances of the vast bulk of Americans. If you're a hawk, the TPP does nothing for you because it's just not a serious tool for containing China. Thus, either way, Congress should just say no to the TPP.

Critical Thinking

1. What are the benefits of free trade?
2. Is the TransPacific Partnership flawed?
3. Describe China's Comprehensive trade strategy?

Internet References

Asian Infrastructure Investment Bank
www.aiib.org/

TransPacific Partnership
https://ustr.gov/tpp/

World Trade Organization
https://www.wto.org/

Article

Prepared by: Robert Weiner, *University of Massachusetts, Boston*

The Three Ways We Get China and Its Neighbors Wrong

Asia Is as It Is, Not as We Wish It To Be

DAN BLUMENTHAL

Learning Outcomes

After reading this article, you will be able to:

- Understand the U.S. grand strategy as applied to Asia.
- Discuss the relationship between European regional order and Asian regional order.

Since the end of World War II, U.S. strategy in Asia has rested upon two pillars. The first has been the need to maintain "preponderant" power, to guard against the rise of a great-power challenger who could dominate the eastern part of the Eurasian landmass and its Asian rimlands. This strategy historically required a continued military presence in locations where the United States had fought during World War II and the early Cold War: Japan, South Korea, and the Philippines. The U.S. military also had to retain the ability to surge its forces into the region at any time and any place they might be needed.

The second element of U.S. strategy in Asia called for building a regional *political* order that would advance American interests and values. To accomplish this, first and foremost the United States helped rebuild Japan and forged an alliance with it. It also aligned itself with other noncommunist forces in North and Southeast Asia to support their development as decolonized independent countries and their growth as liberal democracies.

This two-pronged strategy was successful—both in heading off the Soviet Communist threat to the region and in providing space for Western-aligned Asian countries to develop. By the time of the Soviet Union's collapse in the late 1980s and early 1990s, experts were predicting that the 21st century would be an "Asian century." Lending force to this prediction was the fact that Japan's economy was then booming, and that the so-called Asian Tigers (South Korea, Taiwan, and Singapore) were likewise experiencing high levels of economic growth.

But then came the economic "explosion" of the People's Republic of China (PRC)—according to the *Economist*, the most rapid accumulation of wealth in human history, lifting millions out of poverty and bringing China into the family of nations. At the same time, India, too, appeared poised to unburden its economy of socialist shackles, and the rest of Southeast Asia seemed no less eager to follow northeast Asia's path. But the big story was, and remains, China.

Starting with President's Nixon's "opening" to China in the 1970s, the main thrust of U.S. strategy toward the Middle Kingdom had been focused on integrating it into the liberal world order: the panoply of organizations and institutions that historically enabled the development of peaceful nation-states. But by the early 1990s, it was becoming clear that Asia's continued peace was not preordained.

With Beijing translating its wealth into greater military and political power and making a frank bid for dominance of the region, the strategy of integration, however necessary it remained, was no longer sufficient. In the post–Cold War world, the U.S. needed an answer to China's rise as the potential great-power challenger in Asia whose emergence Washington had long sought to prevent.

The answer—less a departure from than an amendment to the strategy of integration—was to take hedging actions against a more aggressive China. Under Bill Clinton and both Presidents Bush, Washington kept American troops forward-deployed in Japan and South Korea and shored up its extant alliances not only with Japan but also with South Korea, Australia, and the Philippines. In addition, it strove to strengthen non-allied security partnerships with Taiwan and Singapore while searching for new partners in Indonesia, Malaysia, India, and Vietnam. The idea was to continue seeking the benefits of the hoped-for "Asian century" while actively protecting our forward-based security position.

Enter the Obama administration, which from early on brought to Asia policy a new assumption. Obama believed that the United States had been too distracted by the Middle East, and lately by the war on terror, to pay sufficient attention to Asia policy. Indeed, over involvement in the Middle East, so this thinking goes, had been a strategic mistake. While his predecessors' approach to China wasn't altogether wrongheaded, it had been insufficient on both the hedging and the integration sides.

Obama made the case that in a time of allegedly limited resources, the U.S. needed to "rebalance its portfolio," shifting capabilities away from the Middle East and Europe and toward Asia. From this perspective, the 2015 nuclear deal with Iran, coupled with Obama's decision not to intervene in Syria, was not to be seen as (in the former instance) just an agreement on nonproliferation or (in the latter) an expression of caution—but as a way out of the Middle East morass. If that exit could be achieved, it was posited, the U.S. would be in a stronger position to hedge against China's aggressive behavior.

This new attitude implicitly unwound the strategy of American preponderance *across* the critical regions of Eurasia, and also the relative weight given to the hedging and integration aspects of that strategy. When it came to order-building in particular, Obama brought a new verve and vigor to his predecessors' efforts. The administration joined such Asian groupings as the East Asia Summit, signed the Treaty of Amity and Cooperation with the Association of Southeast Asian Nations (ASEAN), and became a more frequent attendee at ASEAN+ meetings. The guiding assumption appeared to be that Asia already had, or could soon have, the kind of liberal order that defines present-day Europe.

And so the next administration will be handed an Asia strategy based on three main assumptions. First, that China will continue to grow more powerful, and the United States must continue to hedge against its aggression. Second, that there already exists a "liberal order" in Asia that simply needs to be strengthened. Third, that the U.S. can achieve success on both fronts even as it retrenches from Europe and the Middle East—regions long thought to be critical elements of a strategy of Eurasian preponderance.

Each of these assumptions is deeply flawed.

Will China Grow More Powerful?

China, it is now agreed, has entered a period of prolonged economic slowdown. Its own reported 2015 numbers showed that the economy grew only 6.9 percent, down from the breakneck double-digit rates of the first decade of the 21st century. But the true number is surely far lower. Economist Derek Scissors argues: "If Xi [Jinping] does not quickly move beyond talk to profound pro-market reform, China will not slow or struggle—it will just stop."

China's economic slowdown has had many causes, including an abysmal demographic situation, high levels of debt, and an inefficient and corrupt financial system. Global demand is shrinking, which means China's export-driven growth model is approaching its end. And the regime's response to the 2008 global financial crisis—namely, government stimulus and the accumulation of massive debt—will continue to cause more problems down the road.

A strategic reassessment is in order, and the first question to be asked is what China's slowdown means for the regime's internal stability and external behavior.

Tough choices clearly lie ahead for President Xi Jinping. Internally, he must find new ways to build legitimacy for a Chinese Communist Party that faced little organized resistance as long as most Chinese living standards were improving. Theoretically, of course, China could reverse direction and implement substantive market reforms. But politically those reforms do not seem to be in the offing—it is simply too risky to let capital leave Chinese banks and flow freely, as was envisioned by the 2013 Communist Party Plenum. So far, the approach of Xi Jinping has instead been a high-profile "anti-corruption" campaign that has helped to further centralize power and featured a crackdown on the media, lawyers, intellectuals, and churches.

Washington has come to expect that continued high levels of economic growth in China would sustain its double-digit increases in annual defense expenditures and impressive military-modernization program, which has produced the world's most active missile-, ship-, and aircraft-building programs. In addition, Beijing deployed a toolkit of economic inducements to purchase the support of countries it has deemed strategically valuable—from the eastern coast of Africa, where it wants naval bases, to the Middle East, where it needs oil, to Asian countries that it wants to keep from allying fully with the United States.

Under worsening economic conditions, can China be expected to sustain its national security policy abroad, and for how long? So far, there have been no signs of abatement in Beijing's policy of militarizing the South China Sea and challenging Japan in the East China Sea.

Nor has the PRC displayed any inclination to deviate from its historical willingness to engage in adventurism during times of trouble, earlier instances of which reach back to Deng's post–Cultural Revolution attack on Vietnam and Mao's pre-Great Leap Forward bombardment of Taiwan's offshore islands.

Meanwhile, won't a slowing economy cause internal trouble that will require Beijing to make further investments to police the large Chinese empire, spanning from Tibet and Xinjiang to Hong Kong?

Washington has not even begun to think about an alternative strategy for this slowing China. There is no consensus about such basic questions as whether the lethargic growth and political decay in China might actually benefit the American interest in keeping Asia free of a hostile hegemonic power.

In that light, a wise strategy would alter the current policy of hoping for greater Chinese economic growth. While it is detrimental to global economic growth, China's slowdown provides an opportunity to better align our economic interest with our security interest. Under such a strategy, for example, the United States could work harder to push its open-markets agenda throughout the rest of South and Southeast Asia, with particular attention granted to countries that we want to befriend. Especially in the case of a China that might be prone to lashing out, the U.S. would benefit from economically successful Asian partners able to invest in a more serious, region-wide hedging policy.

Has the 'Liberal Order' in Asia Been Strengthened?

The purpose of current U.S. diplomacy in Asia has been increased engagement with Asia's multilateral organizations—hence, the Obama administration's signing of the ASEAN Treaty of Amity and Cooperation in 2009, joining the East Asia Summit in 2011, sending high-level officials to the ASEAN Regional Forum, and successfully negotiating the Trans-Pacific Partnership that has become a major sticking point in the U.S. presidential election. But do these institutions really amount to a functioning and coherent Asian political/security order?

The answer is no. When we speak of a functioning regional liberal order, we have one model: post–World War II Europe (granting all of its manifold problems). European politics and economics have long been "ordered" around a security system based on a preponderance of American power tied into a collective alliance, NATO. American primacy within the alliance system provided the time and space for Europe to develop a political-economic system both regional and international in scope.

This is a "liberal" order in the classical sense that it encourages free trade, free markets, and a respect for liberal interpretations of international law and custom. Its institutions include the International Monetary Fund, the World Bank, and the World Trade Organization at the global level, and the EU, OSCE, and NATO at a regional level. And this order was not conjured up out of nothing in 1945; ultimately, it was based on what Walter Russell Mead has described as centuries of an Anglo-American order.

This order was rooted in the pre–World War I British belief that "free trade would promote peace between nations based on common interests, and increasing prosperity," according to Mead. "People-to-people contact, facilitated by international human rights and religious organizations, would remove the misunderstandings that led to war and create bonds of friendship as well."

Before the breakdown that led to the two world wars, Europeans had gone through a "Westphalian" progression in which empires, kingdoms, and religious movements became nation-states with agreed-upon sets of customs, rules, and practices, often referred to as "classical diplomacy." These rules included sovereign equality among nations, respect for territorial boundaries, promises to stay out of one another's domestic politics, and attempts to resolve disputes peacefully while circumscribing the use of force. Most of Europe's constituent countries also shared common political cultures and a common civilizational legacy.

Without those shared civilizational and cultural legacies, statesmen on both sides of the Atlantic could not possibly have been as optimistic as they were about the prospect of rebuilding a liberal system in Europe. American and European leaders spoke in urgent and almost spiritual terms about the need for a Western order. As Undersecretary of State Robert Lovett stated, the "cement" of the NATO treaty "was not the Soviet threat, but the common Western approach and that Western attachment to the worth of the individual."

The postwar European order was animated by a positive vision for what type of order the constituent states wanted to create and a strong sense of the principles that the system would defend. Over time the European nations felt safe enough to abandon much of the Westphalian system and create "post-modern" supranational institutions that weakened the importance of national sovereignty.

This same "European world order" has now become the reference point for discussions of an "Asian order." But that Asian order simply does not exist.

Asia does *not* enjoy a history of a functioning system of independent nation-states. In postwar Asia, the United States arranged its strategy around Japan, the only Asian country that had modernized and industrialized and participated in the international order before the war. The only other candidate for a great-power partner was China—and indeed, before and during the war, Western statesmen saw China as a potentially key part of the new international order. But then it fell to Mao in 1949 and joined the other side.

As for Asia's other countries, none has ever really made the transition into a "Westphalian" nation-state. After the war, Indonesia, India, Burma, and Vietnam struggled for independence from European and American colonization. Malaysia and Singapore were created as nation-states when the British pulled out of Asia. Many Southeast Asian countries did not gain full independence until the 1960s and 1970s.

In Northeast Asia today, Japan, South Korea, and Taiwan do operate within a semblance of a liberal order thanks to the U.S. alliance system and Japan's prewar processes of modernization. Yet, these countries do not have habits of mutual cooperation.

This is why all attempts to arrange Asia into a collective alliance (as in the Southeast Asia Treaty Organization) have fallen apart. Potential member countries simply do not share a common perception of threat, and efforts to build political and regional organizations with teeth are still limited. To be sure, Asian nations have joined such institutions as the WTO, the IMF and World Bank, and UN organizations. They have thereby benefitted from access to international markets and capital, as well as from the free use of the commons.

From time to time, Asian states have also searched for ways to make common cause regionally, as when Chinese and Soviet subversion was spreading and America was beginning to lose the Vietnam War. In 1967, the Southeast Asian nations created ASEAN as a bulwark against Communism; today, its member countries are fitfully learning to play balance-of-power politics in order to maintain their sovereign independence, increase their strategic autonomy, and block new forms of hegemony.

Still, ASEAN is no EU or NATO—and should not have been expected to be so. When it was founded, countries were still fighting for their independence and to define their territorial limits. Until 1966, Indonesia, the largest Southeast Asian nation, was still trying to undermine the formation of Malaysia as an independent country. Mutual suspicion between Indonesia and Vietnam blocked the resolution of the third Indochina war until 1991. Whenever key national-security interests were at stake, ASEAN members went outside the confines of ASEAN to resolve their problems. With little security cohesion among the Asian countries, the Asian security system still today depends on U.S. primacy and in particular on the American commitment to an open maritime order.

Given this brief history, is postwar Europe really a model for Asia and regional integration? The next administration should ask itself that question, and also what can realistically and fruitfully be expected of emergent Asian countries facing the stresses and convulsions of a declining China. The first requirement of a liberal order is constituent liberal nation-states.

Thus, the first order of business for the United States is to continue encouraging the development of strategically autonomous countries developing along liberal lines. Given their short histories as independent nations, the U.S. should be satisfied with a Southeast Asia that can engage in the kind of "classical diplomacy" that resolves territorial and maritime disputes on the basis of sovereign equality and international custom, and that can defend itself against the uncertainties attendant on China's rise and current turmoil. The region must undergo its own modern Westphalian progression before it is urged to jump into postmodern diplomacy.

Can Our Strategy in Asia Succeed While We Retrench Elsewhere?

The most important question to be asked of the new approach to Asia is not about its tactical mistakes but about its misguided strategic conception that the U.S. can "pivot" away from Eurasia's other critical regions.

The Cold War grand strategy of containment had its roots in the view that, to preserve its security and way of life, America could not abide the dominance of a hostile power in *any* critical region in Eurasia. That is still the case. Put in a more positive and activist light, America has an interest in making the liberal order open to all comers, particularly in Europe, the Middle East, and Asia. To do so, it still needs a global maritime strategy that can tip balances in its favor and keep trade functioning well.

Take U.S. commitments to Asian friends and allies. These depend upon security of energy supply and transit from the Middle East and therefore upon the dominant role of the United States in the Persian Gulf. It is doubtful that China or Russia will step into that role, and if they did, Asian allies would be right to be concerned.

On the issue of nonproliferation, Japan and South Korea in particular look to the U.S. global role in stemming the tide of nations anxious to acquire weapons of mass destruction. When Syria broke a longstanding taboo in 2013 by using chemical weapons, and got away with it, the event and its significance did not go unnoticed in Tokyo. America's friends in Asia wait in nervous apprehension as Iran, allegedly in nuclear purgatory, nevertheless continues its slow march toward nuclear weapons. They have seen this happen in North Korea in the face of very similar attempts by American diplomats to prevent it. A nuclear Iran will most certainly change the calculation of Asian countries like Japan and South Korea that have the technological capabilities to acquire their weapons.

Add to these negative developments in the Middle East the festering jihadist threat in Iraq and Syria. The terrorists who have fought in these conflicts are making their way back to Muslim-majority countries in Asia such as Indonesia. And what of Russia, busy annexing Crimea, menacing Europe—and reestablishing its influence in the Middle East? Asian friends pay close attention to territories that are forcefully changed, to the rise of another revisionist power like Iran, and to the potential for unfavorable alignments between Moscow and Beijing.

The Obama administration has had some real accomplishments in Asia, from defense agreements with the Philippines and India, to the successful negotiation of the Trans-Pacific Partnership. But these advances will be fleeting if the next administration gets U.S. global strategy wrong. The good news in Asia is, unlike in 1945, there are now many countries that share Washington's liberal values. Even so, order-building in Southeast Asia should be mindful of the relatively short history these countries have as independent, strategically autonomous, Westphalian countries. They will be jealously protective of their sovereignty, will still work out their territorial boundaries, and will remain skeptical of grand institutionalist schemes. Washington can start to help build a solid foundation for 21st-century Asian states by understanding them better and learning to work with them as they are rather than as we would wish them to be—and, crucially, by helping them to hedge more effectively against a China whose rough road ahead will likely result in more aggressive behavior against them, both economically and militarily.

Critical Thinking

1. Why is China going through an economic slowdown?
2. Is the U.S. pivot to Asia a mistake?

Internet References

Asian Infrastructure Investment Bank
www.aib.org/

One Belt, One Road
https://en.wikipedia.org/wiki/One_Belt_One_Road

The Regional Comprehensive Economic Partnership
https://en.wikipedia.org/wiki/Regional_Comprehensive_Economic-Partnership

Dan Blumenthal is the director of Asian studies at the American Enterprise Institute.

Article

Prepared by: Robert Weiner, *University of Massachusetts, Boston*

The Global Challenge of the Refugee Exodus

GALLYA LAHAV

The horrific terrorist attacks in Paris on November 13, 2015, and those on American soil on September 11, 2001, have much more in common than the involvement of radicalized foreigners and international networks. Both of these events came immediately on the heels—within three days, in fact—of major international agreements to facilitate human mobility between sending and receiving countries. Just as then-President Vicente Fox of Mexico secured a deal for his citizens in the North American labor market days before the attack on the World Trade Center, embattled ministers and heads of states from the European Union met with African leaders in Valetta, Malta, on November 11 to work out a practical redistribution plan for dealing with the mass exodus of refugees trying to reach Europe.

Both of these initiatives to forge international policy cooperation on migration and refugee movements were quickly dashed by seemingly knee–jerk reactions across Western liberal states to temporarily close borders, lock down civil society, suspend rights and privileges, and contravene their own treaties and laws. Labor, trade, and humanitarian considerations were quashed by national security and "public order" exigencies. Fueled by public outrage and political protests, the resurgence of nationalist and populist sentiment following the Paris attacks all but shelved urgent relocation plans for the massive influx of refugees from protracted wars in the Middle East.

Not only did these events cement the link between human mobility and security; they catapulted refugee and migration politics onto the foreign and security policy agenda, prompting a proliferation of intergovernmental and international meetings. The salience of migration in the security agenda was best summarized by German Chancellor Angela Merkel's proclamation that "immigration is the largest problem facing Europe in this decade." The "new security" paradigm has put a spotlight on emerging threats like human mobility, fundamentalism, environmental degradation, smuggling, and terrorism—global issues that cross boundaries.

The movement of impoverished masses making their way to safer shores from regions including the Middle East, sub-Saharan Africa, and Central America, has grown over the past five years, and rose to unprecedented levels in 2014, according to the United Nations High Commissioner for Refugees (UNHCR). The numbers arriving in Europe reached crisis levels in the sweltering political and summer heat of 2015. Personified by the piercing image of a young Syrian boy's corpse retrieved from the surf by a Turkish soldier, the human toll was inescapable.

Yet the humanitarian narrative was quickly overshadowed by the spectacle of Paris, a symbol of liberty, under assault by Islamist radicalism and terrorism inspired from abroad. This sequence of tragic events showed how swiftly politicization of such issues could upset a fragile balance and move the debate from a humanitarian to a security framework. The sudden shift in the discourse surrounding refugee movements underscored the fluid interests and competing trade-offs of refugee politics in the post–Cold War era.

The terrorist suspects who surfaced in Paris embodied the range of threats facing liberal democracies as they deal with refugees and migration. Among the suspected perpetrators, Western officials identified a disguised or bogus asylum seeker of Syrian origin; a Belgian-born and Western-educated middle-class Muslim of Tunisian-Moroccan origin; and a number of other radicalized EU nationals from ethnic minority backgrounds in France and Belgium. These profiles encapsulated the multiple internal and external dimensions of the threats that inform policies on border control, minority integration, and identity politics in the 21st century.

The refugee crises of this decade amount to an emerging global challenge facing almost all industrialized liberal democracies, pitting their humanitarian norms against materialist values of survival and well-being. These crises have acutely tested the delicate immigration and asylum policy consensus that largely prevailed across Western countries throughout the Cold War period. Until the political earthquake of 9/11, this equilibrium was founded on the premise that each dimension of immigration policy could be addressed in relative isolation, and that decisions concerning one dimension did not significantly circumscribe the options for others. Since 2001, subsequent terrorist attacks have suggested to some that open economic borders, humanitarian passage, and immigrant integration policies now conflict with the core responsibility of liberal states to safeguard the physical safety of their citizens.

The new security context of the post–Cold War world poses what I have elsewhere called a political "trilemma" when it comes to balancing markets, rights, and security interests in dealing with human mobility. Liberal democracies have struggled with contradictory goals of maintaining open markets for trade and allowing freedoms for ethnically diverse populations while protecting their borders from the security threats associated with global mobility. How can liberal states in an international system reconcile the need to open borders—for the sake of human mobility, demographic balance, sustainable development, global markets, tourism, and human rights norms—with political, societal, and security pressures to effectively protect their citizens and control their borders?

Cooperation among states on human mobility has been largely based on restrictive policies.

Europe's Challenge

This difficulty has been most evident at the EU regional level, where democratic member states are forced to balance national impulses favoring protectionism with communitarian demands for more cooperation. On what basis might states with different historical exigencies and approaches to migration find their interests merging? Collective action among 28 diverse member states has proved intractable (in instructive ways).

The ongoing European refugee emergency is emblematic of the challenges generated by forced migration. It constitutes an enormous crisis for the vision of European integration. Moreover, assuming that EU integration is representative of the larger globalizing goal of free movement for all four economic factors of production (goods, capital, services, people), its struggles are revealing of the challenges faced by all advanced liberal democracies.

To the extent that achieving the aim of a single market, enshrined in the 1957 Treaty of Rome, rests on the success of freedom of movement, a common immigration and asylum policy is essential to founder Jean Monnet's concept of a frontier-free Europe. The functional rationale for this goal was that the pursuit of economic and social well-being, within a framework of human rights and democratic norms, would create a rational incentive for states to cooperate and further integrate with each other.

We have seen an incremental development of European instruments on the supranational level, such as the Common European Asylum System, the Schengen open borders system, the Dublin system for handling asylum claims, and the cross-border enforcement agency Frontex. This trend has been reinforced by a notable shift since the 2009 Lisbon Treaty toward deferring to the EU on policy decisions. Despite these encouraging signs of cooperation, the current refugee crisis has reopened serious rifts between member states over national borders. As demonstrated by the disputes between French and British authorities over migrants converging on Calais to seek passage to Britain through the Channel Tunnel, and by the abrupt border closures by some eastern and southern European countries, member states have diverged dramatically in fulfilling their obligations.

The disparate perspectives of member states toward humanitarian movements are reflected in the uneven reception of refugees and burden-sharing proclivities among the EU countries. Attitudes have varied widely, from generous Sweden and Germany, Europe's main economic powerhouses in northern Europe, to the economically embattled countries in the south and east (Hungary, Croatia, Greece, Austria) that have been at the geographic forefront of the crisis. Some countries, like Slovakia and Poland, have taken to specifying the types of people in need that they are willing to help—namely, Christian refugees. Other EU nations such as Greece and Italy have been struggling to avoid saddling their citizens, already exasperated by economic hardship, with the potentially catastrophic burden of massive refugee influxes.

The challenge is faced not only by individual member states (especially transit countries in the south and east) with weak migration infrastructures and beaten-down economies. The core challenge confronts the entire European Union, founded on the principles of free mobility and solidarity. The lack of a comprehensive, common asylum policy (including refugee quotas, reception centers, and a common list of safe third countries) that also recognizes member states' capacities and the public mood is a danger to the entire EU enterprise. At the moment, Brussels risks reneging on previous steps it has taken regarding human mobility (including the Dublin Regulations and Schengen Agreements). It also risks losing the support of national publics and even some member states, including key ones such as Britain, which are threatening to exit the union altogether because of migration concerns.

Policy responses that include detention, deportation, or refoulement represent a slow erosion of liberal norms.

Imperiled Principles

The current refugee crisis casts doubts over three major principles inscribed in the conscience of Europeans: free movement of persons, human rights protections, and social harmonization or solidarity. First, a Schengen breakdown, symbolized by the temporary shutdown of national borders (currently allowed for up to 90 days), compromises the principle of a free human mobility zone within the single market. Second, the buck-passing by safe "first-arrival" countries that are sending refugees on to their neighbors represents a serious breach of the Dublin Convention, and thereby jeopardizes the principle of human rights protections. Finally, the nature of a "peoples' Europe," as set out in the Maastricht Treaty of 1992, stressing the universality of human rights throughout the Union, is in flux. Even if the first two challenges are surmountable, the looming question is: What kind of Europe does the EU want to be? What is its identity, and where do Muslims fit within the rapidly redrawn lines between insiders and outsiders?

Lurking behind much of modern European identity building has been the ghost of Christendom; and as in the United States, relations with Muslim minorities have become strained. Amid rising anxiety over security, concern about the cultural impact of migrants and refugees has extended beyond perceived threats to language and customs. Fears of radical, anti-Western political culture in Muslim communities are prevalent. These tensions present a serious crisis for democracies. The religious cleavage of secularism or pluralism versus fundamentalism complicates refugee politics and lends some unwelcome credence to the late political scientist Samuel Huntington's contentious and gloomy prophesy about rivalry between "civilizations" becoming the fault line of future conflicts in world politics.

European integration, like globalization, compounds the challenges that refugee and migration issues have long posed to the exercise of sovereignty by states seeking to control territory, identity, and citizenship. The reinvention of borders has compelled Europeans to rethink fundamental questions of identity—of "us" and "them." The growing tendency toward restrictive and protectionist migration policies across Europe stems less from demographic changes than from the reactions of policymakers and ordinary citizens to migration in the context of changing borders.

Indeed, the rush to control migration seems initially puzzling in a Europe built on the principle of free movement, dependent on global mobility, committed to maintaining a robust welfare system, and facing a serious demographic crisis of aging populations and falling birthrates. It also runs counter to rising public expectations in the EU, especially among the young generations. Eurobarometer polls of European youths between 15 and 24 years old from 2005 to 2015 found that "free movement" ranked higher in importance than any other motivations for regional integration, including the euro, social protection, and peace. The hardening of migration and refugee controls despite the liberalization of borders for other global economic reasons is one of the contradictions of incomplete integration.

Asylum Redefined

Refugees have a sacred and separate space in migration politics. And yet, as was noted by a pair of eminent scholars, the late Aristide Zolberg and Astri Suhrke, their forced movements may be defined in at least three ways: legally (as stipulated in national or international law), politically (as interpreted to meet political exigencies), and sociologically (as reflected in empirical reality). Legally speaking, the modern right to asylum has its roots in the aftermath of World War I and the Russian Revolution. Forced to flee by the Bolsheviks and famine, an unprecedented wave of 1.5 million Russians who had been stripped of their citizenship were resettled by the League of Nations.

The humanitarian system broke down under Nazi aggression and its aftermath, until the United Nations took up the task of rebuilding it at the international level with the establishment of the UNHCR in 1950. Since then, the main pillar of refugee protections has been firmly institutionalized in the Geneva Convention of 1951 and its 1967 protocol. The narrow definition of refugees (broadened only slightly in 1967 to extend beyond the original European refugees of World War II) has remained the standard and template for all other international and regional instruments dealing with forced migrations.

Despite the tenacity of these legal standards, political definitions have shifted dramatically from the Cold War ideological competition to the post–Cold War geopolitical preoccupation with religious and ethnic conflict. Nation-states interpret their legal and humanitarian obligations in the context of shifting political and foreign policy concerns. Cuban and Soviet refugees, for example, are no longer guaranteed asylum in the West.

The extent to which accepted asylum applications to Europe show overrepresentation from countries such as Eritrea, Afghanistan, and Iraq relative to other countries of origin, such as Serbia, Kosovo, Pakistan, and Albania, is striking. The numbers reflect neither legal nor sociological considerations but political affinities. So, too, changes in the types of refugees (which now include, for example, those facing female genital mutilation, environmental calamities, and gang violence) mean that the numbers of people being pushed out involuntarily have increased greatly, belying limited legal definitions and institutional capacities. Clearly, the contemporary refugee crisis stems from the growing incongruence between narrow and anachronistic legal definitions and evolving political and sociological realities.

While the relative size of these flows is not unprecedented, their compositional breakdown is revealing. In 1945, 20 million European refugees were resettled; today, there reportedly are 19.5 million refugees in the world. In contrast to those earlier, mainly European flows, most of today's asylum applicants are fleeing violence and conflict outside Europe. In 2014, the world's largest source of refugees was the Middle East and North Africa. According to UNHCR statistics, one in every five displaced people worldwide came from Syria. More tellingly, the vast majority of refugees in 2014 were from countries in the developing world, such as Ethiopia, Kenya, and Pakistan: nearly 9 out of 10 refugees lived in such countries, compared with 7 out of 10 a decade earlier.

This period has seen the breakdown of countries in the former Soviet Bloc such as Ukraine or Kyrgyzstan, the collapse of states such as Afghanistan, Iraq, Libya, and Yemen, and the reconfiguration of others, such as Sudan, Eritrea, and Somalia. The range of potential candidates for refugee status has been vastly expanded by ethnic and religious conflicts in the Middle East and the wider region (particularly in Syria, Iraq, Afghanistan, and Libya) and in South and Southeast Asia (in Pakistan, Myanmar, and Bangladesh, locus of the Rohingya refugee crisis); war, poverty, and repression in Africa (for instance in Eritrea, Somalia, Nigeria, Ivory Coast, Mali, Burundi, Central African Republic, and the Democratic Republic of Congo); and the fraying of national boundaries elsewhere (prompting migrations of Roma, Kurds, and other ethnic groups).

Absorbing the Flow

The overwhelming displacement of Syrians and Libyans since 2011 has pushed refugees next door, to Jordan, Lebanon, Iraq, Turkey, Egypt, and Tunisia. The absorption of refugees in those neighboring countries, while keeping them as far from Western borders as possible, has also led to further destabilization of already fragile states. However, as these countries have become oversaturated, they have closed their borders, forcing the West to deal with the inevitable diversion of the refugee flow to other regions.

Although in 2015 only 10 percent of Syrians moved to Europe out of the roughly 4 million who have left their homeland, the political challenge now faced by Europe—which has agreed to resettle and distribute 160,000 refugees, according to the last EU relocation plan—is rather minimal by comparative standards. However, while past refugees were settled "temporarily" in close proximity to their country of origin, the current reality is based on long-term projections of permanent absorption. As UNHCR statistics suggest, only 126,800 refugees were resettled in their home countries in 2014—the lowest number in 31 years. This means cost–benefit assessments extend well beyond migration admissions and quotas and must include a consideration of permanent settlement.

Whereas vulnerable people tried to get to the Balkans from Germany during World War II, or fled from Serbian ethnic cleansing in the 1990s to countries such as Hungary, the route today is reversed, as refugees from the Middle East and Africa try to get to Serbia on their way to Austria and Germany. Asylum-seekers headed for Europe often start in Greece, which they can reach via a short boat trip from Turkey. Then, they move on through Macedonia and Serbia and into Hungary, where thousands have been crossing the border every day, crawling over or under a razor-wire fence meant to keep them out. Most go from there to other countries in the EU, sometimes paying smugglers to drive them. The danger of drowning has led migrants to increasingly seek land routes to Europe, especially through the Western Balkans.

The unprecedented number of deaths among people trying to reach Europe, which exceeded 2,500 in 2015 alone, reveals that all routes, by land or sea, have been closing. The reintroduction of archaic border fences by countries such as Spain (in its North African territories of Ceuta and Melilla), Bulgaria (on its border with Turkey), or Hungary (on its border with Serbia) has been designed to keep unwanted flows out, and as far away as possible. As a result, refugees seek alternative, dangerous routes through the Arctic Circle via Russia to the Nordic countries, or through harsh deserts, the Gulf of Aden, or the Red Sea to other unlikely countries that are culturally distant, such as Israel, Ethiopia, and Iran; or to others like Jordan, Malaysia, and Pakistan, which are not signatories of the Geneva Convention. And of course, despite the lukewarm reception in some countries, many still come to Europe. If they are lucky, some make it even farther, to Australia, the United States, and Canada.

Gray Areas

As the empirical and political redefinition of refugees outpaces legal definitions, scholars and policymakers alike are forced to reconsider the old distinctions between voluntary (mostly economically driven) and involuntary (humanitarian) migration. It is increasingly apparent that the refugee crisis is also a migration crisis. The elusiveness of policy categories not only deflects institutional responsibility, it neglects the gray areas which include unaccompanied minors and victims of natural catastrophe, trafficking, female genital mutilation, and other forms of discrimination. An untold number of those people fall through the terminological cracks in definitions of protected status.

How long can legal definitions maintain the differences between those who flee persecution on the basis of race, nationality, religion, or belonging to certain political or social groups, and those who flee other life-threatening events such as food insecurity, gang wars (which have driven unaccompanied children from Central America), or economic displacement? The link between climate change and massive human mobility goes beyond boundaries, as do civil strife, sustainable development issues, and other "new security" threats. Population movements are driven by compounding factors. Among the initial sparks for the imploding ethnic and sectarian conflicts in Syria and in Sudan's Darfur region were severe droughts and other ecological shocks, which aggravated fierce economic competition for scarce resources.

The definition of refugees has expanded in scope and complexity, and so have the potential solutions. Yet legal formulations have not been keeping pace. According to some estimates, there are now approximately 60 million uprooted, forcibly displaced, or stateless persons around the globe (equal in population to some of the larger European countries), most of whom are precluded from seeking the protections of existing legal rights. The scale of this problem obliges states to address the changing notions of refugee status and to align them with empirical realities.

The task for the international community is to uphold and adjust legal standards to meet the times. This involves bridging the enormous gap between generalized threats such as gang warfare, climate displacement, and food insecurity (which are not covered by the Geneva Convention), on the one hand, and narrowly defined forms of persecution, on the other. It also requires attention to the failure to uphold legal principles ratified by the world community. Policy responses that include detention, deportation, or *refoulement* (the return of refugees to a country where they face persecution) represent a slow erosion of liberal norms set out by international and supranational instruments such as the UN Refugee Convention, the Schengen Agreement, the Dublin Convention, and the Convention Against Torture. They also prevent any meaningful policy fixes.

Long-Term Solutions

Most scholars and observers of the post–World War II period have concluded that liberal principles are embedded in the evolution of the contemporary Western world. The principle of free movement of all factors of production has dominated the prevailing discourse. Globalization and regional economic entities such as the European Union have ensured efficient flows across borders. Liberal markets presupposed Adam Smith's "invisible hand," assuming that the international system would neutralize inequalities and find equilibrium if the poor regions could send their impoverished risk-takers to faraway capital-rich ones. Liberal norms for international mobility were institutionalized in the Bretton Woods system, facilitating efficient flows of foreign or guest labor.

Globalization is both a boon and a hindrance to international migration.

In the same spirit of postwar thinking, human rights norms sought to ensure compassionate migration flows. While the Geneva regime instituted refugee protections of non-*refoulement* and nondiscrimination, international human rights instruments guaranteed basic protections to all individuals regardless of citizenship.

But the breakdown of the Cold War system has unleashed new dilemmas for the world, testing the liberal paradigm on which the current migration–asylum equilibrium has rested. The increasing inclination of national governments to view refugee and migration questions through the prism of national security has both compelled and repelled greater bilateral and multilateral cooperation. The security paradigm has disclosed a series of paradoxes and unintended consequences looming in the background of refugee politics.

Before celebrating international cooperation and further integration, we need to recall that unlike other areas of globalization (such as trade), cooperation among states on human mobility has been largely based on restrictive policies. Indeed, with specific exceptions, such as the US Bracero program for Mexican guest workers from the 1940s to the 1960s, cooperation on migration has predominantly existed in the form of prevention. This is also true of refugee policies. In the EU, these have been less about establishing a common European asylum system and more concerned with reducing migration pressures.

Long-term perspectives should factor in lessons learned; international cooperation may be more compatible with national interests than is often presumed. Contrary to conventional theories of globalization and regional integration, cooperation may bolster, not compromise, state sovereignty. International and transnational organizations can serve as an opportunity for increasing, rather than constraining, the regulatory power of nation-states. States may deal more effectively with migration challenges by joining international or supranational institutions like the EU.

The tendency to outsource refugees to other countries that are already crumbling in the Middle East or elsewhere is shortsighted. The presence of 4–6 million homeless and stateless people undermines the goal of helping to stabilize those compromised countries, to which rejected asylum seekers reluctantly return or where they are stranded on the edge of society. It is rather duplicitous to offer development aid and humanitarian assistance, as the European Neighborhood Policy has done, to strategic partners like Ukraine, Libya, Morocco, Tunisia, Egypt, and Turkey—countries hardly known for their civil rights records—to help them monitor migrants and asylum-seekers. Beyond the human toll, the prospective costs, in terms of the further regional destabilization that comes with growing numbers of stateless and displaced persons, are immeasurable.

Finally, the piecemeal attempts to tackle the refugee crisis have belied the externalities of migration policy. The growing interdependence of migration with other policy domains means that outcomes are contingent on developments in other areas, from foreign affairs to welfare policy. Long-term solutions require holistic and comprehensive approaches that include diplomatic and military engagement, social and cultural integration, labor and demographic considerations, development aid, and environmental protections. They also require extending burden sharing (beyond financial assistance) to include more affluent countries in the area such as Saudi Arabia, the United Arab Emirates, and Qatar, and outside the region, such as Japan, Singapore, Russia, and the United States.

An alternative to disengagement in unstable regions of the world is population movement outside of them. In today's world, power is no longer commensurate with military might. Non-state actors such as the Islamic State (ISIS), Al-Qaeda, and Boko Haram, as well as states with poor military infrastructure, can deploy what the political scientist Kelly Greenhill calls "weapons of mass migration." Global strategies therefore need to attend to the insidious psychological trauma among uprooted, suffering, and marginalized peoples. The antidote to jihadist ideology is to prevent extremism and alienation at home.

Short-term fixes to current crises undermine long-term solutions in an increasingly interdependent world. The double-edged sword of globalization is both a boon and a hindrance to international migration. The expanded regulatory apparatus for migration includes global high-tech surveillance, cross-border intelligence, and real-time databases and information systems that are equally available to sophisticated smuggling networks. Desperate refugees may also rely on smartphones and social networks, a striking feature of the current exodus.

The massive flows of Syrians that dominate today's headlines are fleeing ISIS and a brutally oppressive regime at the same time. Ultimately, amid lagging rates of minority integration by the multicultural societies in the West, they may be abandoned to the alienation and hopelessness that feed radicalization and help terrorist organizations recruit. A responsible and holistic approach to integration needs to address threats including growing populist parties, the radicalization of alienated youths, and domestic violence, along with rising economic disparities. When globalization's own weapons are turned against itself, they threaten to undermine its core liberal values.

Critical Thinking

1. Why are some members of the European Union preventing refugees from crossing their borders?
2. What steps should the European Union take to deal with the refugee crisis?

Internet References

Doctors Without Borders
www.doctorswithoutborders.org/

The International Organization for Migration
www.iom.int/

The International Rescue Committee
www.rescue.org

The United Nations High Commissioner for Refugees
www.unhcr.org/en-us

Gallya Lahav is an associate professor of political science at Stony Brook University, the State University of New York.

Article

Prepared by: Robert Weiner, *University of Massachusetts, Boston*

The US–Cuba Thaw and Hemispheric Relations

MICHAEL SHIFTER

Learning Outcomes

After reading this article, you will be able to:

- Discuss why the U.S. reopened diplomatic relations with Cuba.
- Explain how Latin American countries reacted to the restoration of diplomatic relations between the U.S. and Cuba.

For over five decades, no issue so sharply divided the United States from Latin America and the Caribbean as Washington's punitive economic and diplomatic isolation of Cuba. The unexpected announcement made on December 17, 2014, by US President Barack Obama and Cuban President Raúl Castro that they would move toward normalization of diplomatic relations brought the prolonged period of estrangement between Washington and Havana to an end. The policy of isolation, a Cold War relic that had utterly failed in its goal of producing regime change in Havana—and was widely perceived to be a prisoner of US domestic politics, especially the powerful Cuban American community—was gradually dismantled in 2015. The dramatic shift in approach and mindset was best exemplified in August 2015, when John Kerry—the first US secretary of state to visit Cuba since 1945—formally inaugurated the US embassy in Havana, following a corresponding event for the Cuban embassy in Washington.

To be sure, the US trade embargo, adopted in 1962 and reinforced by the 1996 Helms-Burton Act, remains in effect and can only be removed by Congress. For the time being at least, that policy is unlikely to be lifted by a Republican-controlled Congress that has opposed Obama's move toward greater opening. Both Marco Rubio and Ted Cruz, Cuban-American senators and candidates for the 2016 Republican presidential nomination, have roundly rejected the administration's approach, defending the embargo as necessary so long as Cuba remains undemocratic.

In Cuba too, and throughout the region, the embargo is viewed as the chief obstacle to a full normalization of relations (the return of Guantanamo Bay to Cuban sovereignty is another, albeit secondary, issue). Still, there are unmistakable signs that the US–Cuba thaw is irreversible, and that the removal of the embargo is just a matter of time. There is little question about the will and interest on the part of both governments to move toward deeper engagement on all fronts. Polls in both countries consistently reveal overwhelming support for the policy shift, with more than 7 in 10 Americans backing the resumption of diplomatic relations.

The move has been popular even within the Cuban-American community in Miami, which in the past stood in the way of any attempts at rapprochement. Some prominent, previously hardline Cuban Americans, such as Carlos Gutierrez, who served as commerce secretary under President George W. Bush, now openly embrace the Obama administration's approach.

Moreover, a growing coalition of business interests, led by the US Chamber of Commerce, is urging lawmakers to lift the embargo, which could bring important trade and investment opportunities and benefits for American companies. Obama has also called on Congress to remove the embargo, and has signed executive orders and issued regulations that have eased restrictions on travel, remittances, and trade.

Symbolic Stroke

While Obama administration officials and many analysts were surprised by the relatively muted reaction in the United States to the policy breakthrough on Cuba, they fully expected that the announcement would be cheered throughout Latin America and the Caribbean. Washington's anachronistic and punitive Cuba policy had long been one of the main irritants in US–Latin American relations. It is not that Cuba's political and economic model had much appeal in the region but rather that Washington's heavy-handed approach toward the weak and vulnerable Caribbean nation was regarded as conduct more befitting an old-fashioned imperial power than a genuine partner in inter-American affairs.

This concern reached a critical point at the 2012 Summit of the Americas in Cartagena, Colombia, when virtually all Latin American leaders made it clear to Obama that they would not participate in another such summit (a regular assembly of hemispheric heads of state that started in 1994) if Washington continued to insist on excluding Havana. That forced the issue, uniting Latin America in its opposition to the United States. In a historic turn of events, the 2015 meeting in Panama will be remembered for the presence of Raúl Castro and his animated conversation with Obama. Regional leaders unanimously praised the long-awaited US policy shift.

Cuba's presence at the Panama summit marked a moment of renewed goodwill and optimism in hemispheric relations. Washington regained some lost legitimacy and credibility in Latin America and enhanced its soft power. Moreover, few doubt that with this single stroke, Obama secured his legacy in Latin America. To find a historical precedent for a policy decision with such a substantial impact on US–Latin American relations, one would have to go back to Panama Canal treaty of 1977. While Obama's Latin American legacy will ultimately be judged on a variety of issues—including drug and immigration policy, the crisis in Venezuela, relations with Mexico and Brazil, and Central America's deepening security predicament—the breakthrough on Cuba will trump all others. This will be especially visible should Obama choose to visit the island before the end of his term. In a December 2015 interview, Obama said he would "very much" like to travel to Cuba, but also made it clear he wanted to "talk to everybody," including dissidents.

The rapprochement offers a crucial opportunity to pursue a more productive and balanced hemispheric dialogue.

Nonetheless, despite the centrality and symbolic significance of the Cuba issue, Washington's relations with Latin American should be analyzed within a broader frame. On two other matters of unfinished business—drug trafficking and immigration—there has been modest progress at best. To its credit, the Obama administration has shifted its stance on the former issue, jettisoning the "war on drugs" rhetoric and showing more flexibility, tolerance, and openness toward alternative approaches. Still, US antidrug policies continue to be a source of frustration in many countries.

The comprehensive immigration reform that many in Latin America expected, especially given the prominent role of Latinos in US politics, has not materialized. Obama has employed his executive powers to pursue some reforms, but his initiatives have been challenged in the courts. And there is considerable concern about the administration's deportation practices. Setbacks on immigration reform have been compounded by shrill and occasionally racist language on the issue in the primary stage of the 2016 presidential campaign. Until it is fixed, the broken US immigration system will remain a major obstacle to improved relations with Latin America, especially for the nearest neighbors in Mexico, Central America, and the Caribbean.

More fundamentally, there are underlying forces largely unaffected by the shift on Cuba policy that will result in the United States and Latin America continuing to go their separate ways. Latin American nations are more confident and assertive than ever, and have diversified their ties around the globe. China, the nations of the European Union, and many other countries, including India, Russia, South Korea, and Japan, have established a significant and growing presence in the region. Washington's bilateral relations are more varied and complex than ever, in some cases marked by cooperation, in others by conflict. The US profile in South America, in particular, has diminished in recent years. Although the importance of the so-called Bolivarian axis—a leftist bloc led by Venezuela—has waned, suspicions and resentments about US motives persist.

The US–Cuba rapprochement does, however, offer a fresh climate and a crucial opportunity for a more productive and balanced hemispheric dialogue. The United States remains a key economic partner for nearly all Latin American countries, and most acknowledge the importance of maintaining cordial relations with Washington. The difficult economic outlook and rapidly evolving political landscape—marked by the emergence of new leaders in some nations and the decline of some anti-US figures—can set the stage for further cooperation. Although inter-American relations are unlikely to be substantially recast in the near future, the removal of the failed policy of isolating Cuba, which irked even Washington's closest Latin American allies for decades, is a major step forward.

Critical Thinking

1. Why is the US–Cuban thaw irreversible?
2. Has the diplomatic opening secured President Obama's legacy in history?
3. What are the major objections to the reopening of diplomatic relations between the United States and Cuba?

Internet References

Cuban Permanent Mission to the United Nations
Cubadiplomatica.cu

Embassy of Cuba in the United States.
www.embassy.org/embassies/cu.html

US Embassy in Cuba
https://cu.usembassy.gov/

Michael Shifter is president of the Inter-American Dialogue, an adjunct professor at Georgetown University's School of Foreign Service, and a current history contributing editor.

Article Prepared by: Robert Weiner, *University of Massachusetts, Boston*

The Information Revolution and Power

JOSEPH S. NYE JR.

Learning Outcomes

After reading this article, you will be able to:

- Understand the relationship between the information revolution and soft power.
- Understand what two power shifts are occurring in the 21st century.

One of the notable trends of the past century that will likely continue to strongly influence global politics in this century is the current information revolution. And with this information revolution comes an increase in the role of soft power—the ability to obtain preferred outcomes by attraction and persuasion rather than coercion and payment.

Information revolutions are not new—one can think back to the dramatic effects of Gutenberg's printing press in the 16th century. But today's information revolution is changing the nature of power and increasing its diffusion. Sometimes called "the third industrial revolution," the current transformation is based on rapid technological advances in computers and communications that in turn have led to extraordinary declines in the costs of creating, processing, transmitting, and searching for information.

One could date the ongoing information revolution from Intel cofounder Gordon Moore's observation in the 1960s that the number of transistors fitting on an integrated circuit doubles approximately every 2 years. As a result of Moore's Law, computing power has grown enormously, and by the beginning of the 21st century doubling this power cost one-thousandth of what it did in the early 1970s.

Meanwhile, computer-networked communications have spread worldwide. In 1993, there were about 50 websites in the world; by 2000, the number had surpassed 5 million, and a decade later had exceeded 500 million. Today, about a third of the global population is online; by 2020 that share is projected to grow to 60 percent, or 5 billion people, many connected with multiple devices.

The key characteristic of this information revolution is not the *speed* of communications among the wealthy and the powerful; for a century and a half, instantaneous communication by telegraph has been possible between Europe and North America. The crucial change, rather, is the radical and ongoing reduction in the *cost* of transmitting information. If the price of an automobile had declined as rapidly as the price of computing power, one could buy a car today for $10 to 15.

When the price of a technology shrinks so rapidly, it becomes readily accessible and the barriers to entry are reduced. For all practical purposes, transmission costs have become negligible; hence the amount of information that can be transmitted worldwide is effectively infinite.

Winning Stories

In the middle of the 20th century, people feared that the computers and communications of the information revolution would create the central governmental control dramatized in George Orwell's dystopian novel *1984*. Instead, as computing power has decreased in cost and computers have shrunk to the size of smartphones and other portable devices, their decentralizing effects have outweighed their centralizing effects, as WikiLeaks and Edward Snowden have demonstrated.

Power over information is much more widely distributed today than even a few decades ago. Information can often provide a key power resource, and more people have access to more information than ever before. This has led to a diffusion of power away from governments to nonstate actors, ranging from large corporations to nonprofits to informal ad hoc groups.

This does not mean the end of the nation-state. Governments will remain the most powerful actors on the global stage. However, the stage will become more crowded, and many nonstate actors will compete effectively for influence. They will do so mostly in the realm of soft power.

The increasingly important cyber domain provides a good example. A powerful navy is important in controlling sea-lanes; it does not provide much help on the internet. The historian A.J.P. Taylor wrote that in 19th century Europe, the mark of a great power was the ability to prevail in war. Yet, as the American defense analyst John Arquilla has noted, in today's global information age, victory may sometimes depend not on whose army wins, but on whose story wins.

Sources of Power

I first coined the term "soft power" in my 1990 book *Bound to Lead,* which challenged the then-conventional view of the decline of US power. After looking at American military and economic power resources, I felt that something was still missing—the ability to affect others by attraction and persuasion rather than just coercion and payment. I thought of soft power as an analytic concept to fill a deficiency in the way analysts thought about power.

The term was eventually used by European leaders to describe some of their power resources, as well as by other governments, such as Japan and Australia. But I was surprised when President Hu Jintao told the Chinese Communist Party's 17th Party Congress in 2007 that his country needed to increase its soft power.

This is a smart strategy, because as China's hard military and economic power grows, it may frighten its neighbors into balancing coalitions. If China can accompany its rise with an increase in its soft power, it can weaken the incentives for these coalitions. Consequently, the Chinese government has invested billions of dollars in this task, and Chinese journals and papers are filled with hundreds of articles about soft power. But what, precisely, is it?

Power is the ability to affect others to obtain the outcomes you want. You can affect their behavior in three main ways: threats of coercion (sticks), inducements or payments (carrots), and attraction that makes others want what you want. A country may obtain the outcomes it desires in world politics because other countries want to follow it—admiring its values, emulating its example, and aspiring to its level of prosperity and openness.

More people have access to more information than ever before.

In this sense, it is important to set the agenda and attract others in world politics, and not only to force them to change through the threat or use of military or economic weapons. This soft power—getting others to want the outcomes that you want—co-opts countries rather than coerces them.

Soft power rests on the ability to shape the preferences of others. It is not the possession of any one country, nor only of countries. For example, companies invest heavily in their brands, and nongovernmental activists often attack their brands to press them to change their practices. In international politics, a nation's soft power rests primarily on three resources: its culture (in places where it is attractive to others), its political values (when it lives up to them at home and abroad), and its foreign policies (when they are seen as legitimate and having moral authority).

Propaganda Ploys

China is doing well in terms of culture, but is having difficulty with values and policies. The world's most populous country has always had an attractive traditional culture; now it has created hundreds of Confucius Institutes around the world to teach its language and culture. Beijing is also increasing its international radio and television broadcasting. Moreover, China's economic success has attracted others. This attraction was reinforced by China's successful response to the 2008 global financial crisis—maintaining growth while much of the West fell into recession—and by its economic aid and investment in poor countries. In the past decade, it became common to refer to these efforts as "China's charm offensive."

Yet, as the University of Denver's Jing Sun observed in the September 2013 issue of *Current History,* China has not reaped a good return on its investment. This is not because soft power is becoming less important in world politics. It is a result of limitations in China's strategy—a strategy that overly stresses culture while neglecting civil society and the damage done by nationalistic policies.

In 2009, Beijing announced plans to spend huge sums to develop global media giants to compete with Bloomberg, Time Warner, and Viacom, using soft power rather than military might to win friends abroad. As George Washington University's David Shambaugh has documented, China has invested billions in external publicity work, including a 24-hour Xinhua cable news channel.

China's soft power, however, still has a long way to go. A recent BBC poll shows that opinions of China's influence are positive in much of Africa and Latin America, but predominantly negative in the United States and everywhere in Europe, as well as in India, Japan, and South Korea. Similarly, a poll taken in Asia after the 2008 Beijing Olympics found that Beijing's charm offensive had not been effective.

China does not yet have global cultural industries on the scale of Hollywood, and its universities are not yet the equal of America's. But more important, it lacks the many nongovernmental organizations that generate much of America's soft power. Chinese officials seem to think that soft power is generated primarily by government policies and public diplomacy, whereas much of America's soft power is generated by its civil society rather than its government.

Great powers try to use culture and narrative to create soft power that promotes their advantage, but it is not an easy sell when it is inconsistent with their domestic realities. For example, while the 2008 Olympic Games were a great success, Beijing's crackdowns shortly thereafter in Tibet, in Xinjiang, and on human rights activists undercut its soft power gains. The Shanghai Expo in 2010 likewise was judged a success, but it was followed by the jailing of Nobel Peace laureate Liu Xiaobo and the artist Ai Weiwei. In the world of communications theory, this is called "stepping on your own message."

And for all the efforts to turn Xinhua and China Central Television into competitors of CNN and the BBC, there is not much of an international audience for brittle propaganda. As *The Economist* reported, "the party has not bought into Mr. Nye's view that soft power springs largely from individuals, the private sector, and civil society. So the government has taken to promoting ancient cultural icons whom it thinks might have global appeal."

Given a political system that relies on one-party control, it is difficult to tolerate dissent and diversity. Moreover, the Chinese Communist Party has based its legitimacy on high rates of economic growth and appeals to nationalism. The nationalism reduces the universal appeal of "the Chinese Dream" promoted by President Xi Jinping, and encourages policies in the South China Sea and elsewhere that antagonize its neighbors. For example, when Chinese ships drove Philippine fishing boats from the Scarborough Shoal in 2012, China gained control of the remote area, and from a domestic nationalist point of view, the action was a success. However, it came at the cost of reduced Chinese soft power in Manila.

Russian President Vladimir Putin has recently called for an effort to increase his country's soft power, but he might consider lessons from China the next time he locks up dissidents or bullies neighbors such as Georgia or Ukraine. A successful soft power strategy must attend to all three resources: culture, political values, and foreign policies that are seen as legitimate in the eyes of others. Investment in government propaganda is not a successful strategy for increasing a country's soft power.

Positive Sums

The development of soft power need not be a zero-sum game. All countries can gain from finding attraction in each other. Just as the national interests of China and the United States are partly congruent and partly conflicting, their soft powers are reinforcing each other in some issue areas and contradicting each other in others.

This is not something unique to soft power. In general, power relationships can be zero- or positive-sum depending on the objectives of the actors. For example, if two countries both desire stability, a balance of military power in which neither side fears attack by the other can be a positive-sum relationship. Likewise, if China and the United States both become more attractive in each other's eyes, the prospects of damaging conflicts will be reduced. If the rise of China's soft power reduces the likelihood of conflict, it can be part of a positive-sum relationship.

In the long term, there will always be elements of both competition and cooperation in the US-China relationship, but the two countries have more to gain from the cooperative element, and this can be strengthened by the rise in both countries' soft power. Prudent policies would aim to make that a trend in coming decades.

The 21st century is experiencing two great power shifts: a "horizontal" transition among countries from west to east, as Asia recovers its historic proportion of the world economy, and a "vertical" diffusion of power away from states to nongovernmental actors. This diffusion is fueled by the current information revolution, and it is creating an international politics that will involve many more actors than in the several centuries since the Treaty of Westphalia enshrined the norm of sovereignty.

But power diffusion also affects relations among states. It strengthens transnational actors and puts new transnational issues on the agenda, such as terrorism, global financial stability, cyber-conflict, pandemics, and climate change. No government can solve these problems acting on its own. In seeking to organize coalitions and networks to deal with such challenges, governments will need to exercise the powers not only of coercion and payment, but also of attraction and persuasion.

Critical Thinking

1. How is the information revolution contributing to the diffusion of power in the international system?
2. Why is information a key power source?
3. Are states still important in view of the information revolution?

Internet References

Department of Homeland Security
http://www.dhs.gov

International Corporation for Names and Numbers
http://www.icann.org

International Organization on Migration
http://www.iom/nt/cms/en/sites/iom/home

International Telecommunications Union
http://www.itu.int

National Security Agency
http://www.nsa.gov

World Wide Web Consortium
www.w3.org

Joseph S. Nye Jr., a Current History contributing editor, is a professor of political science at Harvard University and the author most recently of *Presidential Leadership and the Creation of the American Era* (Princeton University Press, 2013).

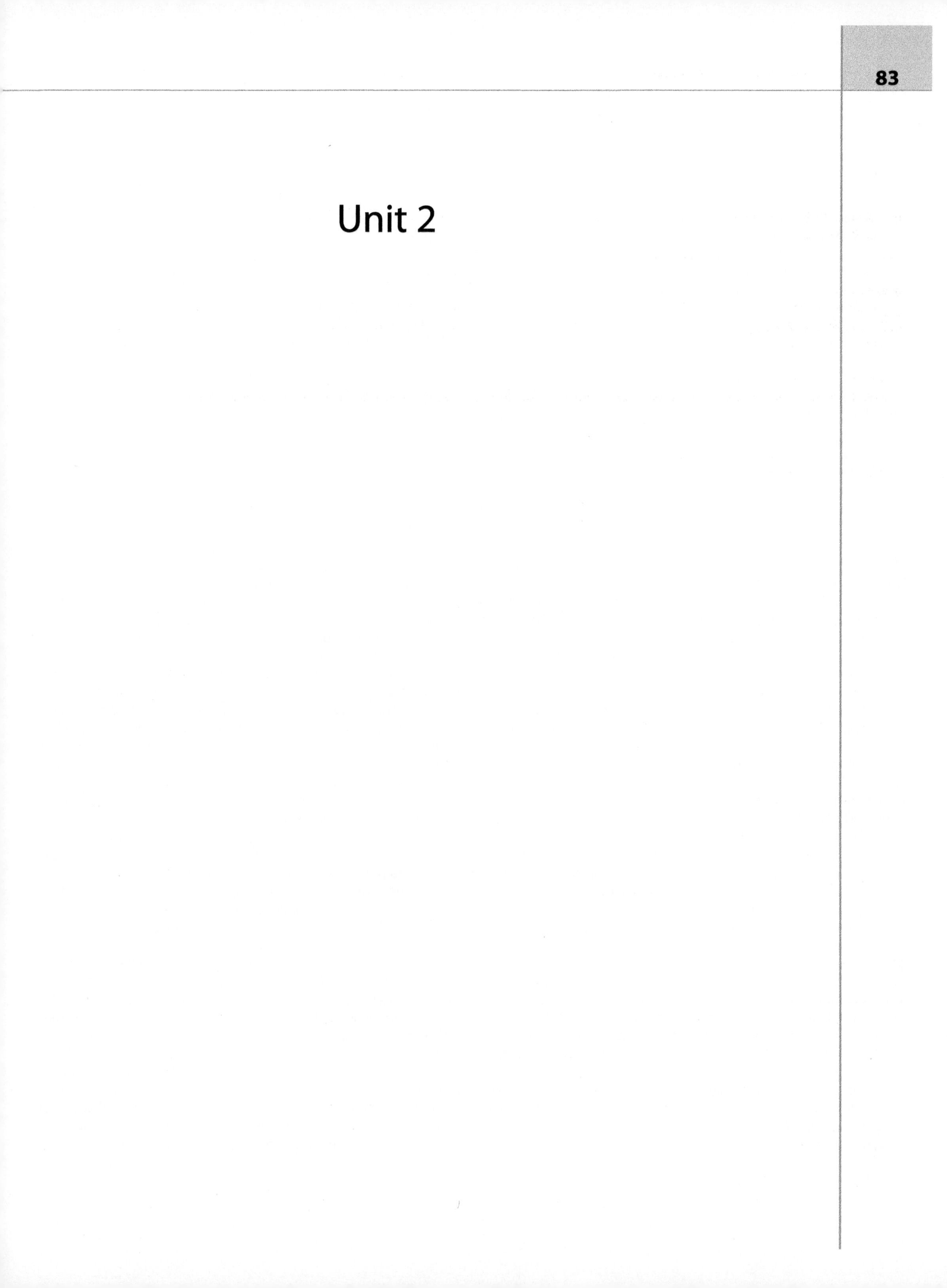

Unit 2

UNIT

Prepared by: Robert Weiner, *University of Massachusetts, Boston*

Population, Natural Resources, and Climate Change

After World War II, the global population reached an estimated two billion people. It had taken 250 years to triple the population to that level. In the six decades following World War II, the population tripled again to six billion. By 2050, or about 100 years after World War II, some analysts forecast that the population will go up to 10–12 billion. While demographers develop various scenarios forecasting population growth, it is important to remember that there are circumstances that could lead not to growth but to significant decline in global population. The spread of AIDS and other infectious diseases like the Ebola virus are cases in point. The lead article in this unit provides an overview of general demographic trends, with a special focus on issues related to aging. Making predictions about the future of the world's population is a complicated task, for there are a variety of forces at work and considerable variation from region to region. The dangers of oversimplification must be overcome if governments and international organizations are going to respond with meaningful policies. Perhaps one could say that this is not a global population challenge, but many population challenges that vary from country to country and region to region.

The increase in population has also put pressure on countries to gain access to vital resources such as oil and natural gas. A recent trend has been for multinational corporations to prospect for oil in more advanced economies such as New Zealand. Multinational corporations find advanced economies more stable and less corrupt than developing countries. Rather than scarcity, advances in energy technology have also resulted in a significant increase in oil and natural gas production in the United States. This is done through the technique of fracking natural gas and oil from shale rocks. Fracking, however, results in the release of toxic chemicals which affect the water supplies of surrounding communities. Analysts predict that the United States has the capacity to become the leading producer of natural gas and oil in the world and that it will surpass Saudi Arabia as the world's leading exporter of energy. This has important geopolitical implications because an increase in the export of US natural gas to Europe could reduce European dependence on Russian natural gas imports. Increased US production of natural gas and oil, and other developments such as reduced demand in Europe, created a glut of oil in the world market in 2015 which continued into 2016. The result was a significant reduction in the benchmark price of Brent crude oil, contributing to a decline in the Russian economy and other major oil producers. The oversupply of oil in the world market has also created a crisis for the major oil cartel, The Organization of Petroleum Exporting Countries (OPEC), which has faced pressure from some of its members to reduce the production of oil in order to keep prices up.

The world's population also faces an existential threat from global warming, according to credible scientists. The main question is whether the international community is willing to reach the consensus that will effectively regulate the emission of greenhouse gases into the atmosphere. In 2014–2015, the United States and China, the two largest emitters of greenhouse gases reached an agreement to reduce the amount released into the atmosphere. China has also promised, in a statement made by the President of China in a state visit to the United States in September 2015, to provide the developing countries with about $3 billion in aid to help reduce the emission of greenhouse gases. Some analysts, however, are rather pessimistic about the prospects for environmental diplomacy that will replace the Kyoto Protocol of 1997 with a new climate treaty at a conference which met in Paris in 2015. The conference was held successfully in Paris in December 2015. In 2016, the "double trigger" of approval of the agreement by 55 percent of the states responsible for 55 percent of the emissions of greenhouse gases was achieved.

Article Prepared by: Robert Weiner, *University of Massachusetts, Boston*

The New Population Bomb: The Four Megatrends That Will Change the World

JACK A. GOLDSTONE

Learning Outcomes

After reading this article, you will be able to:

- Identify the four demographic trends.
- Discuss how international politics is changing due to these trends.
- Summarize the Afghanistan case study.

Forty-two years ago, the biologist Paul Ehrlich warned in The Population Bomb that mass starvation would strike in the 1970s and 1980s, with the world's population growth outpacing the production of food and other critical resources. Thanks to innovations and efforts such as the "green revolution" in farming and the widespread adoption of family planning, Ehrlich's worst fears did not come to pass. In fact, since the 1970s, global economic output has increased and fertility has fallen dramatically, especially in developing countries.

The United Nations Population Division now projects that global population growth will nearly halt by 2050. By that date, the world's population will have stabilized at 9.15 billion people, according to the "medium growth" variant of the UN's authoritative population database World Population Prospects: The 2008 Revision. (Today's global population is 6.83 billion.) Barring a cataclysmic climate crisis or a complete failure to recover from the current economic malaise, global economic output is expected to increase by two to three percent per year, meaning that global income will increase far more than population over the next four decades.

But twenty-first-century international security will depend less on how many people inhabit the world than on how the global population is composed and distributed: where populations are declining and where they are growing, which countries are relatively older and which are more youthful, and how demographics will influence population movements across regions.

These elements are not well recognized or widely understood. A recent article in *The Economist*, for example, cheered the decline in global fertility without noting other vital demographic developments. Indeed, the same UN data cited by *The Economist* reveal four historic shifts that will fundamentally alter the world's population over the next four decades: the relative demographic weight of the world's developed countries will drop by nearly 25 percent, shifting economic power to the developing nations; the developed countries' labor forces will substantially age and decline, constraining economic growth in the developed world and raising the demand for immigrant workers; most of the world's expected population growth will increasingly be concentrated in today's poorest, youngest, and most heavily Muslim countries, which have a dangerous lack of quality education, capital, and employment opportunities; and, for the first time in history, most of the world's population will become urbanized, with the largest urban centers being in the world's poorest countries, where policing, sanitation, and health care are often scarce. Taken together, these trends will pose challenges every bit as alarming as those noted by Ehrlich. Coping with them will require nothing less than a major reconsideration of the world's basic global governance structures.

Europe's Reversal of Fortunes

At the beginning of the eighteenth century, approximately 20 percent of the world's inhabitants lived in Europe (including Russia). Then, with the Industrial Revolution, Europe's population boomed, and streams of European emigrants set off for the Americas. By the eve of World War I, Europe's population had more than quadrupled. In 1913, Europe had more people than China, and the proportion of the world's population living in Europe and the former European colonies of North America had risen to over 33 percent. But this trend reversed after World War I, as basic health care and sanitation began to spread to poorer countries. In Asia, Africa, and Latin America, people began to live longer, and birthrates remained high or fell only slowly. By 2003, the combined populations of Europe, the United States, and Canada accounted for just 17 percent of the global population. In 2050, this figure is expected to be just 12 percent—far less than it was in 1700. (These projections, moreover, might even understate the reality because they reflect the "medium growth" projection of the UN forecasts, which assumes that the fertility rates of developing countries will decline while those of developed countries will increase. In fact, many developed countries show no evidence of increasing fertility rates.) The West's relative decline is even more dramatic if one also considers changes in income. The Industrial Revolution made Europeans not only more numerous than they had been but also considerably richer per capita than others worldwide. According to the economic historian Angus Maddison, Europe, the United States, and Canada together produced about 32 percent of the world's GDP at the beginning of the nineteenth century. By 1950, that proportion had increased to a remarkable 68 percent of the world's total output (adjusted to reflect purchasing power parity).

This trend, too, is headed for a sharp reversal. The proportion of global GDP produced by Europe, the United States, and Canada fell from 68 percent in 1950 to 47 percent in 2003 and will decline even more steeply in the future. If the growth rate of per capita income (again, adjusted for purchasing power parity) between 2003 and 2050 remains as it was between 1973 and 2003—averaging 1.68 percent annually in Europe, the United States, and Canada and 2.47 percent annually in the rest of the world—then the combined GDP of Europe, the United States, and Canada will roughly double by 2050, whereas the GDP of the rest of the world will grow by a factor of five. The portion of global GDP produced by Europe, the United States, and Canada in 2050 will then be less than 30 percent—smaller than it was in 1820.

These figures also imply that an overwhelming proportion of the world's GDP growth between 2003 and 2050—nearly 80 percent—will occur outside of Europe, the United States, and Canada. By the middle of this century, the global middle class—those capable of purchasing durable consumer products, such as cars, appliances, and electronics—will increasingly be found in what is now considered the developing world. The World Bank has predicted that by 2030 the number of middle-class people in the developing world will be 1.2 billion—a rise of 200 percent since 2005. This means that the developing world's middle class alone will be larger than the total populations of Europe, Japan, and the United States combined. From now on, therefore, the main driver of global economic expansion will be the economic growth of newly industrialized countries, such as Brazil, China, India, Indonesia, Mexico, and Turkey.

Aging Pains

Part of the reason developed countries will be less economically dynamic in the coming decades is that their populations will become substantially older. The European countries, Canada, the United States, Japan, South Korea, and even China are aging at unprecedented rates. Today, the proportion of people aged 60 or older in China and South Korea is 12–15 percent. It is 15–22 percent in the European Union, Canada, and the United States and 30 percent in Japan. With baby boomers aging and life expectancy increasing, these numbers will increase dramatically. In 2050, approximately 30 percent of Americans, Canadians, Chinese, and Europeans will be over 60, as will more than 40 percent of Japanese and South Koreans.

Over the next decades, therefore, these countries will have increasingly large proportions of retirees and increasingly small proportions of workers. As workers born during the baby boom of 1945–65 are retiring, they are not being replaced by a new cohort of citizens of prime working age (15–59 years old).

Industrialized countries are experiencing a drop in their working-age populations that is even more severe than the overall slowdown in their population growth. South Korea represents the most extreme example. Even as its total population is projected to decline by almost 9 percent by 2050 (from 48.3 million to 44.1 million), the population of working-age South Koreans is expected to drop by 36 percent (from 32.9 million to 21.1 million), and the number of South Koreans aged 60 and older will increase by almost 150 percent (from 7.3 million to 18 million). By 2050, in other words, the entire working-age population will barely exceed the 60-and-older population. Although South Korea's case is extreme, it represents an increasingly common fate for developed countries. Europe is expected to lose 24 percent of its prime working-age population (about 120 million workers) by 2050, and its 60-and-older population is expected to increase by 47 percent. In the United States, where higher fertility and more immigration are expected than in Europe, the working-age population will grow by 15 percent over the next four decades—a steep decline from its growth of 62 percent between 1950 and 2010. And by 2050, the United States' 60-and-older population is expected to double.

All this will have a dramatic impact on economic growth, health care, and military strength in the developed world. The forces that fueled economic growth in industrialized countries during the second half of the twentieth century—increased productivity due to better education, the movement of women into the labor force, and innovations in technology—will all likely weaken in the coming decades. College enrollment boomed after World War II, a trend that is not likely to recur in the twenty-first century; the extensive movement of women into the labor force also was a one-time social change; and the technological change of the time resulted from innovators who created new products and leading-edge consumers who were willing to try them out—two groups that are thinning out as the industrialized world's population ages.

Overall economic growth will also be hampered by a decline in the number of new consumers and new households. When developed countries' labor forces were growing by 0.5–1.0 percent per year, as they did until 2005, even annual increases in real output per worker of just 1.7 percent meant that annual economic growth totaled 2.2–2.7 percent per year. But with the labor forces of many developed countries (such as Germany, Hungary, Japan, Russia, and the Baltic states) now shrinking by 0.2 percent per year and those of other countries (including Austria, the Czech Republic, Denmark, Greece, and Italy) growing by less than 0.2 percent per year, the same 1.7 percent increase in real output per worker yields only 1.5–1.9 percent annual overall growth. Moreover, developed countries will be lucky to keep productivity growth at even that level; in many developed countries, productivity is more likely to decline as the population ages.

A further strain on industrialized economies will be rising medical costs: as populations age, they will demand more health care for longer periods of time. Public pension schemes for aging populations are already being reformed in various industrialized countries—often prompting heated debate. In theory, at least, pensions might be kept solvent by increasing the retirement age, raising taxes modestly, and phasing out benefits for the wealthy. Regardless, the number of 80- and 90-year-olds—who are unlikely to work and highly likely to require nursing-home and other expensive care—will rise dramatically. And even if 60- and 70-year-olds remain active and employed, they will require procedures and medications—hip replacements, kidney transplants, blood-pressure treatments—to sustain their health in old age.

All this means that just as aging developed countries will have proportionally fewer workers, innovators, and consumerist young households, a large portion of those countries' remaining economic growth will have to be diverted to pay for the medical bills and pensions of their growing elderly populations. Basic services, meanwhile, will be increasingly costly because fewer young workers will be available for strenuous and labor-intensive jobs. Unfortunately, policymakers seldom reckon with these potentially disruptive effects of otherwise welcome developments, such as higher life expectancy.

Youth and Islam in the Developing World

Even as the industrialized countries of Europe, North America, and Northeast Asia will experience unprecedented aging this century, fast-growing countries in Africa, Latin America, the Middle East, and Southeast Asia will have exceptionally youthful populations. Today, roughly nine out of ten children under the age of 15 live in developing countries. And these are the countries that will continue to have the world's highest birthrates. Indeed, over 70 percent of the world's population growth between now and 2050 will occur in 24 countries, all of which are classified by the World Bank as low income or lower-middle income, with an average per capita income of under $3,855 in 2008.

Many developing countries have few ways of providing employment to their young, fast-growing populations. Would-be laborers, therefore, will be increasingly attracted to the labor markets of the aging developed countries of Europe, North America, and Northeast Asia. Youthful immigrants from nearby regions with high unemployment—Central America, North Africa, and Southeast Asia, for example—will be drawn to those vital entry-level and manual-labor jobs that sustain advanced economies: janitors, nursing-home aides, bus drivers, plumbers, security guards, farm workers, and the like. Current levels of immigration from developing to developed countries are paltry compared to those that the forces of supply and demand might soon create across the world.

These forces will act strongly on the Muslim world, where many economically weak countries will continue to experience dramatic population growth in the decades ahead. In 1950, Bangladesh, Egypt, Indonesia, Nigeria, Pakistan, and Turkey had a combined population of 242 million. By 2009, those six countries were the world's most populous Muslim-majority countries and had a combined population of 886 million. Their populations are continuing to grow and indeed are expected to increase by 475 million between now and 2050—during which time, by comparison, the six most populous developed countries are projected to gain only 44 million inhabitants. Worldwide, of the 48 fastest-growing countries today—those with annual population growth of two percent or more—28 are majority Muslim or have Muslim minorities of 33 percent or more.

It is therefore imperative to improve relations between Muslim and Western societies. This will be difficult given that many Muslims live in poor communities vulnerable to radical appeals and many see the West as antagonistic and militaristic. In the 2009 Pew Global Attitudes Project survey, for example, whereas 69 percent of those Indonesians and Nigerians surveyed reported viewing the United States favorably, just 18 percent of those polled in Egypt, Jordan, Pakistan, and Turkey (all U.S.

allies) did. And in 2006, when the Pew survey last asked detailed questions about Muslim-Western relations, more than half of the respondents in Muslim countries characterized those relations as bad and blamed the West for this state of affairs.

But improving relations is all the more important because of the growing demographic weight of poor Muslim countries and the attendant increase in Muslim immigration, especially to Europe from North Africa and the Middle East. (To be sure, forecasts that Muslims will soon dominate Europe are outlandish: Muslims compose just three to ten percent of the population in the major European countries today, and this proportion will at most double by midcentury.) Strategists worldwide must consider that the world's young are becoming concentrated in those countries least prepared to educate and employ them, including some Muslim states. Any resulting poverty, social tension, or ideological radicalization could have disruptive effects in many corners of the world. But this need not be the case; the healthy immigration of workers to the developed world and the movement of capital to the developing world, among other things, could lead to better results.

Urban Sprawl

Exacerbating twenty-first-century risks will be the fact that the world is urbanizing to an unprecedented degree. The year 2010 will likely be the first time in history that a majority of the world's people live in cities rather than in the countryside. Whereas less than 30 percent of the world's population was urban in 1950, according to UN projections, more than 70 percent will be by 2050.

Lower-income countries in Asia and Africa are urbanizing especially rapidly, as agriculture becomes less labor intensive and as employment opportunities shift to the industrial and service sectors. Already, most of the world's urban agglomerations—Mumbai (population 20.1 million), Mexico City (19.5 million), New Delhi (17 million), Shanghai (15.8 million), Calcutta (15.6 million), Karachi (13.1 million), Cairo (12.5 million), Manila (11.7 million), Lagos (10.6 million), Jakarta (9.7 million)—are found in low-income countries. Many of these countries have multiple cities with over one million residents each: Pakistan has eight, Mexico 12, and China more than 100. The UN projects that the urbanized proportion of sub-Saharan Africa will nearly double between 2005 and 2050, from 35 percent (300 million people) to over 67 percent (1 billion). China, which is roughly 40 percent urbanized today, is expected to be 73 percent urbanized by 2050; India, which is less than 30 percent urbanized today, is expected to be 55 percent urbanized by 2050. Overall, the world's urban population is expected to grow by 3 billion people by 2050.

This urbanization may prove destabilizing. Developing countries that urbanize in the twenty-first century will have far lower per capita incomes than did many industrial countries when they first urbanized. The United States, for example, did not reach 65 percent urbanization until 1950, when per capita income was nearly $13,000 (in 2005 dollars). By contrast, Nigeria, Pakistan, and the Philippines, which are approaching similar levels of urbanization, currently have per capita incomes of just $1,800–$4,000 (in 2005 dollars).

According to the research of Richard Cincotta and other political demographers, countries with younger populations are especially prone to civil unrest and are less able to create or sustain democratic institutions. And the more heavily urbanized, the more such countries are likely to experience Dickensian poverty and anarchic violence. In good times, a thriving economy might keep urban residents employed and governments flush with sufficient resources to meet their needs. More often, however, sprawling and impoverished cities are vulnerable to crime lords, gangs, and petty rebellions. Thus, the rapid urbanization of the developing world in the decades ahead might bring, in exaggerated form, problems similar to those that urbanization brought to nineteenth-century Europe. Back then, cyclical employment, inadequate policing, and limited sanitation and education often spawned widespread labor strife, periodic violence, and sometimes—as in the 1820s, the 1830s, and 1848—even revolutions.

International terrorism might also originate in fast-urbanizing developing countries (even more than it already does). With their neighborhood networks, access to the Internet and digital communications technology, and concentration of valuable targets, sprawling cities offer excellent opportunities for recruiting, maintaining, and hiding terrorist networks.

Defusing the Bomb

Averting this century's potential dangers will require sweeping measures. Three major global efforts defused the population bomb of Ehrlich's day: a commitment by governments and nongovernmental organizations to control reproduction rates; agricultural advances, such as the green revolution and the spread of new technology; and a vast increase in international trade, which globalized markets and thus allowed developing countries to export foodstuffs in exchange for seeds, fertilizers, and machinery, which in turn helped them boost production. But today's population bomb is the product less of absolute growth in the world's population than of changes in its age and distribution. Policymakers must therefore adapt today's global governance institutions to the new realities of the aging of the industrialized world, the concentration of the world's economic and population growth in developing countries, and the increase in international immigration.

During the Cold War, Western strategists divided the world into a "First World," of democratic industrialized countries;

a "Second World," of communist industrialized countries; and a "Third World," of developing countries. These strategists focused chiefly on deterring or managing conflict between the First and the Second Worlds and on launching proxy wars and diplomatic initiatives to attract Third World countries into the First World's camp. Since the end of the Cold War, strategists have largely abandoned this three-group division and have tended to believe either that the United States, as the sole superpower, would maintain a Pax Americana or that the world would become multipolar, with the United States, Europe, and China playing major roles.

Unfortunately, because they ignore current global demographic trends, these views will be obsolete within a few decades. A better approach would be to consider a different three-world order, with a new First World of the aging industrialized nations of North America, Europe, and Asia's Pacific Rim (including Japan, Singapore, South Korea, and Taiwan, as well as China after 2030, by which point the one-child policy will have produced significant aging); a Second World comprising fast-growing and economically dynamic countries with a healthy mix of young and old inhabitants (such as Brazil, Iran, Mexico, Thailand, Turkey, and Vietnam, as well as China until 2030); and a Third World of fast-growing, very young, and increasingly urbanized countries with poorer economies and often weak governments. To cope with the instability that will likely arise from the new Third World's urbanization, economic strife, lawlessness, and potential terrorist activity, the aging industrialized nations of the new First World must build effective alliances with the growing powers of the new Second World and together reach out to Third World nations. Second World powers will be pivotal in the twenty-first century not just because they will drive economic growth and consume technologies and other products engineered in the First World; they will also be central to international security and cooperation. The realities of religion, culture, and geographic proximity mean that any peaceful and productive engagement by the First World of Third World countries will have to include the open cooperation of Second World countries.

Strategists, therefore, must fundamentally reconsider the structure of various current global institutions. The G-8, for example, will likely become obsolete as a body for making global economic policy. The G-20 is already becoming increasingly important, and this is less a short-term consequence of the ongoing global financial crisis than the beginning of the necessary recognition that Brazil, China, India, Indonesia, Mexico, Turkey, and others are becoming global economic powers. International institutions will not retain their legitimacy if they exclude the world's fastest-growing and most economically dynamic countries. It is essential, therefore, despite European concerns about the potential effects on immigration, to take steps such as admitting Turkey into the European Union. This would add youth and economic dynamism to the EU—and would prove that Muslims are welcome to join Europeans as equals in shaping a free and prosperous future. On the other hand, excluding Turkey from the EU could lead to hostility not only on the part of Turkish citizens, who are expected to number 100 million by 2050, but also on the part of Muslim populations worldwide.

NATO must also adapt. The alliance today is composed almost entirely of countries with aging, shrinking populations and relatively slow-growing economies. It is oriented toward the Northern Hemisphere and holds on to a Cold War structure that cannot adequately respond to contemporary threats. The young and increasingly populous countries of Africa, the Middle East, Central Asia, and South Asia could mobilize insurgents much more easily than NATO could mobilize the troops it would need if it were called on to stabilize those countries. Long-standing NATO members should, therefore—although it would require atypical creativity and flexibility—consider the logistical and demographic advantages of inviting into the alliance countries such as Brazil and Morocco, rather than countries such as Albania. That this seems far-fetched does not minimize the imperative that First World countries begin including large and strategic Second and Third World powers in formal international alliances.

The case of Afghanistan—a country whose population is growing fast and where NATO is currently engaged—illustrates the importance of building effective global institutions. Today, there are 28 million Afghans; by 2025, there will be 45 million; and by 2050, there will be close to 75 million. As nearly 20 million additional Afghans are born over the next 15 years, NATO will have an opportunity to help Afghanistan become reasonably stable, self-governing, and prosperous. If NATO's efforts fail and the Afghans judge that NATO intervention harmed their interests, tens of millions of young Afghans will become more hostile to the West. But if they come to think that NATO's involvement benefited their society, the West will have tens of millions of new friends. The example might then motivate the approximately one billion other young Muslims growing up in low-income countries over the next four decades to look more kindly on relations between their countries and the countries of the industrialized West.

Creative Reforms at Home

The aging industrialized countries can also take various steps at home to promote stability in light of the coming demographic trends. First, they should encourage families to have more children. France and Sweden have had success providing child care, generous leave time, and financial allowances to families with young children. Yet there is no consensus among policymakers—and certainly not among demographers—about what policies best encourage fertility.

More important than unproven tactics for increasing family size is immigration. Correctly managed, population movement can benefit developed and developing countries alike. Given the dangers of young, underemployed, and unstable populations in developing countries, immigration to developed countries can provide economic opportunities for the ambitious and serve as a safety valve for all. Countries that embrace immigrants, such as the United States, gain economically by having willing laborers and greater entrepreneurial spirit. And countries with high levels of emigration (but not so much that they experience so-called brain drains) also benefit because emigrants often send remittances home or return to their native countries with valuable education and work experience.

One somewhat daring approach to immigration would be to encourage a reverse flow of older immigrants from developed to developing countries. If older residents of developed countries took their retirements along the southern coast of the Mediterranean or in Latin America or Africa, it would greatly reduce the strain on their home countries' public entitlement systems. The developing countries involved, meanwhile, would benefit because caring for the elderly and providing retirement and leisure services is highly labor intensive. Relocating a portion of these activities to developing countries would provide employment and valuable training to the young, growing populations of the Second and Third Worlds.

This would require developing residential and medical facilities of First World quality in Second and Third World countries. Yet even this difficult task would be preferable to the status quo, by which low wages and poor facilities lead to a steady drain of medical and nursing talent from developing to developed countries. Many residents of developed countries who desire cheaper medical procedures already practice medical tourism today, with India, Singapore, and Thailand being the most common destinations. (For example, the international consulting firm Deloitte estimated that 750,000 Americans traveled abroad for care in 2008.)

Never since 1800 has a majority of the world's economic growth occurred outside of Europe, the United States, and Canada. Never have so many people in those regions been over 60 years old. And never have low-income countries' populations been so young and so urbanized. But such will be the world's demography in the twenty-first century. The strategic and economic policies of the twentieth century are obsolete, and it is time to find new ones.

Reference

Goldstone, Jack A. "The new population bomb: the four megatrends that will change the world." *Foreign Affairs* 89.1 (2010): 31. *General OneFile*. Web. 23 Jan. 2010. http://0-find.galegroup.com.www.consuls.org/gps/start.do?proId5IPS&userGroupName5a30wc.

Critical Thinking

1. Using the websites below, develop a comparison of demographic trends between two different regions of the world.
2. Compare the projected demographic makeup of the United States in 2050 with China, Mexico, Pakistan, and Russia.

Internet References

INED (French Institute for Demographic Studies)
www.ined.fr/en/everything_about_population

PRB World Population
www.prb.org/Publications/Datasheets/2011/world-population-data-sheet/world-map.aspx#/map/population

Worldmapper
www.worldmapper.org

Article

Prepared by: Robert Weiner, *University of Massachusetts, Boston*

Fighting Water Wars: Regional Environmental Cooperation as a Roadmap for Peace

LEONARDO ORLANDO

Learning Outcomes

After reading this article, you will be able to:

- Discuss the factors that contribute to freshwater scarcity
- Understand the importance of taking a transdisciplinary approach to the issue of freshwater cooperation and conflict

Introduction

There is a general consensus among experts that water scarcity in the 21st century will seriously threaten world peace. As it is essential for virtually all domestic, agricultural and industrial uses, as well as for maintaining ecological balance, water is the fundamental pillar of socioeconomic, cultural, political, and environmental development. Water's deepening scarcity, the thinking goes, could thus lead to an upsurge in interstate armed clashes. However, management of transboundary waters has historically led to a process of cooperation rather than confrontation—a trend that is greatly enhanced where institutions are able to properly cultivate environmental governance, and when civil society can deploy transnational ties.

In other words, water has not lead to major interstate clashes, at least not *yet*. The question now is how should we cope with increasing water stress so as to continue fostering cooperation? This article intends to explore this issue, starting with a brief overview of global water resources, followed by a presentation of the main paradigms regarding environmental scarcity and conflict, and ending with a case study of conflict containment through water cooperation.

The World Water Situation: A Sword of Damocles

Less than 3 percent of the world's water is fresh—the remaining is seawater and undrinkable. Of this 3 percent, over 2.5 percent is frozen, locked up in Antarctica, the Arctic, and other glaciers. Thus, humanity must rely on this 0.5 percent for all its fresh water needs. Currently, more than 780 million people lack access to an improved water source.[1] More than 3.4 million people die each year from poor water sanitation and hygiene related causes,[2] and nearly all of these deaths—99 percent—take place in the developing world, making unclean water a greater threat to human security than violent conflict in these countries.[3]

Moreover, analysts see a grim future for the world's water resources, for several reasons. First, massive pollution of rivers, lakes, and aquifers across the world, along with the alteration of watercourses (a result of the construction of dams, mainly for the production of hydropower), are progressively reducing earth's water reserves. Secondly, the excessive use of water for industrial and agricultural purposes is cutting off essential access to water for numerous peoples around the world. The rapid growth of the population in recent decades will, if the same level of consumption is maintained, further reduce the availability of water resources. Third, as it was established by the Intergovernmental Panel on Climate Change (IPCC), global warming and climate change will have a strongly negative impact on access to and demand for water in most countries, due to changes in the hydrological cycle and levels of precipitation.

Confronting Paradigms: Water Wars versus Hydric Peace

Based on current scientific projections, political scientists and international relations analysts believe that in the coming decades we should see increased interstate hostilities over water scarcity. Water shortages can quickly become a cause of conflict, as upsurges in violent clashes threaten local and international security.[4] The alarm bells first sounded in 1991, with an article published in Foreign Affairs by Joyce Starr titled "Water Wars."[5] Then, in August 1995, the Vice President for Environmentally and Socially Sustainable Development at the World Bank, Ismail Serigaildein, warned that "if the wars of this century were fought over oil, the wars of the next century will be fought over water—unless we change our approach to managing this precious and vital resource."[6]

However, in the early 2000s the study of environmental conflicts, regarding water in particular, shifted to question the relationship between environmental factors and conflict. This change stressed the plurality of factors that lead to conflict. Indeed, the outbreak of conflict cannot be reduced to a single cause but must be understood in relation to the social, economic, and political context that shapes its particular characteristics. Within this framework, a team led by Professor Aaron T. Wolf, from Oregon State University, conducted a three-year study that collected information on water interactions among States from 1950 to 2000. During those 50 years, there were over 1,800 of these "hydric interactions" documented, of which only 37 led to serious conflicts, none of which involved war.[7] In fact, the last known war over water occurred more than 4,500 years ago between the city-states of Lagash and Umma in the Tigris-Euphrates basin.[8]

This research concluded that the management of transboundary waters drives a process of cooperation instead of confrontation, a trend that is greatly enhanced in watersheds where there are existing political relationships and institutions that can shape environmental governance. Moreover, the resolution of water conflicts can prompt regional integration that changes how environmental governance is discussed. By fostering civil society opportunities for collaboration and peacebuilding, conflict can be avoided. Indeed, nearly 450 agreements on international waters were signed between 1820 and 2007, demonstrating the potential for transboundary waters to become a source of cooperation rather than conflict.

Thus, some studies on the prevention and resolution of water-related conflicts are skeptical of the supposed inevitability of an ecological disaster, and instead point to careful management of transboundary waters as a way to foster global environmental governance. In this sense, water is seen as a catalyst for cooperation: the total interdependence of all parties concerned in a water system requires actors to find agreements concerning its administration, under penalty of a disruption that would make all of them victims.[9]

The hydropolitic conflicts studied by Wolf exhibit certain peculiarities of great interest: the displacement of governments as the only actors in transboundary water management, and the establishment of civil society ties that go beyond the borders of the nation-state.[10] Continuing with this perspective, Ken Conca points out that rivers are a source of livelihood, pillars of culture and community, and a key component in development strategies. As such, they must not be taken in isolation, but as elements of larger and more complex socioecological systems.[11] He considers three main elements of river management: human needs, the requirements of agro-industrial growth, and the uses of ecosystems. These elements cannot be confined to state borders. Following these clues, this article focuses on the role played by the regional institutional framework of water governance as a key actor for containing conflict and fostering cooperation.

Containing Conflict and Building Peace: Regional Water Cooperation

Lake Victoria is an example of how peace can be achieved in spite of water scarcity—a site where conflict is not only contained, but where the institutional framework used for this purpose serves to foster regional cooperation.[12] Lake Victoria, whose water is used for drinking as well as for household and industrial use, supports the livelihoods of people living along its shores, mainly in the fishing industry. It is the habitat of tilapia (*Oreochromis niloticus*) and Nile perch (*Lates niloticus*), two types of fish that are in great demand. However, the serious human and environmental situation at the Lake Victoria basin are turning it into a "sick giant."[13] People living in its surroundings languish in abject poverty and are devastated by health threats such as HIV/AIDs and water borne diseases like bilharzias and diarrhea. Chemical runoff from industry and agriculture as well as human waste runoff create severe pollution and health hazards. The lake is also infested with deadly water hyacinth, which is depriving the lake of oxygen. Moreover, within the next 25 years the region's population is expected to double, adding to the demand generated by growth in industry and agriculture. The constant threat of droughts adds to the urgency of the problem.[14]

However, in spite of the difficult environmental and human problems around Lake Victoria, there is cooperation rather than conflict in the basin. How could this be possible? In order to answer this question, I undertook three months of fieldwork research in the area of Lake Victoria.[15] From March to June 2014, I conducted dozens of semistructured interviews with international and regional organizations, regional and national officials, local representatives, NGOs, and scholars along eleven cities in Kenya, Tanzania, and Uganda.[16]

I also performed extensive interviews of nearly one hundred fishermen, individually or in small groups, in their places of work and residence, and along relevant sites of the three shores of the lake: Dunga Beach in Kisumu, Kenya; Kigungu in Entebbe, Uganda; and Igombe in Mwanza, Tanzania. These interviews were supplemented by observations of various fishing activities performed both on land and on the water. These activities involved fishing of the lake's predilected specimens; treatment of the fishing nets and other gear after use; sale of the day's catch to the brokers who resell it in different local markets; measurement and verification of fishing catch for eventual export; and health controls of the catch and compliance to fisheries legislation performed by local and national inspectors.

These interviews and observations aimed to paint a picture of the everyday experiences of the inhabitants of the areas affected by some type of water conflict. I recorded their collective and individual reactions, partnerships, or cross-border conflict between neighbors, and the actions of the participating NGOs to conserve the environment or to assist local populations. I also noted the perceptions of local governments in tackling these problems, the considerations of potential threats to national security, and the different means of addressing the problems. Fishing has been chosen as the leading activity for our research because of its crucial importance for managing the resources of Lake Victoria: it is its principal economic activity, and one of the largest in the region. From this starting point, we have reached a number of preliminary results that support the following four points.

1. ***Regional integration processes are a key factor for peacebuilding in transboundary waters***
   Almost every actor interviewed, at the national, regional or international levels claims that the situation is one of both conflict and cooperation. Conflict, which is sometimes latent, is present to the extent that the lake resources are becoming scarce, particularly the fishing stocks. However, officials—as well as activists of nongovernmental organizations and the fishermen themselves—are convinced that conflict containment and cooperation are a result of an ad hoc regional institutional framework created for the management of the lake's resources. Most of the interviewed actors believe the following three organizations play a principal role in the prevention and resolution of conflicts: Lake Victoria Basin Commission (LVBC), Lake Victoria Fisheries Organization (LVFO), and Lake Victoria Environmental Management Project Phase II (LVEMP II). Thus, they logically look to the East African Community to foster this cooperation, as the responsibility to effectively manage the lake's resources rests on its shoulders. All parties agree that strengthening the East African Community's involvement in the governance of Lake Victoria is a key priority.

2. ***States are unable to manage transboundary water basins on their own***
   The main objective of the regional organizations created to prevent conflict and strengthen cooperation around Lake Victoria is to harmonize the various regulations and regulatory inconsistencies applied by states to the lake. There are often different laws enforced among the riparian countries regarding matters such as the type of fishing nets permitted (this being a crucial factor to the extent that it determines the size of the fish captured, and with it, its maturity—the capture of immature fish can be disastrous for the renewal of fish species) and the homogenization of fishing restriction periods, which, when practiced at different times, pushed fishermen to venture beyond their national borders of the lake and thus encouraged transnational disputes. To reconcile these differences, it is necessary to include all stakeholders involved in the management of transboundary waters; or, in other words: any lack of cooperation inevitably entails conflict.

3. ***Transboundary civil society linkages entail a transnational community***
   One of the main objectives of interviewing local actors was to understand how they perceive their cross-border neighbors, with whom they share the scarce resources of the lake. Based on these interviews, one is able to confidently state that, through transnational civil-societal linkages, the lake's community transcends national identity.[17] It is in fact possible to imagine the existence of a single community in which all of the lake's inhabitants feel a part of. Indeed, the conflicts that actors report are explained in terms of altercations between fishermen and enforcement officials (notably between Kenyan residents and the Ugandan coastal police), or between "good" and "bad" people.

   This indicates that the concerned parties perceive the conflict not in terms of citizenship, but on moral parameters applicable to the whole community of the lake. This conclusion can be supported by the strong mobility of fishermen, who circulate among different coastal villages to undertake their activities, always integrating in local communities, but also redistributing economic gains to their original places of provenance (through regular visits or remittances). This means that incomes from fishing do not stay where they are made, but circulate around the whole area of the watershed. Moreover, strong family ties are generated from marriages between citizens of different countries, creating lasting transboundary linkages. As a corollary to these elements, actors request a deepening of the management of the lake's resources at the regional level, to the detriment of national policies that might be incompatible among them and therefore contrary to the general interest of the inhabitants of the lake as a whole.

4. ***Transdisciplinarity as the pillar of transboundary water management***

Last but not least, the inquiry reveals a dimension that is not properly analyzed in academic and political analyses of water governance: transdisciplinarity. Transdisciplinarity describes a new integrative approach that transcends existing research boundaries.[18] Problems that involve an interface of human and natural systems are complex and multidimensional,[19] hence the foolishness of addressing them in the isolation of any single discipline. Indeed, technical solutions without knowledge of social sciences to accompany them are empty, and water policies that lose sight of the scientific dimension are blind. The management of water resources has *sui generis* characteristics that require, for its proper approach, to go beyond these two approaches. Actually, if transboundary basins constitute complex socialecological systems, the reality is that the science of the engineer, the political scientist, or the international analyst cannot alone give a full account of the situation, much less propose plausible solutions.

A fertile hydropolitical approach must include both social anthropology and the sociology of cross-border linkages, cooperation, and conflict. These disciplines allow us to understand both the social dynamics and the particularities of the eco-human systems on which the policies operate. Moreover, a deeper perspective that aspires to a more thorough understanding of water systems and to long term policies requires the intervention of political ecology, which is understood not only as the political analysis of human-environmental dynamics, but also as the process of bringing to the political arena the scientific comprehension of nature itself.[20] Simultaneously, the spiritual dimension that water encompasses demands philosophical and religious study to understand the *hydropolitical* consequences of certain decisions. Thus, it is only through extensive interdisciplinary studies that we can weave governance of natural resources, conflict management, cooperation, and peacebuilding together, and deepen the institutional framework of the processes of regional integration.

Conclusion

The Stoic philosopher Epictetus explains the idea that, if while sailing in the open sea I feel a coming storm, there are two paths I can take: the terror that leads to error in the assessment of both reality and in the proper actions to undertake, or impassibility, which comes from the comprehensive and objective representation of what it is in front. The latter is the only way to act according to the truth.[21] Increasing world water scarcity is a storm whose clamor will most likely will fall upon us with rage. This storm will certainly generate fear, undermine trust, and enhance the possibility of conflict between nations. If we are to address potential conflicts over water only through a security lens, it will be what the sociologist Robert Merton termed a "self-fulfilling prophecy,"[22] effectively bringing to war what could have been a roadmap to peace.

What we can do then is follow the road toward peace that some case studies, such as Lake Victoria, show us is possible. Though there are cultural, political, and historical differences in each case, it is possible to look for similarities in how a regional institutional framework of water management works. Because, as for the alchemist who thinks of all metals as potential gold, for water management, conflict over transboundary basins lays out an actual process of peace, waiting to be implemented.

Notes

1. World Health Organization. "Progress on Drinking-Water and Sanitation—2012 Update," March 6, 2012.
2. Annette Prüss-üstün, Robert Bos, Fiona Gore, and Jamie Bartram, *Safer water, better health: Costs, benefits and sustainability of interventions to protect and promote health*, World Health Organization, 2008.
3. Kevin Watkins, *Human Development Report 2006-Beyond scarcity: Power, poverty and the global water crisis*, United Nations Development Programme, 2006.
4. Thomas F. Homer-Dixon, *Environment, Scarcity, and Violence* (Princeton: Princeton University Press, 2010).
5. Joyce Starr, "Water wars," *Foreign Policy* (1991): 17–36.
6. Ismail Serageldin's personal homepage, www.serageldin.com/water.htm.
7. Jerome Delli Priscoli and Aaron T. Wolf, *Managing and transforming water conflicts* (New York: Cambridge University Press, 2009).
8. Jerome Delli Priscoli and Aaron T. Wolf, *Managing and transforming water conflicts* (New York: Cambridge University Press, 2009), 12.
9. Shafiqul Islam and Lawrence Susskind, *Water Diplomacy: A Negotiated Approach to Managing Complex Water Networks* (New York: Routledge, 2012).
10. Priscoli and Wolf.
11. Ken Conca, *Governing Water: Contentious Transnational Politics and Global Institution Building* (Cambridge: MIT Press, 2006): 123–124.
12. Marielle J. Canter and Stephen N. Ndegwa, "Environmental Scarcity and Conflict: A Contrary Case from Lake Victoria," *Global Environmental Politics* 2 (3) (2002): 40–62.
13. Joseph L. Awange and Obiero Ong'ang'a, *Lake Victoria: Ecology, Resources, Environment* (The Netherlands: Springer Science & Business Media, 2006), 2.

14. Patricia Kameri-Mbote, "From Conflict to Cooperation in the Management of Transboundary Waters: The Nile Experience," Heinrich Boell Foundation, Washington D.C. (2005), 1-2; Joseph L. Awange and Obiero Ong'ang'a, *Lake Victoria: Ecology, Resources, Environment* (The Netherlands: Springer Science & Business Media, 2006), 1–3.
15. Ongoing PhD dissertation at Sciences Po Paris, under supervision of Professor Ariel Colonomos, "De l'interdépendance à l'intégration: gouvernance environnementale et construction de la paix en eaux transfrontières. Les cas du fleuve Uruguay et du lac Victoria."
16. The following actors were interviewed: International and Regional Organizations: United Nations Environment Programme (UNEP), United Nations Settlements Programme (UN-HABITAT); Regional Officials: East African Community (EAC), Lake Victoria Basin Commission (LVBC), Lake Victoria Fisheries Organization (LVFO), Lake Victoria Environmental Management Project Phase II (LVEMP II), The Nile Basin Initiative (NBI); National Officials: Ministry of Agriculture, Livestock and Fisheries of the Republic of Kenya, Ministry of Agriculture, Animal Industry and Fisheries of the Republic of Uganda, Ministry of Water of the United Republic of Tanzania, Ministry of Livestock and Fisheries Development of the United Republic of Tanzania; Local political representatives: leaders of the Beach Management Units (BMU) from Dunga Beach (Kenya), Kigungu (Uganda) and Igombe (Tanzania); Representatives of NGOs: Sustainable Development Environment Watch (SUSWATCH Kenya), Lake Victoria Center for Research and Development (OSIENALA), Uganda Coalition for Sustainable Development (UCSD), The Environmental Management and Economic Development Organisation (EMEDO), The East African Civil Society Organizations Forum (EACSOF); Scholars: University of Nairobi, University of Dar es Salaam, Institut de recherche pour le développement (IRD), Institut français de recherché en Afrique (IFRI-Nairobi), National Fisheries Resources Research Institute of Uganda (NaFIRRI).
17. Conca, Ken, and Geoffrey D. Dabelko, eds., "Environmental peacemaking," Woodrow Wilson Center Press, 2002, 226–229.
18. Costanza, Robert, and Sven Erik Jorgensen, eds., "Understanding and solving environmental problems in the 21st century: toward a new, integrated hard problem science," Gulf Professional Publishing, 2002, 1.
19. Wickson, Fern, Anna L. Carew, and A. W. Russell, "Transdisciplinary research: characteristics, quandaries and quality," *Futures* 38, (9) (2006): 1048.
20. Bruno Latour, *Politiques de la nature: comment faire entrer les sciences en démocratie* (Paris: La découverte, 1999).
21. Epictetus, *Discourses, Fragments, Handbook* (Oxford: Oxford University Press, 2014), Book II, Chapter XVIII.
22. Robert K. Merton, "The self-fulfilling prophecy," *The Antioch Review* 8 (2) (Summer 1948): 193–210.

Critical Thinking

1. What role do regional organizations play in promoting water cooperation in the Lake Victoria Basin?
2. What role does civil society play in promoting water cooperation in the Basin?
3. How does a transdisciplinary approach contribute to transboundary water management?

Internet References

Lake Victoria Basin Commission

Internationalwatercooperation.org/tbwaters/?actor=40

The International Water Events Data Base

https://www.transboundarywaters.orst.edu/database/event_bat_scale.html

Leonardo Orlando is a PhD candidate in Political Science and International Relations at Sciences Po Paris. He is a junior research fellow at The French Institute for Research in Africa (IFRA-Nairobi), Doctoral Research Fellow at Consejo Nacional de Investigaciones Científicas y Técnicas (CONICET, Argentina), and an assistant professor at Universidad de Ciencias Empresariales y Sociales (UCES, Argentina) and Universidad Argentina de la Empresa (UADE, Argentina). During the academic year 2014–2015, he was a visiting research scholar at The Fletcher School of Law and Diplomacy at Tufts University.

Article

Prepared by: Robert Weiner, *University of Massachusetts, Boston*

Welcome to the Revolution: Why Shale Is the Next Shale

Edward L. Morse

Learning Outcomes

After reading this article, you will be able to:

- Understand what the shale revolution is all about.
- Gain an insight into the relationship between the shale revolution and the global oil and natural gas situation.

Despite its doubters and haters, the shale revolution in oil and gas production is here to stay. In the second half of this decade, moreover, it is likely to spread globally more quickly than most think. And all of that is, on balance, a good thing for the world.

The recent surge of U.S. oil and natural gas production has been nothing short of astonishing. For the past 3 years, the United States has been the world's fastest-growing hydrocarbon producer, and the trend is not likely to stop anytime soon. U.S. natural gas production has risen by 25 percent since 2010, and the only reason it has temporarily stalled is that investments are required to facilitate further growth. Having already outstripped Russia as the world's largest gas producer, by the end of the decade, the United States will become one of the world's largest gas exporters, fundamentally changing pricing and trade patterns in global energy markets. U.S. oil production, meanwhile, has grown by 60 percent since 2008, climbing by three million barrels a day to more than eight million barrels a day. Within a couple of years, it will exceed its old record level of almost 10 million barrels a day as the United States overtakes Russia and Saudi Arabia and becomes the world's largest oil producer. And U.S. production of natural gas liquids, such as propane and butane, has already grown by one million barrels per day and should grow by another million soon.

What is unfolding in reaction is nothing less than a paradigm shift in thinking about hydrocarbons. A decade ago, there was a near-global consensus that U.S. (and, for that matter, non-OPEC) production was in inexorable decline. Today, most serious analysts are confident that it will continue to grow. The growth is occurring, to boot, at a time when U.S. oil consumption is falling. (Forget peak oil production; given a combination of efficiency gains, environmental concerns, and substitution by natural gas, what is foreseeable is peak oil demand.) And to cap things off, the costs of finding and producing oil and gas in shale and tight rock formations are steadily going down and will drop even more in the years to come.

The evidence from what has been happening is now overwhelming. Efficiency gains in the shale sector have been large and accelerating and are now hovering at around 25 percent per year, meaning that increases in capital expenditures are triggering even more potential production growth. It is clear that vast amounts of hydrocarbons have migrated from their original source rock and become trapped in shale and tight rock, and the extent of these rock formations, like the extent of the original source rock, is enormous—containing resources far in excess of total global conventional proven oil reserves, which are 1.5 trillion barrels. And there are already signs that the technology involved in extracting these resources is transferable outside the United States, so that its international spread is inevitable.

In short, it now looks as though the first few decades of the 21st century will see an extension of the trend that has persisted for the past few millennia: the availability of plentiful energy at ever-lower cost and with ever-greater efficiency, enabling major advances in global economic growth.

Why the Past Is Prologue

The shale revolution has been very much a "made in America" phenomenon. In no other country can landowners also own mineral rights. In only a few other countries (such as Australia,

Canada, and the United Kingdom) is there a tradition of an energy sector featuring many independent entrepreneurial companies, as opposed to a few major companies or national champions. And in still fewer countries are there capital markets able and willing to support financially risky exploration and production.

This powerful combination of indigenous factors will continue to drive U.S. efforts. A further 30 percent increase in U.S. natural gas production is plausible before 2020, and from then on, it should be possible to maintain a constant or even higher level of production for decades to come. As for oil, given the research and development now under way, it is likely that U.S. production could rise to 12 million barrels per day or more in a few years and be sustained there for a long time. (And that figure does not include additional potential output from deepwater drilling, which is also seeing a renaissance in investment.)

Two factors, meanwhile, should bring prices down for a long time to come. The first is declining production costs, a consequence of efficiency gains from the application of new and growing technologies. And the second is the spread of shale gas and tight oil production globally. Together, these suggest a sustainable price of around $5.50 per 1,000 cubic feet for natural gas in the United States and a trading range of $70–90 per barrel for oil globally by the end of this decade.

These trends will provide a significant boost to the U.S. economy. Households could save close to $30 billion annually in electricity costs by 2020, compared to the U.S. Energy Information Administration's current forecast. Gasoline costs could fall from an average of 5 to 3 percent of real disposable personal income. The price of gasoline could drop by 30 percent, increasing annual disposable income by $750, on average, per driving household. The oil and gas boom could add about 2.8 percent in cumulative GDP growth by 2020 and bolster employment by some three million jobs.

Beyond the United States, the spread of shale gas and tight oil exploitation should have geopolitically profound implications. There is no longer any doubt about the sheer abundance of this new accessible resource base, and that recognition is leading many governments to accelerate the delineation and development of commercially available resources. Countries' motivations are diverse and clear. For Saudi Arabia, which is already developing its first power plant using indigenous shale gas, the exploitation of its shale resources can free up more oil for exports, increasing revenues for the country as a whole. For Russia, with an estimated 75 billion barrels of recoverable tight oil (50 percent more than the United States), production growth spells more government revenue. And for a host of other countries, the motivations range from reducing dependence on imports to increasing export earnings to enabling domestic economic development.

Risky Business?

Skeptics point to three problems that could lead the fruits of the revolution to be left to wither on the vine: environmental regulation, declining rates of production, and drilling economics. But none is likely to be catastrophic.

Hydraulic fracturing, or "fracking"—the process of injecting sand, water, and chemicals into shale rocks to crack them open and release the hydrocarbons trapped inside—poses potential environmental risks, such as the draining or polluting of underground aquifers, the spurring of seismic activity, and the spilling of waste products during their aboveground transport. All these risks can be mitigated, and they are in fact being addressed in the industry's evolving set of best practices. But that message needs to be delivered more clearly, and best practices need to be implemented across the board, in order to head off local bans or restrictive regulation that would slow the revolution's spread or minimize its impact.

As for declining rates of production, fracking creates a surge in production at the beginning of a well's operation and a rapid drop later on, and critics argue that this means that the revolution's purported gains will be illusory. But there are two good reasons to think that high production will continue for decades rather than years. First, the accumulation of fracked wells with a long tail of production is building up a durable base of flows that will continue over time, and second, the economics of drilling work in favor of drilling at a high and sustained rate of production.

Finally, some criticize the economics of fracking, but these concerns have been exaggerated. It is true that through 2013, the upstream sector of the U.S. oil and gas industry has been massively cash-flow negative. In 2012, for example, the industry spent about $60 billion more than it earned, and some analysts believe that such trends will continue. But the costs were driven by the need to acquire land for exploration and to pursue unproductive drilling in order to hold the acreage. Now that the land-grab days are almost over, the industry's cash flow should be increasingly positive.

It is also true that traditional finding and development costs indicate that natural gas prices need to be above $4 per 1,000 cubic feet and oil prices above $70 per barrel for the economics of drilling to work—which suggests that abundant production might drive prices down below what is profitable. But as demand grows for natural gas—for industry, residential and commercial space heating, the export market, power generation, and transportation—prices should rise to a level that can sustain increased drilling: the $5–6 range, which is about where prices were this past winter. Efficiency gains stemming from new technology, meanwhile, are driving down break-even drilling costs. In the oil sector, most drilling now brings an adequate return on investment at prices below $50 per barrel, and within a few years, that level could be under $40 per barrel.

Think Globally

Since shale resources are found around the globe, many countries are trying to duplicate the United States' success in the sector, and it is likely that some, and perhaps many, will succeed. U.S. recoverable shale resources constitute only about 15 percent of the global total, and so if the true extent and duration of even the U.S. windfall are not yet measurable, the same applies even more so for the rest of the world. Many countries are already taking early steps to develop their shale resources, and in several, the results look promising. It is highly likely that Australia, China, Mexico, Russia, Saudi Arabia, and the United Kingdom will see meaningful production before the end of this decade. As a result, global trade in energy will be dramatically disrupted.

A few years ago, hydrocarbon exports from the United States were negligible. But by the start of 2013, oil, natural gas, and petrochemicals had become the single largest category of U.S. exports, surpassing agricultural products, transportation equipment, and capital goods. The shift in the U.S. trade balance for petroleum products has been stunning. In 2008, the United States was a net importer of petroleum products, taking in about two million barrels per day; by the end of 2013, it was a net exporter, with an outflow of more than two million barrels per day. By the end of 2014, the United States should overtake Russia as the largest exporter of diesel, jet fuel, and other energy products, and by 2015, it should overtake Saudi Arabia as the largest exporter of petrochemical feedstocks. The U.S. trade balance for oil, which in 2011 was −$354 billion, should flip to +$5 billion by 2020.

By then, the United States will be a net exporter of natural gas, on a scale potentially rivaling both Qatar and Russia, and the consequences will be enormous. The U.S. gas trade balance should shift from −$8 billion in 2013 to +$14 billion by 2020. U.S. pipeline exports to Mexico and eastern Canada are likely to grow by 400 percent, to eight billion cubic feet per day, by 2018, and perhaps to 10 billion by 2020. U.S. exports of liquefied natural gas (lng) look likely to reach nine billion cubic feet per day by 2020.

Sheer volume is important, but not as much as two other factors: the pricing basis and the amount of natural gas that can be sold in a spot market. Most LNG trade links the price of natural gas to the price of oil. But the shale gas revolution has delinked these two prices in the United States, where the traditional 7:1 ratio between oil and gas prices has exploded to more than 20:1. That makes lng exports from the United States competitive with LNG exports from Qatar or Russia, eroding the oil link in LNG pricing. What's more, traditional LNG contracts are tied to specific destinations and prohibit trading. U.S. LNG (and likely also new LNG from Australia and Canada) will not come with anticompetitive trade restrictions, and so a spot market should emerge quickly. And U.S. LNG exports to Europe should erode the Russian state oil company Gazprom's pricing hold on the continent, just as they should bring down prices of natural gas around the world.

In the geopolitics of energy, there are always winners and losers. OPEC will be among the latter, as the United States moves from having had a net hydrocarbon trade deficit of some nine million barrels per day in 2007, to having one of under six million barrels today, to enjoying a net positive position by 2020. Lost market share and lower prices could pose a devastating challenge to oil producers dependent on exports for government revenue. Growing populations and declining per capita incomes are already playing a central role in triggering domestic upheaval in Iraq, Libya, Nigeria, and Venezuela, and in that regard, the years ahead do not look promising for those countries.

At the same time, the U.S. economy might actually start approaching energy independence. And the shale revolution should also lead to the prevalence of market forces in international energy pricing, putting an end to OPEC's 40-year dominance, during which producers were able to band together to raise prices well above production costs, with negative consequences for the world economy. When it comes to oil and natural gas, we now know that though much is taken, much abides—and the shale revolution is only just getting started.

Critical Thinking

1. Will the shale revolution allow the United States to become energy independent?
2. What are some of the environmental drawbacks associated with the mining of shale?
3. What will be the effect of a glut of oil on the energy market as a result of the shale revolution?

Internet References

Fracking's Future
http://harvardmagazine.com/2013/01/frackings.future

Hydraulic Fracturing
http://www.dangersoffracking.com

National Renewable Energy Laboratory
http://www.nrel.gov

International Energy Agency
http://www.iea.org

Organization of Petroleum Exporting Countries
http://www.opec_web/en

EDWARD L. MORSE is Global Head of Commodities Research at Citi.

Article

Prepared by: Robert Weiner, *University of Massachusetts, Boston*

Think Again: Climate Treaties

David Shorr

Learning Outcomes

After reading this article, you will be able to:

- Explain what is meant by climate change.
- Understand what causes climate change.

Time is running short for the international community to tackle climate change.

Pressure to act comes from rising temperatures and sea levels, superstorms, brutal droughts, and diminishing food crops. It also comes from fears that these problems are going to get worse. Modern economies have already boosted the concentration of carbon dioxide (CO_2) in the atmosphere by 40 percent since the Industrial Revolution. If the world stays on its current course, CO_2 levels could double by century's end, potentially raising global temperatures several more degrees. (The last time the planet's CO_2 levels were so high was 15 million years ago, when temperatures were 5 to 10 °F higher than they are today.)

Another source of pressure, however, is self-imposed. Under the auspices of the United Nations, the next global climate treaty—to be negotiated among some 200 countries, with the central goal of cutting greenhouse gas emissions—should be enacted in 2015, to replace the now-outmoded 1997 Kyoto Protocol. (Once passed by state parties, the new treaty would actually go into effect in 2020.)

The race against both nature and the diplomatic clock is stressful. But in the rush to do something, the international community—most notably, and ironically, those individuals and organizations most fervent about combating global warming—is often doing the wrong thing. It has become fixated on the notion of consensus codified in international law.

The U.N. process for climate diplomacy has been in place for more than two decades, punctuated since 1995 by annual meetings at which countries assess global progress in protecting the environment and negotiate treaties and other agreements to keep the ball rolling. Kyoto was finalized at the third such conference. A milestone, it established targets for country-based emissions cuts. Its signal failure, however, was leaving the world's three largest emitters of greenhouse gases unconstrained, two of them by design. Kyoto gave developing countries, including China and India, a blanket exemption from cutting emissions. Meanwhile, the United States bristled at its obligations—particularly in light of the free pass given to China and India—and refused to ratify the treaty.

Still, Kyoto was lauded by many because it was a legally binding accord, a high bar to clear in international diplomacy. The agreement's provisions were compulsory for countries that ratified it; violating them would invite a stigma—a reputation for weaseling out of promises deemed essential to saving the planet.

Today, the principle of "if you sign it, you stick to it" continues to guide a lot of conventional thinking about climate diplomacy, particularly among the political left and international NGOs, which have been driving forces of U.N. climate negotiations, and among leaders of developing countries that are not yet major polluters but are profoundly affected by global warming. For instance, in the lead-up to the last annual U.N. climate conference—held in Warsaw, Poland, in November 2013, Oxfam International's executive director, Winnie Byanyima, said the world should not accept a successor agreement to Kyoto that has anything less than the force of international law: "Of course not . . . . If it's not legally binding, then what is it?" Ultimately, Byanyima and other civil society leaders walked out of the conference to protest what they viewed as a failure to take steps toward a new, ironclad treaty.

The frustration in Warsaw showed an ongoing failure among many staunch advocates of climate diplomacy to learn the key lesson of Kyoto: Legal force is the wrong litmus test for judging an international framework. Idealized multilateralism has become a trap. It only leads to countries agreeing to the lowest common denominator—or balking altogether.

Evidence shows that a drive for the tightest possible treaty obligations has the perverse effect of provoking resistance. In a seminal 2011 study of climate diplomacy, David Victor of the University of California, San Diego, concluded, "The very attributes that made targets and timetables so attractive to environmentalists—that they set clear, binding goals without much attention to cost—made the Kyoto treaty brittle because countries that discovered they could not honor their commitments had few options but to exit."

This argument may sound like one made by many political conservatives, who opposed Kyoto and have long been wary of treaties in general. But the point is not that international efforts are useless. It is that global agreements are most useful when they include a healthy measure of realism in the demands that they make of countries. Instead of insisting on a binding agreement, diplomats must identify what governments and other actors, like the private sector, are willing to do to combat global warming and develop mechanisms to choreograph, incentivize, and monitor them as they do it. Otherwise, U.N. talks will remain a dialogue of the deaf, as the Earth keeps cooking.

To explain multilateralism's recent failures, from the Kyoto Protocol to the Warsaw conference, its most fervent advocates often take aim at the same purported stumbling block: the spinelessness of politicians. Fainthearted presidents and prime ministers shy away from commitments to protect the planet because it is more politically expedient to focus on economic growth, no matter the environmental consequences.

Thanks to this conventional wisdom, "political will" has become a loaded term. If a leader doesn't sign on to a tough, legally binding treaty, he or she must be morally bankrupt. Mary Robinson, a former president of Ireland who now runs a foundation dedicated to climate change issues, has called the "legal character" of climate agreements "an expression of or an extension of political will." Meanwhile, Kumi Naidoo, executive director of Greenpeace International, has written that he hopes governments will "find the political will to act beyond short-sighted electoral cycles and the corrupting influence of some business elites."

The fallacy of the political will argument, however, is that it assumes everyone already agrees on the steps necessary to address climate change and that the only remaining task is follow-through. It is true that the weight of scientific evidence tells us humanity can only spew so many more gigatons of CO_2 into the air before subjecting the planet and its inhabitants to dire consequences. But the only guidance this gives policymakers is that they must transition to low-carbon economies, stat. It does not tell them how they should do this or how they can do it most efficiently, with the least cost incurred. As a result, advocates of strict climate treaties hammer home the imperative for environmental action without providing for discussion about how countries can actually transform their economies in practice.

Consider environmental author and activist Bill McKibben's comments in early 2013 praising Germany for using more renewable energy: "There were days last summer when Germany generated more than half the power it used from solar panels within its borders. What does that tell you about the relative role of technological prowess and political will in solving this?"

Unfortunately, it tells us very little. It doesn't tell us what it would take to stretch the reliance on solar energy beyond some sunny German days or the subsidy levels required to make solar power a more widely used energy source. It also tells us nothing about how we could translate Germany's accomplishments to countries with very different political and economic circumstances. And it doesn't explain what would induce those diverse countries to accept a multilateral arrangement boosting the global use of renewable energy. All McKibben's factoid tells us is that the myth of political will is quite powerful.

Certainly, economic imperatives should not override environmental ones. Yet the standard for climate diplomacy should not be broad appeals for boldness that ask policymakers to deny trade-offs rather than wrestle with them—particularly in the countries that the world needs most in the fight against global warming. Last fall, after the Warsaw meeting, many experts and pundits were quick to place blame for the gathering's tumult. "The India Problem: Why is it thwarting every international climate agreement?" a headline on Slate demanded. Other observers scorned India and China for saying they would not make "commitments" to greenhouse gas cuts in the 2015 climate agreement. (The meeting's attendees ultimately settled on the word "contributions.")

These complaints, however, are increasingly out of date.

It's true that, throughout most of the 2000s, China and India clung to the exemption that the Kyoto Protocol had granted them, arguing that the industrialized world had caused global warming and that developing countries shouldn't be deprived of their own chance to prosper. This has induced great anxiety because, since 2005, China's annual share of CO_2 emissions has grown from around 16 percent to more than 25 percent, while India has emerged as the world's third-largest carbon emitter. In short, without China and India, progress on climate change will be virtually impossible.

By 2010, however, Beijing and New Delhi had begun to change their stance. A desire to save face diplomatically, combined with increasing pollution at home and domestic need for energy efficiency, have made China and India more willing to cut emissions than ever before.

Chinese leaders in particular are eager to recast their country as an environmental paragon, rather than a pariah. Some analysts attribute this shift to China's aspirations to global prominence. Playing off the popular idea of the "Chinese century," Robert Stavins, director of the Harvard Project on Climate Agreements, has said, "If it's your century, you don't obstruct—you lead." Recently, China has taken significant steps forward with green energy, mimicking many of the regulations and mandates that have helped the United States achieve environmental progress. Wind, solar, and hydroelectric power now provide one-quarter of China's electricity-generating capacity. More energy is being added to China's grid each year from clean sources than from fossil fuels. And in a show of its willingness to step up to the diplomatic plate, China signed an accord with the United States in 2013 that scales down emissions of hydrofluorocarbons, which are so-called super-greenhouse gases.

Yet these changes have not substantially bent the curve of China's total emissions. According to Chris Nielsen and Mun Ho of Harvard University's China Project, this is largely because the country's rapid economic growth makes the tools that have slowed emissions in other economies less effective in China: "[T]he unprecedented pace of China's economic transformation makes improving China's air quality a moving target." Ultimately, Nielsen and Ho argue, the only way for China to rein in emissions will be to attach a price to carbon, through either a tax or a cap-and-trade system. As if on cue, China is now setting up municipal and provincial markets in which polluters can trade emissions credits, with the goal of creating a national market by 2016.

The point here is that the leaders of countries with rapidly developing economies cannot predict environmental payoffs with any real confidence. Tools that work well for others may not for them. That's why China and India are hesitant to sign legally binding treaties, which would put them on the hook to hit targets that could prove much harder to reach than anticipated. They don't want to undertake costly reforms that might not have the predicted benefits, and they do not want to risk the hefty criticism that failure to abide by a treaty would surely bring.

Chinese and Indian leaders realize they'll be judged by their contributions to a cleaner environment, and they embrace the challenge. (Recently in India, more than 20 major industry players launched an initiative to cut emissions.) And they are apt to be less guarded on the international stage if a new climate agreement functions as a measuring stick, not a bludgeon—much like the 2009 Copenhagen accord has done.

In December 2008, the U.N.'s annual cunate conference, hosted in Copenhagen, produced an agreement that is still roundly condemned by environmentalists, the leaders of developing countries, and political liberals alike. Unlike the Kyoto Protocol, the agreement let countries voluntarily set their own targets for emissions cuts over 10 years. "The city of Copenhagen is a crime scene tonight," the executive director of Greenpeace U.K. declared when the deal was reached. Lumumba Di-Aping, the chief negotiator for a group of developing countries known as the G-77, which had wanted major polluters like the United States to take greater responsibility for global warming, said the agreement had "the lowest level of ambition you can imagine."

In reality, however, the conference wasn't a fiasco. It offered the basis for a promising, more flexible regime for climate action that could be a model for the 2015 agreement.

The Copenhagen agreement had a number of advantages. It didn't have to be ratified by governments, which can delay implementation by years. Moreover, in an important new benchmark for climate negotiations, the agreement set the goal of preventing a global average temperature rise of more than 2 °C, with all countries' emission cuts to be gauged against that objective. This provision went to the heart of climate diplomacy's collective-action problem: Apportioning responsibility for cutting emissions among countries is always tricky, but the 2° target creates a shared definition of success.

Most importantly, however, the shift to voluntary pledges showed the first glimmers of lessons learned from the most common mistakes of climate negotiations. In the U.N. process, countries usually operate by consensus: They must all agree on each other's respective climate goals, a surefire recipe for dysfunction. (In 2010, the chair of annual climate talks refused to let a single delegation—Bolivia—block consensus, which counts as a daring move at U.N. conferences.)

Under Copenhagen, by contrast, countries can pledge to do their share while remaining within their comfort zones as dictated by circumstances back home. For instance, faced with economic imperatives to continue delivering high growth, China and India pledged at Copenhagen to reach targets pegged relative to carbon intensity (emissions per unit of economic output) rather than absolute levels of greenhouse gases. This was as far as they were willing to go—but it was further than they'd ever gone before.

Admittedly, the Copenhagen conference wasn't perfect. The deal was struck on the conference's tail end, after U.S. President Barack Obama barged in on a meeting already under way among the leaders of China, India, Brazil, and South Africa. Many of the other delegates registered outrage that the five leaders had negotiated a deal in private by having the conference merely "take note" of the accord.

But the following year's U.N. conference fleshed out the Copenhagen framework, and it has since gained enough legitimacy that 114 countries have agreed to the accord and another

27 have expressed their intention to agree. Taken together, this includes the world's 17 largest emitters, responsible for 80 percent of carbon-based pollution.

The Copenhagen accord will expire as the Kyoto successor agreement takes effect in 2020. But it shouldn't be viewed as just a stopgap. In giving governments more flexibility, Copenhagen offers the chance to build more confidence—and ambition—where historically there has only been uncertainty and rancor. Any future climate agreement should do the same.

"Countries Will Never Keep Mere Promises to Cut Emissions."

Never say never

The most obvious criticism of Copenhagen's system, of course, is that, while it is nice for countries to set voluntary goals, they will never meet them unless they are legally compelled to do so. That is why, just after the Copenhagen deal was reached, then-British Prime Minister Gordon Brown hastily said, "I know what we really need is a legally binding treaty as quickly as possible."

To date, there has been progress on meeting targets set under Copenhagen. The United States and the European Union, for instance, are all within reach of meeting their 10-year goals, perhaps even ahead of schedule. Meanwhile, China's pledge to cut carbon intensity, based on 2005 levels, has become the framework for the country's new emissions-trading markets.

But the most important reason to have confidence in the Copenhagen deal lies in its provisions for measurement, reporting, and verification. If done right, these so-called MRV mechanisms will alert the world as to how countries are (or are not) reducing greenhouse gases, while also pushing states to keep pace toward pledged cuts.

MRVs rely on peer pressure. Countries report to and monitor one another, tracking and urging progress. This kind of system has already proved effective in a variety of international policy areas. For instance, the Mutual Assessment Process of the G-20 and International Monetary Fund brings together the major economic powers to discuss whether their respective policies are helping to maximize global economic growth or are instead widening imbalances between export and consumer-based economies. The process is fairly new, but already, it is widely credited with prodding China long reluctant to discuss these issues in multilateral forums (sound familiar?)—to let its currency appreciate and to make boosting domestic consumption a main plank of its 5-year (2011–2015) plan.

MRVs have also proved valuable in narrower climate regimes, such as the European Union's cap-and-trade mechanism. As a 2012 Environmental Defense Fund report explained, "[B]ecause EU governments based the system's initial caps and emissions allowance allocation on estimates of regulated entities' emissions . . . governments issued too many emissions allowances (over-allocation). Now, however, caps are established on the basis of measured and verified past emissions and best-practices benchmarks, so over-allocation is less of a problem." In other words, MRVs have helped the European Union tighten market standards, correcting an earlier miscalculation and actually heightening the system's ambition.

The Copenhagen agreement enhanced the utility of global, climate-related MRVs by requiring greater transparency from developing countries. Under Kyoto, these countries were only required to provide a summary of their emissions for 2 years: a choice of either 1990 or 1994, and 2000. Copenhagen, by contrast, committed developing countries to report on their emissions biennially—the first reports are due in December—narrowing the gap with the requirement for annual reports that Kyoto imposed on developed countries.

Copenhagen's MRVs are not yet as strong as they could be. For instance, they should require annual reports from all countries, no matter their stages of development. These reports should also include a breakdown of information according to economic subsectors and different greenhouse gases, along with supporting details about data-collection methods. In addition, the process of reviewing reports needs to be fleshed out, taking cues from other strong MRVs that already exist, and wealthier countries should help underwrite the cost to developing countries of preparing comprehensive reports.

The good news is that, given the ongoing nature of U.N. climate diplomacy, it's still possible to strengthen Copenhagen's MRVs. Important new principles and guidelines for peer review have been established in negotiations since 2009, and those involved in climate diplomacy should now buckle down to finish the job. Robust MRVs would guarantee that the world makes the most of the next few years and draws on that experience to chart a new phase of climate action anchored in a 2015 agreement.

"Forget treaties. Solutions will come from the bottom up."

Don't get carried away

Some critics of the U.N. process, hailing from conservative political ranks, the private sector, and other areas, have lost all patience and think that a top-down process, particularly one negotiated in an international forum, is the wrong way to go. They point out that, while national leaders negotiated the Copenhagen deal, actual progress toward its goals is being cobbled together by actors at lower levels—in cities, states, markets, and industries. They are choosing which energy will generate electricity, honing farming practices, improving industrial efficiency, and the like.

Indeed, some policymakers and climate analysts point to the influence of local authorities as a game-changer for climate action. After all, Chinese cities and provinces have begun building emissions-trading markets, and California has passed a law establishing one of the most robust such markets in the world. Meanwhile, leaders of the world's megacities have banded together to cut emissions in what's known as the C40 group, established in 2005. As C40 chair and Rio de Janeiro Mayor Eduardo Paes has put it, "C40's networks and efforts on measurement and reporting are accelerating city-led action at a transformative scale around the world."

Given this sort of local progress, it is certainly worth asking whether diplomats and national policymakers should just get out of the way. Maybe a thoroughly bottom-up approach would be better for the planet than an international climate regime, no matter how flexible. David Hodgkinson, a law professor and executive director of the nonprofit EcoCarbon, which focuses on market solutions for reducing emissions, has argued that such an approach has "more substance" and "probably holds out more hope than a top-down UN deal."

Ultimately, however, this view is misguided. There is no substitute for high-level diplomacy in getting everyone to do their utmost and in keeping track of their efforts. In particular, as Copenhagen reminded the world, the value of the agenda setting, peer pressure, and leverage unique to international diplomacy shouldn't be overlooked. Moreover, we've seen in other policy spheres how the international community can first establish fundamental principles, which then sharpen over time with the aid of global coordinating bodies and more localized initiatives. For instance, the nonbinding 1948 Universal Declaration of Human Rights established a framework for a host of subsequent international treaties, U.N. agencies, regional charters and courts, national policies, and, more recently, corporate responsibility efforts.

Practically speaking, it would also be shortsighted to rely on an assortment of subnational actors to tackle a global problem like climate change. Determining how the work of these actors intersects, what it adds up to, and who monitors that sum are critical matters best managed from the top-down. As the goal of preventing a global average temperature rise of 2 °C reminds us, it is the aggregate of countries' reduced emissions that will be the ultimate test of success.

Even so, the status quo of climate talks, focused on badgering countries to join another legally binding treaty, represents diplomatic overreach. This hasn't worked in the past, and it won't in the future. The international community should give up the quest to sign a legally binding treaty in 2015. Stop fretting about political will and acknowledge the various pressures different countries face. Focus on fully implementing Copenhagen's pledge-and-review system and use that as a model for the successor to Kyoto. Then, allow that new pact to be what steers action and innovation.

Interest in this approach is slowly mounting, including in the U.S. government. Todd Stern, the State Department's special envoy for climate change, said in a 2013 speech, "An agreement that is animated by the progressive development of norms and expectations rather than by the hard edge of law, compliance, and penalty has a much better chance of working." Still, there's a long way to go before the all-or-nothing attitude that has dominated climate diplomacy for so long disappears for good.

In the meantime, the environmental clock keeps ticking.

Critical Thinking

1. Do you think that climate treaties can deal with the problem? Why or why not?
2. Why is it so difficult to persuade developing countries to reduce emissions of greenhouse gases into the atmosphere?
3. Is climate warming a danger now or in the future?

Internet References

Arctic Council
http://www.arctic-council.org/index.php/en

Center for Governance and Sustainability
http://www.umb.edu/cgs

Climate Summit 2014
http://www.un.org/climatechange/summit

UN Framework Convention on Climate Change
http://unfccc.int/2860.php

David Shorr has been analyzing multilateral affairs for over 25 years. He has worked with a range of international organizations and participates in Think20, a global meeting of leading think-tank representatives.

Article

Prepared by: Robert Weiner, *University of Massachusetts, Boston*

The Clean Energy Revolution

Fighting Climate Change with Innovation

VARUN SIVARAM AND TERYN NORRIS

Learning Outcomes

After reading this article, you will be able to:

- Understand the importance of R&D for innovative clean energy technology
- Explain the advantage of clean energy technology as compared to fossil fuels

As the UN Climate Change Conference in Paris came to a close in December 2015, foreign ministers from around the world raised their arms in triumph. Indeed, there was more to celebrate in Paris than at any prior climate summit. Before the conference, over 180 countries had submitted detailed plans to curb their greenhouse gas emissions. And after two weeks of intense negotiation, 195 countries agreed to submit new, stronger plans every five years.

But without major advances in clean energy technology, the Paris agreement might lead countries to offer only modest improvements in their future climate plans. That will not be enough. Even if they fulfill their existing pledges, the earth will likely warm by some 2.7°C to 3.5°C—risking planetary catastrophe. And cutting emissions much more is a political nonstarter, especially in developing countries such as India, where policymakers must choose between powering economic growth and phasing out dirty fossil fuels. As long as this trade-off persists, diplomats will come to climate conferences with their hands tied.

It was only on the sidelines of the summit, in fact, that Paris delivered good news on the technology front. Bill Gates unveiled the Breakthrough Energy Coalition, a group of more than two dozen wealthy sponsors that plan to pool investments in early stage clean energy technology companies. And U.S. President Barack Obama announced Mission Innovation, an agreement among 20 countries—including the world's top three emitters, China, the United States, and India—to double public funding for clean energy R&D to $20 billion annually by 2020. Washington will make or break this pledge, since over half of the target will come from doubling the U.S. government's current $6.4 billion yearly budget.

Fighting climate change successfully will certainly require sensible government policies to level the economic playing field between clean and dirty energy, such as putting a price on carbon dioxide emissions. But it will also require policies that encourage investment in new clean energy technology, which even a level playing field may not generate on its own. That will take leadership from the United States, the only country with the requisite innovative capacity. In the past, the United States has seen investment in clean energy innovation surge forward, only to collapse afterward. To prevent this from happening again, the government should dramatically ramp up its support for private and public R&D at home and abroad. The task is daunting, to be sure, but so are the risks of inaction.

Don't Stop Thinking about Tomorrow

The key to a low-carbon future lies in electric power. Improvements in that sector are important not just because electric power accounts for the largest share of carbon dioxide emissions but also because reaping the benefits of innovations downstream—such as electric vehicles—requires a clean electricity supply upstream. Fossil-fueled power plants now account for nearly 70 percent of electricity globally. But by 2050, the International Energy Agency has warned, this figure must plummet to seven percent just to give the world a 50 percent chance of limiting global warming to two degrees Celsius. More fossil-fueled power is acceptable only if the carbon emissions can be captured and stored underground. And zero-carbon power sources, such as solar, wind, hydroelectric, and nuclear power, will need to grow rapidly, to the point where they supply most of the world's electricity by the middle of the century.

The problem, however, is that the clean technologies now making progress on the margins of the fossil-fueled world may not suffice in a world dominated by clean energy. The costs of solar and wind power, for example, are falling closer to those of natural gas and coal in the United States, but this has been possible because of flexible fossil fuel generators, which smooth out the highly variable power produced by the sun and wind. Ramping up the supply of these intermittent sources will oversupply the electrical grid at certain times, making renewable power less valuable and requiring extreme swings in the dwindling output of fossil fuel generators. Nuclear and hydroelectric power, for their part, are more reliable, but both have run into stiff environmental opposition. As a result, trying to create a zero-carbon power grid with only existing technologies would be expensive, complicated, and unpopular.

Similarly, cleaning up the transportation sector will require great technological leaps forward. Alternative fuels are barely competitive when oil prices are high, and in the coming decades, if climate policies succeed in reducing the demand for oil, its price will fall, making it even harder for alternative fuels to compete. The recent plunge in oil prices may offer a mere foretaste of problems to come: it has already put biofuel companies out of business and lured consumers away from electric vehicles.

All of this means that a clean, affordable, and reliable global energy system will require a diverse portfolio of low-carbon technologies superior to existing options. Nuclear, coal, and natural-gas generators will still be necessary to supply predictable power. But new reactor designs could make nuclear meltdowns physically impossible, and nanoengineered membranes could block carbon emissions in fossil-fueled power plants. Solar coatings as cheap as wallpaper could enable buildings to generate more power than they consume. And advanced storage technologies—from energy-dense batteries to catalysts that harness sunlight to split water and create hydrogen fuel—could stabilize grids and power vehicles. The wish list goes on: new ways to tap previously inaccessible reservoirs of geothermal energy, biofuels that don't compete with food crops, and ultra-efficient equipment to heat and cool buildings.

Trying to create a zerocarbon power grid with only existing technologies would be expensive, complicated, and unpopular.

Every one of those advances is possible, but most need a fundamental breakthrough in the lab or a first-of-its-kind demonstration project in the field. For example, the quest for the ideal catalyst to use sunlight to split water still hasn't produced a winning chemical, and an efficient solar power coating called "perovskite" still isn't ready for widespread use. So it is alarming that from 2007 to 2014, even as global financial flows to deploy mature clean energy doubled to $288 billion, private investment in early stage companies sank by nearly 50 percent, to less than $2.6 billion. But the United States can reverse that trend.

Third Time's the Charm?

Since the development of civilian nuclear power after World War II, the United States has experienced two booms in clean energy innovation, followed by two busts. The first boom, a response to the oil shocks of the 1970s, was driven by public investment. From 1973 to 1980, the federal government quadrupled investment in energy R&D, funding major improvements in both renewable and fossil fuel energy sources. But when the price of oil collapsed in the 1980s, the administration of President Ronald Reagan urged Congress to leave energy investment decisions to market forces. Congress acquiesced, slashing energy R&D funding by more than 50 percent over Reagan's two terms.

The second wave of investment in clean energy innovation began with the private sector. Soon after the turn of the millennium, venture capital investors began pumping money into U.S. clean energy start-ups. Venture capital investment in the sector grew tenfold, from roughly $460 million per year in 2001 to over $5 billion by 2010. Thanks to Obama's stimulus package, federal funding soon followed, and from 2009 to 2011, the government plowed over $100 billion into the sector through a mix of grants, loans, and tax incentives (although most of this influx subsidized the deployment of existing technologies). Some of the start-ups from this period became successful publicly traded companies, including the electric car maker Tesla, the solar-panel installer SolarCity, and the software provider Opower.

But the vast majority failed, and the surviving ones returned too little to make up for the losses. Indeed, of the $36 billion that venture capital firms invested from 2004 to 2014, up to half may ultimately be lost. The gold rush ended abruptly: from 2010 to 2014, venture capital firms cut their clean energy investment portfolios by 75 percent. And the federal government, reeling from political blowback over the bankruptcies of some recipients of federal loan guarantees (most famously, the solar-panel manufacturer Solyndra), pared back its support for risky ventures, too.

Yet all was not lost, for the failures of these two waves offer lessons for how to make sure the next one proves more enduring. First, they revealed just how important government funding is: after the drop in federal energy R&D in the 1980s, patent filings involving solar, wind, and nuclear power plunged. Today, although the United States is the largest funder of energy R&D in the world, it chronically underspends compared with its investments in other national research priorities. Its $6.4 billion clean energy R&D budget is just a fraction of the amount spent on space exploration ($13 billion), medicine ($31 billion), and defense ($78 billion). Given the gap, Congress should follow through on the Mission Innovation pledge and at least double funding for clean energy R&D. Already, Congress increased spending on applied energy R&D by 10 percent in its 2016 budget, more than it increased spending on any other major R&D agency or program. But starting in 2017, doubling the budget in five years will require annual increases of at least 15 percent.

The second lesson is that the government should fund not only basic research but applied research and demonstration projects, too. Washington's bias goes back decades. In his seminal 1945 report, *Science, the Endless Frontier*, Vannevar Bush, President Franklin Roosevelt's top science adviser, urged the government to focus on basic research, which would generate insights that the private sector was supposed to translate into commercial technologies. Successive administrations mostly heeded his advice, and Reagan doubled down on it, slashing nearly all funding for applied energy R&D. By the late 1990s, basic research would account for 60 percent of all federal spending on energy R&D. Instead of creating space for the private sector to pick up where the government left off, however, the budget cuts scared it away. Private investment shrank by half from 1985 to 1995, stranding public investments in alternative fuels, solar photovoltaic panels, and advanced nuclear reactors.

A similar story unfolded at the end of the second boom in clean energy innovation. When one-time stimulus funding expired after 2011, public funding for demonstration projects—which prove whether new technologies work in real-world conditions—fell by over 90 percent. Private investors had expected to share the risk of such projects with the federal government, but when government funding evaporated, investors pulled their money out—canceling, among others, several projects to capture and store carbon emissions from coal power plants.

Thus, policymakers should increase the kind of public investment that attracts private capital. To that end, the first priority should be to restore public funding for demonstration projects. The last redoubt of support for these projects can be found in the Department of Energy's politically embattled loan guarantee program. To insulate funding from political caprice, the American Energy Innovation Council, a group of business leaders, has proposed an independent, federally chartered corporation that would finance demonstration projects. Others have proposed empowering states or regions to fund their own projects, with matching federal grants. If they make it past Congress, both proposals could unlock considerable private investment.

The Department of Energy has made more progress in supporting technologies not yet mature enough for demonstration. In 2009, with inspiration from the Defense Advanced Research Projects Agency (DARPA), the U.S. military's incubator for high-risk technologies, it created the Advanced Research Projects Agency–Energy (ARPA-E). Several ARPA-E projects have already attracted follow-on investment from the private sector. In 2013, for example, Google acquired Makani Power, a start-up that is developing a kite that converts high-altitude wind energy into power. The department has also curated public–private partnerships among the government, academics, and companies—dubbed "innovation hubs"—to develop advanced technologies. Obama has advocated tripling ARPA-E's budget to $1 billion by 2021 and creating ten new public–private research centers around the country. Congress should approve these proposals.

The Department of Energy should expand its support for one type of public–private partnership in particular: industrial consortia that pool resources to pursue shared research priorities. Once again, DARPA provides a model. In the 1980s, it helped fund a consortium of computer chip manufacturers called SEMATECH, through which the industry invested in shared R&D and technical standards. By the next decade, the United States had regained market leadership from Japan. Clean energy innovation, by contrast, suffers from corporate apathy. From 2006 to 2014, U.S. firms spent a paltry $3 billion per year on in-house clean energy R&D. They were also reluctant to outsource their energy R&D, acquiring clean energy start-ups only half as often as they did biomedical start-ups.

Public–private partnerships should help diversify the set of private investors funding clean energy innovation. Indeed, venture capitalists alone are insufficient, since clean energy investments require capital for periods longer than venture capitalists generally favor. The Breakthrough Energy Coalition may help solve that problem by infusing the sector with more patient capital. Gates has explained that he and his fellow investors would be willing to wait for years, even decades, for returns on their investments. But his vision depends on the government also ramping up support.

Past failures offer a third and final lesson for policymakers: the need to level the playing field on which emerging clean energy technologies compete against existing ones. In the electricity sector in particular, innovative start-ups are at a disadvantage, since they lack early adopters willing to pay a premium for new products. The biggest customers, electric utilities, tend to be highly regulated territorial monopolies that have little tolerance for risk and spend extremely little on R&D (usually 0.1 percent of total revenues). New York and California are reforming their regulations to encourage utilities to adopt new technologies faster; the federal government should support these efforts financially or, at the very least, get out of the way.

Indeed, government intervention can sometimes be counterproductive. Many current clean energy policies, such as state mandates for utilities to obtain a certain percentage of their power from renewable energy and federal tax credits for solar and wind power installations, implicitly support already-mature technologies. Better policies might carve out allotments or offer prizes for emerging technologies that cost more now but could deliver lower costs and higher performance later. The government could even become a customer itself. The military, for example, might buy early stage technologies such as flexible solar panels, energy-dense batteries, or small modular nuclear reactors.

Innovating Abroad

Clean energy innovation at the international level suffers from similar problems. Like Washington, other governments spend too little on R&D, with the share of all publicly funded R&D in clean energy falling from 11 percent in the early 1980s to four percent in 2015. Thanks to Mission Innovation, that trend could soon be reversed. But if spending rises in an uncoordinated way, governments may duplicate some areas of research and omit others.

Since governments prize their autonomy, the wrong way to solve this problem would be through a centralized, top–down process to direct each country's research priorities. Instead, an existing institution should coordinate spending through a bottom-up approach. The most logical body for that task is the Clean Energy Ministerial, a global forum conceived by the Obama administration that brings together energy officials from nearly every Mission Innovation country. Yet the CEM has no permanent staff, and without support from the next U.S. administration, it might disband. The Obama administration should therefore act quickly to convince its Mission Innovation partners to help fund a permanent secretariat and operating budget for the CEM. Once that happens, the body could issue an annual report of each member's R&D expenditures, which countries could use to hold their peers accountable for their pledges to double funding. The CEM could also convene officials to share trends about the frontiers of applied research, gleaned from grant applications submitted to national funding bodies.

Then there is the problem of foreign companies' aversion to investing in innovation. Producers of everything from solar panels to batteries, mostly in Asia, have focused instead on ruthless cost cutting and in many cases have taken advantage of government assistance to build up massive manufacturing capacity to churn out well-understood technologies. Today, over 2/3 of solar panels are produced in China, where most firms spend less than one percent of their revenue on R&D. (In fact, it was largely the influx of cheap, cookie-cutter solar panels from China that caused U.S. solar start-ups to go bankrupt at the beginning of this decade.)

Not only does this global race to the bottom stunt clean energy innovation; it also matches up poorly with the United States' competitive strengths. In other industries, leading U.S. firms generate economic gains both at home and abroad by investing heavily in R&D. In the electronics, semiconductor, and biomedical industries, for instance, U.S. companies reinvest up to 20 percent of their revenues in R&D.

To encourage foreign companies to invest more in clean energy R&D, the United States should embrace public–private collaboration. A good model is the U.S.–China Clean Energy Research Center (CERC), which was set up in 2009 and is funded by the U.S. and Chinese governments, academic institutions, and private corporations. Notably, CERC removes a major obstacle to international collaboration: intellectual property theft. Participants are bound by clear rules about the ownership and licensing of technologies invented through CERC. And unless they agree otherwise, they must submit disputes to international arbitration governed by UN rules. More than 100 firms have signed on, and in 2014, China and the United States enthusiastically extended the partnership. It's time for the United States to apply CERC's intellectual property framework to collaborations with other countries, such as India, with which it has no such agreement.

The Next Revolution

By investing at home and leading a technology push abroad, the United States would give clean energy innovation a badly needed boost. Energy executives would at last rub elbows with top academics at technology conferences. Industrial consortia would offer road maps for dramatic technological improvements that forecast future breakthroughs. And institutional investors would bet on start-ups and agree to wait a decade or more before seeing a return.

To many in Washington, this sounds like an expensive fantasy. And indeed, transforming the energy sector into an innovative powerhouse would prove even harder and costlier than the Manhattan Project or the Apollo mission. In both cases, the government spent billions of dollars on a specific goal, whereas success in clean energy innovation requires both public and private investment in a wide range of technologies.

Yet the United States has achieved similar transformations before. Take the biomedical industry. Like clean energy start-ups, biomedical start-ups endured boom-and-bust investment cycles in the 1980s and 1990s. But today, partly thanks to high and sustained public funding, the private sector invests extensively in biomedical innovation. One might object that the biomedical industry's high profit margins, in contrast to the slim ones that characterize the clean energy industry, allow it to invest more in R&D. But the clean energy sector need not be condemned to permanently small profits: innovative firms could earn higher margins than today's commodity producers by developing new products that serve unmet demands.

With clean energy, the stakes could hardly be higher. If the world is to avoid climate calamity, it needs to reduce its carbon emissions by 80 percent by the middle of this century—a target that is simply out of reach with existing technology. But armed with a more potent low-carbon arsenal, countries could make pledges to cut emissions that were both ambitious and realistic. Emerging economies would no longer face tradeoffs between curbing noxious fossil fuels and lifting their populations out of energy poverty. And the United States would place itself at the forefront of the next technological revolution.

Critical Thinking

1. Why did the first two waves of investment in clean energy technology fail?
2. What are the advantages of private/public partnership in the development of clean energy?
3. Why should the U.S. encourage companies to invest more in clean energy?

Internet References

Climate links
www.climatelinks.org

Climate summit
http://www.un/org/climatechange/summit

UN Framework Convention on Climate Change
http://unfcc.int/2860.php

Varun Sivaram is a Douglas Dillon fellow at the Council on Foreign Relations.

Teryn Norris is a former special adviser at the U.S. Department of Energy.

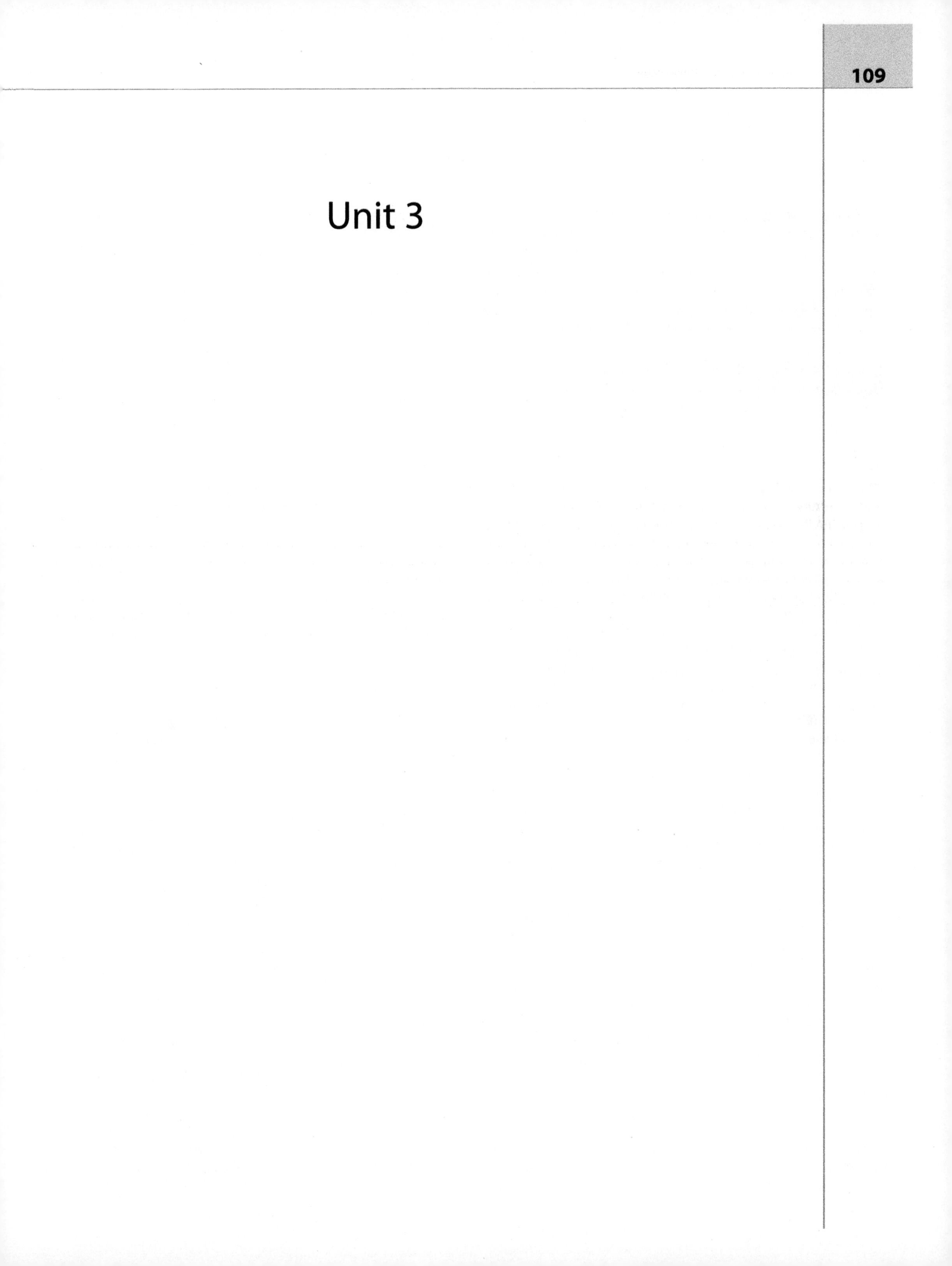

Unit 3

UNIT

Prepared by: Robert Weiner, *University of Massachusetts, Boston*

The Global Political Economy

A defining characteristic of the twentieth century was an intense struggle between proponents of two economic ideologies. At the heart of the conflict was the question of what role government should play in the management of a country's economy. For some, the dominant capitalist economic system appeared to be organized primarily for the benefit of a few wealthy people. From their perspective, the masses were trapped in poverty providing cheap labor to further enrich the privileged elite. These critics argued that the capitalist system could be changed only by giving control of the political system to the state and having the state own the means of production. In striking contrast to this perspective, others argued that the best way to create wealth and eliminate poverty was through the profit motive. The profit motive encouraged entrepreneurs to create products and businesses at the cutting edge of new technologies. An example of this is "The Internet of Things," which can be seen as a new industrial revolution in which devices are connected to the Internet. An open and competitive marketplace, from this point of view, minimized government interference and was the best system for making decisions about production, wages, and the distribution of goods and services.

Conflict at times characterized the contest between capitalism and communism/socialism. The Russian and Chinese revolutions ended the old social order and created radical changes in the political and economic systems in these two important countries. The political structures that were created to support new systems of agricultural and industrial production, along with the centralized planning of virtually all aspects of economic activity eliminated most private ownership of property. These two revolutions were, in short, unparalleled experiments in social engineering.

The economic collapse of the Soviet Union and the dramatic market reforms in China have recast the debate about how to best structure contemporary economic systems. Some believe that with the end of communism and the resulting participation of hundreds of millions of consumers in the global market, an unprecedented era has begun.

Many have noted that this process of globalization is being accelerated by an evolution in communications and computer technology such as the Internet of Things and the use of mobile devices. Proponents of this view argue that a new global economy will ultimately eliminate national economic systems. Others are less optimistic about the process of globalization. They argue that the creation of a single economic system where there are no boundaries to impede the flow of capital and goods and services does not mean a closing of the gap between the world's rich and poor. Rather, they argue that multinational corporations and global financial institutions will have fewer legal constraints on their behavior, and this will lead to not only increased risks of periodic financial crises (such as the Great Recession) but also greater expectations of workers and accelerated destruction of the environment. Other analysts of globalization argue that economic development is resulting in the emergence of a global middle class that is closing the economic gap between nations, while the income gap within states is increasing.

The use of the term political economy in this unit recognizes that economic and political systems are not separate. All economic systems have some type of marketplace, where goods and services are bought and sold. Governments, whether national or international, regulate these transactions to some degree: that is, government sets the rules that regulate the marketplace. One of the most important concepts in assessing the contemporary political economy is development. Developed economies, such as the members of the European Union, are characterized by a profile that includes lower infant mortality rates, longer life expectancies, lower disease rates, higher rates of literacy, and healthier sanitation systems. As the process of globalization proceeds, the question is whether control of the economy and financial system of the world by the West is hindering the development of third world countries. Also, rather than a political revolution taking place on a global scale as envisioned by Marx, economic development around the globe has resulted in the emergence of a global middle class that could find itself in conflict with global elites who control the world's economic and financial system. This development was underscored by the rise of populism as a structural factor that influenced the U.S. presidential election in 2016. The phenomenal rise of the Republican candidate for president was fueled by the discontent of the "Forgotten Man" with the economic stagnation that had characterized the country since 2008. By 2016, it was clear that globalization had resulted in closing the income gap which existed between developed and developing nations (due mainly to the economic gains made by China and India) but had also resulted in growing inequality within such countries as the United States. The resentment of blue-collar workers toward an economic system which resulted in economic stagnation also resulted in the unwillingness of the Democratic and Republican candidates for President in 2016 to support the Obama administration's proposal for a free trade agreement known as the TransPacific Partnership. Free trade agreements have been seen by liberal internationalists as a means of avoiding the "beggar thy neighbor" policies which had resulted in the catastrophe of the Second World War. However, the populist appeal to the industrial base of workers in the US rested on assumptions that free trade agreements such as NAFT (the North American Free Trade Agreement) had resulted in the hollowing out of the manufacturing sector of the U.S.

There is an article in this unit that deal with the European Union. The European movement, which emerged after World War II, had the goal of creating a working peace system on the continent. It was based on the idea, beginning with the European Coal and Steel Community, that economic and technical cooperation between France and Germany would spill over into political cooperation. Since then, what is now called the European Union has expanded from its original six members to include 28 states, some drawn from the former communist world in Eastern Europe. However, by 2015, questions had been raised about the future of the European project. The Eurozone continued to be plagued by the financial difficulties of some of its members which were particularly acute in the southern periphery of the economic organization. In 2015, virtually bankrupt members like Greece had to be rescued by a third bailout by an infusion of financial aid from the European Central Bank, the European Commission, and the International Monetary Fund. The process of European economic and political integration that had been proceeding in fits and starts over the past 60 years faced a further challenge when the United Kingdom voted in June 2016 to withdraw from the organization altogether. The United Kingdom had already had an ambivalent relationship with the European movement and did not join the organization until 1973. There existed a strong current of Euro skepticism in Britain, aptly illustrated by the emergence of the UK Independence Party, which, running on a platform of withdrawal from the European Union, reportedly enjoyed over 20 percent support among the British public.

One article in this unit is more optimistic about the prospects of the European Union, while another is somewhat more skeptical. The British general elections were held in May 2015. To the surprise of analysts and pollsters, the incumbent Conservative Party, led by Prime Minister David Cameron, did much better than anticipated. The Conservatives won a majority in the House of Commons, although not a very large one. However, the margin of victory was sufficient for the Conservatives to govern by themselves, without the support of the Liberal Democrats, ending the coalition which had existed for the preceding five years. In the general elections, the UK Independence Party garnered about 12 percent of the vote rather than 20 percent. Nonetheless, the way was still left open for a referendum in June 2016 as to whether the United Kingdom would stay in the European Union or withdraw from it. In a stunning development, the British voted for a Brexit. The actual exit from the European Union would take about two years to negotiate.

Although globalization has contributed to the emergence of new economic centers in the world economic system, such as the BRICS (Brazil, Russia, India, China, and South Africa), poverty still affects the bottom one billion members of world society. The drive of developing countries in what is still called the third world, to achieve the goal of economic modernization; in a number of cases still has a considerable way to go. The collapse of the Western colonial empires, especially since the end of World War II, found a number of colonies enjoying nominal political independence, but still continuing to exist in a condition of economic dependency on the former metropolitan or imperial powers and the advanced economies of the industrialized core of world society. In a number of cases, the developing countries were relegated to the periphery of the world economic system as exporters of commodities and raw materials to the industrialized sector. Developing countries were caught in an economic bind in which the prices that they received for their commodities and raw materials did not keep pace with the prices they had to pay for the importation of manufactured and semi-manufactured goods from the economically advanced states. Multinational corporations were also able to take advantage of the process of globalization to exploit the resources and labor of the developing countries.

By the 1960s and 1970s, the developing countries constituted a majority of the membership of the United Nations. The third world viewed the United Nations, which had been created primarily as a political organization in 1945 (some 70 years ago) as an institution that could be used to mobilize the international community to promote economic justice by redistributing wealth from the industrialized core of world society to countries on the periphery and semi-periphery of the system. In 1964, the developing countries founded the Group of 77 to function as a "poor man's" lobbying group within the framework of the United Nations Conference on Trade and Development (UNCTAD). The Group of 77, which consisted of states drawn from Africa, Asia, Latin America, and the Middle East, was designed to be used as a bargaining tool to extract economic concessions from the richer states in the world economic system. The Group of 77 focused on such issues as better prices for commodities and raw materials exported by the developing countries, a reduction in the prices of manufactured and semi-manufactured goods, access to the technology of the West to be transferred to them on easy terms, a code of conduct for multinational corporations, and the forgiveness or easier terms for the repayment of loans from Western banks and international financial institutions. With the passage of time, in 2014, as UNCTAD observed its 50th anniversary, it was clear that the Group of 77 was not a cohesive body. For example, some of its members had moved into the ranks of the Newly Industrialized Countries (NICS), the petroleum exporting countries, or the "Asian Tigers."

In the 1970s, the developing countries demanded the creation of a New International Economic Order (NIEO). The purpose of the NIEO was to transfer wealth from the developed economies to the developing states in the third world. Among other things, the drive for the new International Economic Order was based on a philosophy of economic justice, that the former imperial powers and the developed world owed reparations to their former colonies for centuries of economic exploitation. Advocates of the NIEO argued that the economic development of the former metropolitan powers would not have been possible without the exploitation of their former colonies. The mobilization of the developing countries to build a NIEO was based also on the success of OPEC (Organization of Petroleum Exporting Countries) in raising the price of a barrel of oil fourfold in 1973. Some of the developing countries advocated the creation of similar commodity-based cartels to extract wealth from the industrialized states. The Group of 77 used its majority of votes in the United Nations General assembly to push through omnibus resolutions demanding the creation of the NIEO. This could be viewed as an effort to rewrite the Charter of the United Nations to focus on the Economic Rights and Duties of States, such as the recognition of a state's sovereign right of ownership of its natural resources. However, the drive for a New International Economic Order failed, due to the unwillingness of the industrialized countries to recognize an obligation to provide the finances necessary to implement the NIEO.

The 1980s were dubbed a "lost decade" as a number of developing countries found themselves unable to repay the loans that they had received from private banks, which were awash with petrodollars that had been recycled to them by the petroleum exporting states. Eventually, through a combination of default (in the case of Mexico), debt forgiveness, and repayment of loans on far less than the value of the loan itself, the external debt crisis was contained.

The next major phase in the international political economy shifted to the concept of sustainable development. Sustainable development means that economic development should be implemented based on the protection of the resources of the environment for future generations. Developing countries, however, had a tendency to balk at the benchmarks that were set by the developed countries as a hindrance to their efforts to move ahead in economic development.

In spite of the global economic meltdown that occurred in 2008, one of the surprising economic success stories in the third world has been the economic progress that has been made in sub-Saharan Africa. The economic renaissance of Africa has been stimulated by such factors as a commodity boom, increased foreign direct investment, and the widespread use of new communications technology, such as cell phones for microfinancial purposes such as banking. However, the African states still need to further develop the institutional structure necessary to promote economic development. The Group of 77 has also called for a reform of the Bretton Woods institutions' voting system on a more equitable basis to reflect the realities of the globalized economy of the twenty-first century.

Article

Prepared by: Robert Weiner, *University of Massachusetts, Boston*

Trade, Development, and Inequality

URI DADUSH

Learning Outcomes

After reading this article, you will be able to:

- Discuss the relationship between trade, labor-saving technology, and income inequality.
- Identify the reasons behind the major mega-regional trade deals with the TransPacific Partnership and the Transatlantic Trade and Investment Partnership.

To trade or not to trade? Judging by the narrow vote by the US Congress in June 2015 to grant President Barack Obama fast-track negotiating authority for trade agreements, the answer today remains in the affirmative, as it has for decades—but resistance is on the rise. According to the many opponents of such deals, trade is a bitter medicine to be taken only in small doses while guarding carefully against its dangerous side effects. If new trade deals are to go ahead at all, the critics say, they should include strict safeguards, such as provisions to uphold environmental and labor standards, and penalties for currency manipulation. Although the trade debate in the United States, the architect of the postwar global trading system, tends to draw the spotlight, the hand-wringing over trade is even more intense in the developing world. Since developing countries protect their home markets more comprehensively than the United States does, the stakes in their trade debates are higher.

Yet this is an era of hyperglobalization in which consumers have become accustomed to searching online for the best-priced merchandise from all over the world. Trade has surged from 25 percent to 60 percent of world GDP in the past 50 years. Why is it still so controversial? The unemployment and dislocation caused by the global financial crisis provide only part of the explanation.

In the United States, the great crisis of 2008–2009 came on the heels of 30 years of stagnant incomes for the vast majority of households, a period that also saw nearly all of the nation's very considerable income gains accrue at the top of the income and wealth pyramid. Trade, especially with China and other low-income countries, is often blamed for the very high and rising inequality in the United States. Such high inequality contributes to a number of ills, including extremely limited opportunity for the children of poor families, bad health outcomes, crime, capture of the legislative process and of government agencies by moneyed interests, and profound political divisions that impede the formulation and execution of economic reforms.

Rising income inequality is not only an American problem. With few exceptions, it has been a common trend around the world, in both advanced and developing countries—most notably in many of the largest developing countries such as China and India. A recent International Monetary Fund report found that over the past 30 years, inequality has risen in every region of the world except Latin America, which nonetheless still has several countries with levels of inequality surpassed only in South Africa.

There is broad agreement among economists that unskilled labor-saving technologies, not trade, have played the central role in increasing inequality. Many believe that the ongoing information and communications technology revolution all but guarantees that this trend will continue. I share these views. I also believe, however, that trade, interacting in a mutually reinforcing fashion with these technologies, has significantly contributed to the inequality trend in both advanced and developing countries. And yet, since technology and trade also lie at the root of the unprecedented postwar advance in average living standards around the world, the sensible policy response is not to try to suppress or reverse trade (or technology, for that matter), even if that were possible, but to adapt to it and to mitigate its effects on the most vulnerable.

Dozens of trade deals are being negotiated around the world today, including giant "mega-regional" arrangements such as the Trans-Pacific Partnership (TPP) and the Transatlantic Trade and Investment Partnership (TTIP). A TPP deal has just been struck, but it needs to be ratified by national legislatures. These new trade deals are as necessary to sustaining economic growth as previous trade deals were in the past. They may or may not lead to even more inequality, depending on the way they are configured, on other reforms that accompany them, and on the specific circumstances of each country or bloc.

The United States and the European Union are unlikely to see much additional effect on inequality from trade deals, simply because they are already very open economies. Trade deals are likely to have a bigger impact on inequality in developing countries, especially those that have the highest trade barriers, such as India and Brazil. Yet these are also the countries that are most likely to see the highest growth dividends from new trade openings, and where unskilled workers are more likely to be net gainers from trade, even if they are losers relative to their most affluent compatriots.

An enormous unfinished trade agenda lies within national borders.

Ancient Arguments

The debate over trade is not new: Aristotle and Plato might have been in opposing camps. According to the late economic historian Murray Rothbard, "Aristotle, in the Greek tradition, was scornful of moneymaking and scarcely a partisan of laissez-faire . . . [yet] he denounced Plato's goal of the perfect unity of the state . . . pointing out that such extreme unity runs against the diversity of mankind, and against the reciprocal advantage that everyone reaps through market exchange."

Two thousand years later, Adam Smith and David Ricardo, writing at a time when Britain had established its commercial preeminence and was leading the Industrial Revolution, conducted systematic analyses of the gains from trade. Smith's analysis of the welfare-enhancing "invisible hand" of markets and his arguments in favor of the international division of labor and the economies of scale to be gained in world markets, together with Ricardo's advocacy of specialization along the lines of comparative advantage, laid the foundations of modern economics.

Their profound insights made little impression in the United States at the time. The great emerging nation of the era comprehensively protected its infant manufacturing sector while relentlessly copying European technology, a policy that it pursued quite consistently throughout the nineteenth century and well into the twentieth. This protectionism culminated in the trade-suffocating Smoot-Hawley tariff hike during the worst of the Great Depression.

It was only as World War II drew to a close that the United States, having achieved a dominant position as the world's leading industrial power, took up the banner of free trade. At the Bretton Woods conference in 1944, the United States insisted that Britain dismantle its system of imperial preferences, a demand the British strenuously resisted. The disagreement over imperialism was a major reason behind their failure to agree on launching an international trade organization along with the World Bank and International Monetary Fund. The trade wing of what came to be called the Bretton Woods system emerged later, first in the shape of the General Agreement on Tariffs and Trade, which took effect in 1948, and then as the World Trade Organization (WTO), launched in Marrakesh at the end of 1994.

Boom Times

In the immediate postwar years, it was the turn of the newly independent developing countries to become the champions of import substitution, to resist protection of intellectual property, and to adopt industrial policies designed to pick market winners and protect the politically powerful—precisely the policies to which the United States had resorted during its developmental phase. Those policies may or may not have worked well for the young giant—we have no way to be sure whether America might have grown even faster under free trade. What we do know, however, is that import substitution did not yield the desired results in developing countries. Soon enough, a turn toward exports and much freer imports ensued.

That big shift toward more outward-looking economic regimes was initially inspired by the extraordinary export-fueled growth of a small number of developing economies in Asia. This trend gained momentum in the wake of the oil shocks of the 1970s, when countries had to look for ways to cover their surging energy bills. Subsequently, the Latin American debt crises of the 1980s discredited import substitution, and the belief in central planning was undermined by stagnating living standards and lagging technologies across the communist bloc. As many observers predicted at the time, the fall of the Berlin Wall heralded the mother of all trade booms, lasting through the 1990s and early 2000s. The intensification of globalization was accompanied by spectacular growth in many developing countries, led by China, which surpassed the United States and Germany to become the world's largest exporter.

During this period, many economists became convinced that, in addition to the efficiency effects stressed by Smith and Ricardo, trade brought potentially even greater benefits, especially to developing nations, by inducing backward firms and economies to learn from those on the technological frontier. This thinking may have been formalized first by Alexander Gerschenkron in a seminal 1951 essay entitled "Economic Backwardness in Historical Perspective."

The focus on learning encouraged development agencies such as the United Nations Conference on Trade and Development (UNCTAD) and the World Bank to view foreign direct investment (FDI) more as a source of new techniques than as a finance vehicle. FDI, which grew even faster than trade, distributed the value chains of the most advanced manufacturers and service providers to cheaper locations or closer to the largest markets, and in the process created millions of jobs in the developing world. According to UNCTAD, the sales of foreign subsidiaries of multinational enterprises in their host countries today exceed world exports by a wide margin. The techniques and methods employed by these state-of-the-art overseas factories and service centers are systematically emulated or copied by less productive local enterprises. Not surprisingly, countries compete fiercely to attract FDI.

Falling Tariffs

The advance toward an open and predictable trading system has been nothing short of remarkable, and it has been matched by the rise of average incomes around the world. Real average per capita income in the United States has more than tripled since 1950, and incomes in developing countries have grown much faster. Over a billion people have been lifted out of absolute poverty in the past 15 years.

At the same time, high-tariff structures have been dismantled. The average tariff in the advanced countries is now around 2 percent–3 percent, and cannot be raised without violating WTO rules. Moreover, countries must apply the most favored nation (MFN) clause, meaning that all WTO members—which account for 97 percent of world trade—must accord each other at least the same tariff treatment, or treat them more favorably under certain specified circumstances. Quotas and subsidies have been outlawed, except (mainly at the insistence of advanced countries) in agriculture, which remains a heavily protected sector across the world.

MFN tariffs in developing countries are on average near 10 percent, much higher than in the advanced countries but about ⅓ their level during the height of import substitution. For the most part, however, these tariffs are not limited by the WTO or are subject to limits set at very high levels. Therefore, unless they are bound by a bilateral or regional agreement, most developing countries have plenty of room to legally raise their MFN tariffs should they decide to do so. (China is a notable exception on account of its demanding WTO accession protocol.)

According to a recent WTO/OECD paper, the movement toward freer trade has continued despite the ill-fated multilateral trade negotiations launched at Doha in 2001. Over the past 20 years, the applied tariffs of WTO members have declined by 15 percent on average, and the share of developing-country exports that now enter advanced countries duty-free has increased from 55 percent to 80 percent. Unilateral trade liberalization, more generous preference regimes, regional trade deals, and the delayed effects of past multilateral trade rounds have all played a role.

Behind the Border

There is still a long way to go before we secure world free trade, by which I mean zero tariffs on all goods, no quotas or subsidies, and complete freedom of entry in service sectors across the world, as well as equal treatment for foreign investors and suppliers, all bound by international treaty in the WTO. In addition to this admittedly distant or even utopian vision, an enormous unfinished trade agenda lies within national borders: reforming domestic regulations and practices that have the effect, sometimes intended but more often unintended, of restricting trade.

The cost of complying with these regulations, together with the cost of transportation and customs duties, and of distribution through wholesalers and retailers, adds up to "trade costs," which, it is estimated, can easily amount to one or two times the price of the product at the factory door. Economists have identified excessive trade costs (due to inappropriate regulations, inadequate transportation infrastructure, inefficient customs, bribes, and so forth) as a much more important barrier to trade today than tariffs. Numerous ongoing bilateral and regional trade negotiations are designed to address them. The Bali Trade Facilitation Agreement, which still requires ratification by ⅔ of members to take effect under the WTO, deals with a relatively narrow set of these behind-the-border issues, namely customs and regulations affecting international transportation and logistics, but arguably these are the issues of most immediate concern to exporters and importers alike.

Trade both stimulates economic growth and increases income inequality.

Naturally, developing and advanced countries have different agendas for addressing the largest remaining impediments to trade. Within each group there is a wide spectrum of interests that often cross over the dividing line. Developing countries aim to reduce the hugely distorting tariffs, quotas, and subsidies in advanced countries that limit their agricultural exports. They are also looking to limit the relatively high tariffs that advanced countries apply to labor-intensive manufactures, such as garments and shoes. In the context of north–south regional deals, such as the Central American Free Trade Agreement or the EU's Mediterranean agreements, developing countries are the parties most interested in less restrictive rules of origin, which are designed to guard against simple transshipment of goods from third parties but are often so complex and restrictive that exporters prefer to pay the full duty rather than try to document their right to preferential treatment.

Advanced countries, for their part, seek to limit tariffs in developing countries across the manufacturing sector. Many of them also wish to improve access for their (subsidized) agricultural and processed food exports, and to secure access to markets in services such as retailing, finance, and insurance. In addition, advanced countries are the ones most concerned with behind-the-border impediments to trade, such as subsidies or licenses accorded to state-owned enterprises, discrimination in government procurement and treatment of foreign investment, lax protection of intellectual property, and slow or unfair settlement of judicial disputes.

Mega-regional Deals

The United States, together with the European Union, spearheaded the adoption of WTO rules through most of the postwar period, but it has recently taken a very different turn. A decade ago, the twin thrusts of American trade policy consisted of a quest for a comprehensive WTO-driven multilateral trade round in the shape of the Doha agenda and the pursuit of a number of relatively minor bilateral trade agreements intended to spur "competitive liberalization." The idea was to induce countries to engage in bilateral deals and the Doha process to avoid being left out.

Today, American trade policy has pretty much written Doha off on account of what Washington perceives as unbridgeable differences between advanced and developing countries. Instead of small bilateral deals, the Obama administration is pursuing two so-called mega-regionals, the TPP for the Pacific and the TTIP with the EU. These deals explicitly aim to rewrite trade rules for the twenty-first century, effectively bypassing the unwieldy WTO. The TTIP has a long way to go, while a deal on the TPP was struck in October.

Given the number of partners involved and its comprehensive scope—it will cover about 40 percent of global GDP—the TPP is one of the most complex free trade agreements ever negotiated. Eleven countries initially joined the negotiations: Australia, Brunei, Canada, Chile, Malaysia, Mexico, New Zealand, Peru, Singapore, the United States, and Vietnam. Japan is a more recent and hugely important addition, and South Korea is a possible partner in the future. As in the case of the TTIP, trade is already largely free among this group; the aim is a high-standard agreement that will go deep behind the border to enhance trade and investment prospects across the board.

The United States already has established free trade agreements with six of the other eleven countries negotiating the TPP, and since it has a very open economy, the new trade liberalization that will be required of it under the deal is minimal. Accordingly, formal studies of the gains that the United States is likely to derive from tariff reductions under the TPP have come up with very small numbers—0.1 percent or 0.2 percent of GDP. Gains from removing nontariff barriers are potentially much larger, but very difficult to quantify.

The TPP, which excludes China, is motivated by political and security concerns as well as by economics. Yet despite the powerful political motivation behind the TPP (it is seen as a key part of the Obama administration's "pivot to Asia"), the diversity of interests among its prospective members has resulted in lengthy delays. The original deadline for completing the deal—the end of 2013—proved wildly optimistic. Although many expect the TPP to be concluded now that Obama has been granted fast-track trade promotion authority, it faces a tough ratification fight in the US Congress.

Trade deals are likely to have a bigger impact on inequality in developing countries.

The resistance to the TPP in the US Congress and the virulent criticisms leveled against it by civil society groups are difficult for its proponents to understand. They see only the deal's strategic importance and the advantage that the United States must commit to very little new liberalization, while its negotiating partners will have to do most of the hard work to reduce their tariff and nontariff barriers. But this is missing the point: At its core, the powerful resistance to new trade agreements in any shape or form is driven by the conviction that they hurt workers and benefit only the privileged few.

Technology and Inequality

The traditional view of trade is that it is triggered by variations in endowments of factors (resources such as land, labor, and capital). In a standard model with two factors, labor and capital, traditional trade theory predicts that a labor-abundant country will export labor-intensive products and see inequality decline as wages rise, while a capital-rich country will see inequality increase as it exports products that are capital-intensive and the return to its capital increases. This traditional model was adequately descriptive of trade in past centuries, when, as in Ricardo's famous example, England exported clothing and Portugal exported wine. Capital was scarcely mobile across borders, and technologies spread gradually.

Today, though, the traditional theory fails the empirical test. Contrary to its predictions, what we observe is that trade takes place predominantly between economies with similar endowments, namely advanced countries exporting to each other highly differentiated products in the same industry, such as cars and machine tools. Crucially, increased trade has been associated with higher inequality not just in advanced but also in developing countries.

Prompted to explain this reality, economists have come up with a number of alternative narratives in recent years, only some of which have been tested econometrically or using case studies. Economists now broadly agree that the most powerful underlying force driving increased inequality is not trade by itself but skill-biased technological change—that is, machines and methods that reduce the need for unskilled labor and boost demand for more specialized and skilled workers. Economists have also shown definitively that changes in aggregate or sectoral exports and imports are far too small relative to the size of the economy to account for the large shifts in industrial structure, employment, relative wages, and inequality that we observe. However, even though trade on its own cannot account for these changes in economic structure and inequality, the new stories tell us that the mutually reinforcing effects of trade on skill-biased technological change can increase inequality.

Take, for example, the case of an advanced country that opens up trade with a large low-wage economy. Firms in the advanced countries that compete in international trade are heterogeneous—they may operate in the same industry but they produce diverse products and vary greatly in their efficiency. Trade quickly kills the least efficient firms in sectors where the low-wage economy has an advantage, namely those that produce standardized products and which are highly labor-intensive. Those firms that survive do so on the basis of three complementary strategies: they automate so as to save on labor, they outsource their most labor-intensive activities to the low-wage economy, and they move upmarket into highly differentiated or technologically advanced niches.

Under all these scenarios, the demand for unskilled labor declines and the demand for skilled labor and capital increases. The dislocation of unskilled labor that results is larger than could be anticipated from the traditional static two-factor model, since trade encourages reduced employment of unskilled labor over time through multiple channels. Outsourcing of unskilled-labor-intensive activities results in less investment and growth in those activities in the future. While the sectors that compete with imports or engage in exporting lead in automation, those techniques are likely to spread throughout the economy, reducing the demand for unskilled labor even further. The fall in their wages prompts increased demand for unskilled workers in the nontraded service sector, but not enough to compensate.

Skills in Demand

What about the effect of trade on inequality in a low-wage or developing economy? As predicted by the traditional models, the demand for unskilled labor caused by opening up trade with high-wage economies will tend to raise the wages of the unskilled. However, there are three influences that can offset this effect and cause inequality to rise anyway.

First, as argued by the late development economist Arthur Lewis, an abundance of excess rural labor with a low reservation wage (the lowest wage at which a worker would accept a job) can slow the rise in wages of unskilled workers, especially at a time when hundreds of millions of unskilled workers are joining the global economy.

Second, as in an advanced economy, the opening of trade will favor a developing country's most efficient firms and those most able to adopt the higher standards demanded by world markets, with the capacity to meet precise specifications and timely delivery schedules. These requirements will often prompt the hiring of specialized workers and the purchase of sophisticated machines. According to the International Federation of Robotics, China is by far the world's fastest-growing market for industrial robots; its installations grew at a rate of about 25 percent a year between 2005 and 2012. Indeed, the import of such machines from advanced countries is the most direct channel through which trade spreads technology. Multinational enterprises from advanced countries invariably bring these advanced techniques with them as part of their outsourcing strategy.

Third, as in advanced countries, trade and foreign investment will stimulate the adoption of advanced techniques throughout the economy—not only in the traded sector. The combination of these effects leads to a sharp rise in the demand for skilled labor, which is relatively scarce in developing countries, as well as a rise in the demand for capital. Even though the demand for unskilled workers also rises, they may remain in plentiful supply for a long time, their wages rising relatively slowly, resulting in increased inequality.

The connecting thread in all these stories is that trade and, more broadly, international exchange prompt the spread of the most advanced technologies and encourage every firm exposed to increased competition, whether in advanced or developing countries, to become more efficient, thus raising the demand for skilled labor and for sophisticated capital goods. Moreover, since capital and, to a lesser extent, the most highly skilled professionals are more internationally mobile than unskilled workers, their rewards will tend to increase to match the best opportunities available anywhere in the world.

The Trade Dilemma

The modern theory of how trade both stimulates economic growth and increases income inequality applies in both advanced and developing countries. There is, however, an important difference between the two groups. Not only are developing countries catching up technologically and growing much faster; in labor-abundant developing countries the wages of the unskilled are likely sooner or later to rise with increased trade, even if they lose ground in relative terms to the skilled cohort. In contrast, unskilled labor in advanced countries could be a net loser in both relative and absolute terms. The theory is consistent with what we have observed: rising wages in developing countries, stagnant wages of unskilled workers in advanced countries, and rising inequality in both groups.

If the theory is correct, it presents an acute dilemma. Should countries pursue trade deals and grow more rapidly or should they eschew them, grow more slowly, and avoid their inequality-intensifying effects? The dilemma is sharper for advanced countries whose unskilled laborers may lose outright.

The answer goes back to the minimum concept of efficiency developed at the turn of the twentieth century by the Italian economist Vilfredo Pareto: Countries should pursue the efficient solution (in this case, open trade), making the pie bigger, and then redivide it in favor of the losers so that no one is worse off. Politically, this is easier said than done, but the necessary economic policies are familiar and the instruments are well honed. As the IMF has stressed in a recent analysis, compensating the losers from labor-saving technology or trade need not result in a loss of efficiency. Investments in education, health, and infrastructure that boost the incomes of unskilled workers and level the playing field for their children are likely to enhance economic growth. More progressive income and wealth taxes can be achieved by closing the many tax loopholes and inequities that distort economic incentives. Cuts to subsidies that favor rich farmers, purchasers of large homes, or drivers of gas-guzzling vehicles are likely to both increase efficiency and reduce inequality.

Developing countries are least prepared to execute these policies because they have limited taxation and administrative capacity, but they—and their unskilled workers—are the most likely to benefit from new trade deals even if they increase inequality. Advanced countries such as the United States are already largely open and have little to fear from new trade deals, which can consolidate their export interests without causing an unacceptable further rise in inequality.

However, the United States is also the country with the most pressing need to help unskilled workers cope with the effects of advances in labor-saving technology and their mutually reinforcing interaction with globalization. It has all the tools to respond. Its failure to confront rising inequality presents a threat both to its continued economic growth and to its leadership of the open global trading system that Washington played such a large role in creating.

Critical Thinking

1. What are the reasons for the opposition to the mega-regional trade deals?
2. How does the TransPacific Partnership fit into the "pivot" to Asia?
3. Who will benefit the most from the TransPacific Partnership?

Internet References

Transatlantic Trade and Investment Partnership
https://ustr.gov/ttip

TransPacific Partnership
https./ustr.gov/tpp/

Uri Dadush is a senior associate in the international economics program at the Carnegie Endowment for International Peace and a Current History contributing editor.

Article

Prepared by: Robert Weiner, *University of Massachusetts, Boston*

The Truth about Trade

What Critics Get Wrong about the Global Economy

DOUGLAS A. IRWIN

Learning Outcomes

After reading this article, you will be able to:

- Discuss the real cause of the loss of jobs in the U.S.
- Understand the effects of trade with China on the U.S. economy

Just because a U.S. presidential candidate bashes free trade on the campaign trail does not mean that he or she cannot embrace it once elected. After all, Barack Obama voted against the Central American Free Trade Agreement as a U.S. senator and disparaged the North American Free Trade Agreement (NAFTA) as a presidential candidate. In office, however, he came to champion the Trans-Pacific Partnership (TPP), a giant trade deal with 11 other Pacific Rim countries.

Yet in the current election cycle, the rhetorical attacks on U.S. trade policy have grown so fiery that it is difficult to imagine similar transformations. The Democratic candidate Bernie Sanders has railed against "disastrous" trade agreements, which he claims have cost jobs and hurt the middle class. The Republican Donald Trump complains that China, Japan, and Mexico are "killing" the United States on trade thanks to the bad deals struck by "stupid" negotiators. Even Hillary Clinton, the expected Democratic nominee, who favored the TPP as secretary of state, has been forced to join the chorus and now says she opposes that agreement.

Blaming other countries for the United States' economic woes is an age-old tradition in American politics; if truth is the first casualty of war, then support for free trade is often an early casualty of an election campaign. But the bipartisan bombardment has been so intense this time, and has been so unopposed, that it raises real questions about the future of U.S. global economic leadership.

The antitrade rhetoric paints a grossly distorted picture of trade's role in the U.S. economy. Trade still benefits the United States enormously, and striking back at other countries by imposing new barriers or ripping up existing agreements would be self-destructive. The badmouthing of trade agreements has even jeopardized the ratification of the TPP in Congress. Backing out of that deal would signal a major U.S. retreat from Asia and mark a historic error.

Still, it would be a mistake to dismiss all of the antitrade talk as ill-informed bombast. Today's electorate harbors legitimate, deep-seated frustrations about the state of the U.S. economy and labor markets in particular, and addressing these complaints will require changing government policies. The solution, however, lies not in turning away from trade promotion but in strengthening worker protections.

By and large, the United States has no major difficulties with respect to trade, nor does it suffer from problems that could be solved by trade barriers. What it does face, however, is a much larger problem, one that lies at the root of anxieties over trade: the economic ladder that allowed previous generations of lower-skilled Americans to reach the middle class is broken.

Scapegoating Trade

Campaign attacks on trade leave an unfortunate impression on the American public and the world at large. In saying that some countries "win" and other countries "lose" as a result of trade, for example, Trump portrays it as a zero-sum game. That's an understandable perspective for a casino owner and businessman: gambling is the quintessential zero-sum game, and competition is a win-lose proposition for firms (if not for their customers). But it is dead wrong as a way to think about the role of trade in an economy. Trade is actually a two-way street—the exchange of exports for imports—that makes efficient use of a country's resources to increase its material welfare. The United States sells to other countries the goods and services that it produces relatively efficiently (from aircraft to soybeans to legal advice) and buys those goods and services that other countries produce relatively efficiently (from T-shirts to bananas to electronics assembly). In the aggregate, both sides benefit.

To make their case that trade isn't working for the United States, critics invoke long-discredited indicators, such as the country's negative balance of trade. "Our trade deficit with China is like having a business that continues to lose money every single year," Trump once said. "Who would do business like that?" In fact, a nation's trade balance is nothing like a firm's bottom line. Whereas a company cannot lose money indefinitely, a country—particularly one, such as the United States, with a reserve currency—can run a trade deficit indefinitely without compromising its well-being. Australia has run current account deficits even longer than the United States has, and its economy is flourishing.

One way to define a country's trade balance is the difference between its domestic savings and its domestic investment. The United States has run a deficit in its current account—the broadest measure of trade in goods and services—every year except one since 1981. Why? Because as a low-saving, high-consuming country, the United States has long been the recipient of capital inflows from abroad. Reducing the current account deficit would require foreigners to purchase fewer U.S. assets. That, in turn, would require increasing domestic savings or, to put it in less popular terms, reducing consumption. One way to accomplish that would be to change the tax system—for example, by instituting a consumption tax. But discouraging spending and rewarding savings is not easy, and critics of the trade deficit do not fully appreciate the difficulty involved in reversing it. (And if a current account surplus were to appear, critics would no doubt complain, as they did in the 1960s, that the United States was investing too much abroad and not enough at home.)

Critics also point to the trade deficit to suggest that the United States is losing more jobs as a result of imports than it gains due to exports. In fact, the trade deficit usually increases when the economy is growing and creating jobs and decreases when it is contracting and losing jobs. The U.S. current account deficit shrank from 5.8 percent of GDP in 2006 to 2.7 percent in 2009, but that didn't stop the economy from hemorrhaging jobs. And if there is any doubt that a current account surplus is no economic panacea, one need only look at Japan, which has endured three decades of economic stagnation despite running consistent current account surpluses.

And yet these basic fallacies—many of which Adam Smith debunked more than two centuries ago—have found a new life in contemporary American politics. In some ways, it is odd that antitrade sentiment has blossomed in 2016, of all years. For one thing, although the postrecession recovery has been disappointing, it has hardly been awful: the U.S. economy has experienced seven years of slow but steady growth, and the unemployment rate has fallen to just 5 percent. For another thing, imports have not swamped the country and caused problems for domestic producers and their workers; over the past seven years, the current account deficit has remained roughly unchanged at about two to three percent of GDP, much lower than its level from 2000 to 2007. The pace of globalization, meanwhile, has slowed in recent years. The World Trade Organization (WTO) forecasts that the volume of world trade will grow by just 2.8 percent in 2016, the fifth consecutive year that it has grown by less than three percent, down significantly from previous decades.

What's more, despite what one might infer from the crowds at campaign rallies, Americans actually support foreign trade in general and even trade agreements such as the TPP in particular. After a decade of viewing trade with skepticism, since 2013, Americans have seen it positively. A February 2016 Gallup poll found that 58 percent of Americans consider foreign trade an opportunity for economic growth, and only 34 percent viewed it as a threat.

The View from the Bottom

So why has trade come under such strident attack now? The most important reason is that workers are still suffering from the aftermath of the Great Recession, which left many unemployed and indebted. Between 2007 and 2009, the United States lost nearly nine million jobs, pushing the unemployment rate up to ten percent. Seven years later, the economy is still recovering from this devastating blow. Many workers have left the labor force, reducing the employment-to-population ratio sharply. Real wages have remained flat. For many Americans, the recession isn't over.

Thus, even as trade commands broad public support, a significant minority of the electorate—about a third, according to various polls—decidedly opposes it. These critics come from both sides of the political divide, but they tend to be lower-income, blue-collar workers, who are the most vulnerable to economic change. They believe that economic elites and the political establishment have looked out only for themselves over the past few decades. As they see it, the government bailed out banks during the financial crisis, but no one came to their aid.

Trade still benefits the United States enormously.

For these workers, neither political party has taken their concerns seriously, and both parties have struck trade deals that the workers think have cost jobs. Labor unions that support the Democrats still feel betrayed by President Bill Clinton, who, over their strong objections, secured congressional passage of NAFTA in 1993 and normalized trade relations with China in 2000. Blue-collar Republican voters, for their part, supported the anti-NAFTA presidential campaigns of Pat Buchanan and Ross Perot in 1992. They felt betrayed by President George W. Bush, who pushed Congress to pass many bilateral trade agreements. Today, they back Trump.

Among this demographic, a narrative has taken hold that trade has cost Americans their jobs, squeezed the middle class, and kept wages low. The truth is more complicated. Although imports have put some people out of work, trade is far from the most important factor behind the loss of manufacturing jobs. The main culprit is technology. Automation and other technologies have enabled vast productivity and efficiency improvements, but they have also made many blue-collar jobs obsolete. One representative study, by the Center for Business and Economic Research at Ball State University, found that productivity growth accounted for more than 85 percent of the job loss in manufacturing between 2000 and 2010, a period when employment in that sector fell by 5.6 million. Just 13 percent of the overall job loss resulted from trade, although in two sectors, apparel and furniture, it accounted for 40 percent.

This finding is consistent with research by the economists David Autor, David Dorn, and Gordon Hanson, who have estimated that imports from China displaced as many as 982,000 workers in manufacturing from 2000 to 2007. These layoffs also depressed local labor markets in communities that produced goods facing Chinese competition, such as textiles, apparel, and furniture. The number of jobs lost is large, but it should be put in perspective: while Chinese imports may have cost nearly one million manufacturing jobs over almost a decade, the normal churn of U.S. labor markets results in roughly 1.7 million layoffs every month.

Research into the effect of Chinese imports on U.S. employment has been widely misinterpreted to imply that the United States has gotten a raw deal from trade with China. In fact, such studies do not evaluate the gains from trade, since they make no attempt to quantify the benefits to consumers from lower-priced goods. Rather, they serve as a reminder that a rapid increase in imports can harm communities that produce substitute goods—as happened in the U.S. automotive and steel sectors in the 1980s.

Furthermore, the shock of Chinese goods was a one-time event that occurred under special circumstances. Imports from China increased from 1.0 percent of U.S. GDP in 2000 to 2.6 percent in 2011, but for the past five years, the share has stayed roughly constant. There is no reason to believe it will rise further. China's once-rapid economic growth has slowed. Its working-age population has begun to shrink, and the migration of its rural workers to coastal urban manufacturing areas has largely run its course.

The influx of Chinese imports was also unusual in that much of it occurred from 2001 to 2007, when China's current account surplus soared, reaching ten percent of GDP in 2007. The country's export boom was partly facilitated by China's policy of preventing the appreciation of the yuan, which lowered the price of Chinese goods. Beginning around 2000, the Chinese central bank engaged in a large-scale, persistent, and one-way intervention in the foreign exchange market—buying dollars and selling yuan. As a result, its foreign exchange reserves rose from less than $300 million in 2000 to $3.25 trillion in 2011. Critics rightly groused that this effort constituted currency manipulation and violated International Monetary Fund rules. Yet such complaints are now moot: over the past year, China's foreign exchange reserves have fallen rapidly as its central bank has sought to prop up the value of the yuan. Punishing China for past bad behavior would accomplish nothing.

The Right—and Wrong—Solutions

The real problem is not trade but diminished domestic opportunity and social mobility. Although the United States boasts a highly skilled work force and a solid technological base, it is still the case that only one in three American adults has a college education. In past decades, the ⅔ of Americans with no postsecondary degree often found work in manufacturing, construction, or the armed forces. These parts of the economy stood ready to absorb large numbers of people with limited education, give them productive work, and help them build skills. Over time, however, these opportunities have disappeared. Technology has shrunk manufacturing as a source of large-scale employment: even though US manufacturing output continues to grow, it does so with many fewer workers than in the past. Construction work has not recovered from the bursting of the housing bubble. And the military turns away 80 percent of applicants due to stringent fitness and intelligence requirements. There are no comparable sectors of the economy that can employ large numbers of high school educated workers.

This is a deep problem for American society. The unemployment rate for college-educated workers is 2.4 percent, but it is more than 7.4 percent for those without a high school diploma—and even higher when counting discouraged workers who have left the labor force but wish to work. These are the people who have been left behind in the 21st century economy—again, not primarily because of trade but because of structural changes in the economy. Helping these workers and ensuring that the economy delivers benefits to everyone should rank as urgent priorities.

But here is where the focus on trade is a diversion. Since trade is not the underlying problem in terms of job loss, neither is protectionism a solution. While the gains from trade can seem abstract, the costs of trade restrictions are concrete. For example, the United States has some 135,000 workers employed in the apparel industry, but there are more than 45 million Americans who live below the poverty line, stretching every dollar they have. Can one really justify increasing the price of clothing for 45 million low-income Americans (and everyone else as well) in an effort to save the jobs of just some of the 135,000 low-wage workers in the apparel industry?

Like undoing trade agreements, imposing selective import duties to punish specific countries would also fail. If the United States were to slap 45 percent tariffs on imports from China, as Trump has proposed, U.S. companies would not start producing more apparel and footwear in the United States, nor would they start assembling consumer electronics domestically. Instead, production would shift from China to other low-wage developing countries in Asia, such as Vietnam. That's the lesson of past trade sanctions directed against China alone: in 2009, when the Obama administration imposed duties on automobile tires from China in an effort to save American jobs, other suppliers, principally Indonesia and Thailand, filled the void, resulting in little impact on U.S. production or jobs.

For many Americans, the recession isn't over.

And if restrictions were levied against all foreign imports to prevent such trade diversion, those barriers would hit innocent bystanders: Canada, Japan, Mexico, the EU, and many others. Any number of these would use WTO procedures to retaliate against the United States, threatening the livelihoods of the millions of Americans with jobs that depend on exports of manufactured goods. Trade wars produce no winners. There are good reasons why the very mention of the 1930 Smoot-Hawley Tariff Act still conjures up memories of the Great Depression.

If protectionism is an ineffectual and counterproductive response to the economic problems of much of the work force, so, too, are existing programs designed to help workers displaced by trade. The standard package of Trade Adjustment Assistance, a federal program begun in the 1960s, consists of extended unemployment compensation and retraining programs. But because these benefits are limited to workers who lost their jobs due to trade, they miss the millions more who are unemployed on account of technological change. Furthermore, the program is fraught with bad incentives. Extended unemployment compensation pays workers for prolonged periods of joblessness, but their job prospects usually deteriorate the longer they stay out of the labor force, since they have lost experience in the interim.

And although the idea behind retraining is a good one—helping laid-off textile or steel workers become nurses or technicians—the actual program is a failure. A 2012 external review commissioned by the Department of Labor found that the government retraining programs were a net loss for society, to the tune of about $54,000 per participant. Half of that fell on the participants themselves, who, on average, earned $27,000 less over the four years of the study than similar workers who did not find jobs through the program, and half fell on the government, which footed the bill for the program. Sadly, these programs appear to do more harm than good.

A better way to help all low-income workers would be to expand the Earned Income Tax Credit. The EITC supplements the incomes of workers in all low-income households, not just those the Department of Labor designates as having been adversely affected by trade. What's more, the EITC is tied to employment, thereby rewarding work and keeping people in the labor market, where they can gain experience and build skills. A large enough EITC could ensure that every American was able to earn the equivalent of $15 or more per hour. And it could do so without any of the job loss that a minimum-wage hike can cause. Of all the potential assistance programs, the EITC also enjoys the most bipartisan support, having been endorsed by both the Obama administration and Paul Ryan, the Republican Speaker of the House. A higher EITC would not be a cure-all, but it would provide income security for those seeking to climb the ladder to the middle class.

The main complaint about expanding the EITC concerns the cost. Yet taxpayers are already bearing the burden of supporting workers who leave the labor force, many of whom start receiving disability payments. On disability, people are paid—permanently—to drop out of the labor force and not work. In lieu of this federal program, the cost of which has surged in recent years, it would be better to help people remain in the work force through the EITC, in the hope that they can eventually become taxpayers themselves.

The Future of Free Trade

Despite all the evidence of the benefits of trade, many of this year's crop of presidential candidates have still invoked it as a bogeyman. Sanders deplores past agreements but has yet to clarify whether he believes that better ones could have been negotiated or no such agreements should be reached at all. His vote against the U.S.-Australian free-trade agreement in 2004 suggests that he opposes all trade deals, even one with a country that has high labor standards and with which the United States runs a sizable balance of trade surplus. Trump professes to believe in free trade, but he insists that the United States has been outnegotiated by its trade partners, hence his threat to impose 45 percent tariffs on imports from China to get "a better deal"—whatever that means. He has attacked Japan's barriers against imports of U.S. agricultural goods, even though that is exactly the type of protectionism the TPP has tried to undo. Meanwhile, Clinton's position against the TPP has hardened as the campaign has gone on.

The anti-trade rhetoric of the campaign has made it difficult for even pro-trade members of Congress to support new agreements.

The response from economists has tended to be either meek defenses of trade or outright silence, with some even criticizing parts of the TPP. It's time for supporters of free trade to engage in a full-throated championing of the many achievements of U.S. trade agreements. Indeed, because other countries' trade barriers tend to be higher than those of the United States, trade agreements open foreign markets to U.S. exports more than they open the U.S. market to foreign imports.

That was true of NAFTA, which remains a favored punching bag on the campaign trail. In fact, NAFTA has been a big economic and foreign policy success. Since the agreement entered into force in 1994, bilateral trade between the United States and Mexico has boomed. For all the fear about Mexican imports flooding the US market, it is worth noting that about 40 percent of the value of imports from Mexico consists of content originally made in the United States—for example, auto parts produced in the United States but assembled in Mexico. It is precisely such trade in component parts that makes standard measures of bilateral trade balances so misleading.

NAFTA has also furthered the United States' long-term political, diplomatic, and economic interest in a flourishing, democratic Mexico, which not only reduces immigration pressures on border states but also increases Mexican demand for U.S. goods and services. Far from exploiting Third World labor, as critics have charged, NAFTA has promoted the growth of a middle class in Mexico that now includes nearly half of all households. And since 2009, more Mexicans have left the United States than have come in. In the two decades since NAFTA went into effect, Mexico has been transformed from a clientelistic one-party state with widespread anti-American sentiment into a functional multiparty democracy with a generally pro-American public. Although it has suffered from drug wars in recent years (a spillover effect from problems that are largely made in America), the overall story is one of rising prosperity thanks in part to NAFTA.

Ripping up NAFTA would do immense damage. In its foreign relations, the United States would prove itself to be an unreliable partner. And economically, getting rid of the agreement would disrupt production chains across North America, harming both Mexico and the United States. It would add to border tensions while shifting trade to Asia without bringing back any U.S. manufacturing jobs. The American public seems to understand this: in an October 2015 Gallup poll, only 18 percent of respondents agreed that leaving NAFTA or the Central American Free Trade Agreement would be very effective in helping the economy.

A more moderate option would be for the United States to take a pause and simply stop negotiating any more trade agreements, as Obama did during his first term. The problem with this approach, however, is that the rest of the world would continue to reach trade agreements without the United States, and so U.S. exporters would find themselves at a disadvantage compared with their foreign competitors. Glimpses of that future can already be seen. In 2012, the car manufacturer Audi chose southeastern Mexico over Tennessee for the site of a new plant because it could save thousands of dollars per car exported thanks to Mexico's many more free-trade agreements, including one with the EU. Australia has reached trade deals with China and Japan that give Australian farmers preferential access in those markets, cutting into US beef exports.

If Washington opted out of the TPP, it would forgo an opportunity to shape the rules of international trade in the 21st century. The Uruguay Round, the last round of international trade negotiations completed by the General Agreement on Tariffs and Trade, ended in 1994, before the Internet had fully emerged. Now, the United States' high-tech firms and other exporters face foreign regulations that are not transparent and impede market access. Meanwhile, other countries are already moving ahead with their own trade agreements, increasingly taking market share from U.S. exporters in the dynamic Asia-Pacific region. Staying out of the TPP would not lead to the creation of good jobs in the United States. And despite populist claims to the contrary, the TPP's provisions for settling disputes between investors and governments and dealing with intellectual property rights are reasonable. (In the early 1990s, similar fears about such provisions in the WTO were just as exaggerated and ultimately proved baseless.)

The United States should proceed with passage of the TPP and continue to negotiate other deals with its trading partners. So-called plurilateral trade agreements, that is, deals among relatively small numbers of like-minded countries, offer the only viable way to pick up more gains from reducing trade barriers. The current climate on Capitol Hill means that the era of small bilateral agreements, such as those pursued during the George W. Bush administration, has ended. And the collapse of the Doha Round at the WTO likely marks the end of giant multilateral trade negotiations.

Free trade has always been a hard sell. But the antitrade rhetoric of the 2016 campaign has made it difficult for even pro-trade members of Congress to support new agreements. Past experience suggests that Washington will lead the charge for reducing trade barriers only when there is a major trade problem to be solved—namely, when U.S. exporters face severe discrimination in foreign markets. Such was the case when the United States helped form the General Agreement on Tariffs and Trade in 1947, when it started the Kennedy Round of trade negotiations in the 1960s, and when it initiated the Uruguay Round in the 1980s. Until the United States feels the pain of getting cut out of major foreign markets, its leadership on global trade may wane. That would represent just one casualty of the current campaign.

Critical Thinking

1. Why do the 2016 Presidential candidates oppose free-trade agreements?
2. What is the relationship between technology and the loss of jobs in the U.S.?
3. What has been the effect of trade agreements with China on the U.S. economy?

Internet References

The Transatlantic Trade and Investment Partnership
https://ustr.gov/ttip

The World Trade Organization
https://www.wto.org/

Douglas A. Irwin is John Sloan Dickey third century professor in the Social Sciences in the Department of Economics at Dartmouth College and the author of *Free Trade Under Fire*. Follow him on Twitter @D_A_Irwin.

Article

Prepared by: Robert Weiner, *University of Massachusetts, Boston*

Inequality and Globalization

How the Rich Get Richer as the Poor Catch Up

FRANÇOIS BOURGUIGNON

Learning Outcomes

After reading this article, you will be able to:

- Discuss inequality between and within states.
- Understand the relationship between inequality and the international financial system.

When it comes to wealth and income, people tend to compare themselves to the people they see around them rather than to those who live on the other side of the world. The average Frenchman, for example, probably does not care how many Chinese exceed his own standard of living, but that Frenchman surely would pay attention if he started lagging behind his fellow citizens. Yet when thinking about inequality, it also makes sense to approach the world as a single community: accounting, for example, not only for the differences in living standards within France but also for those between rich French people and poor Chinese (and poor French and rich Chinese).

When looking at the world through this lens, some notable trends stand out. The first is that global inequality greatly exceeds inequality within any individual country. This observation should come as no surprise, since global inequality reflects the enormous differences in wealth between the world's richest and the world's poorest countries, not just the differences within them. Much more striking is the fact that, in a dramatic reversal of the trend that prevailed for most of the 20th century, global inequality has declined markedly since 2000 (following a slower decline during the 1990s). This trend has been due in large part to the rising fortunes of the developing world, particularly China and India. And as the economies of these countries continue to converge with those of the developed world, global inequality will continue to fall for some time.

Even as global inequality has declined, however, inequality within individual countries has crept upward. There is some disagreement about the size of this increase among economists, largely owing to the underrepresentation of wealthy people in national income surveys. But whatever its extent, increased inequality within individual countries has partially offset the decline in inequality among countries. To counteract this trend, states should pursue policies aimed at redistributing income, strengthen the regulation of the labor and financial markets, and develop international arrangements that prevent firms from avoiding taxes by shifting their assets or operations overseas.

The Great Substitution

Economists typically measure income inequality using the Gini coefficient, which ranges from zero in cases of perfect equality (a theoretical country in which everyone earns the same income) to one in cases of perfect inequality (a state in which a single individual earns all the income and everyone else gets nothing). In continental Europe, Gini coefficients tend to fall between 0.25 and 0.30. In the United States, the figure is around 0.40. And in the world's most unequal countries, such as South Africa, it exceeds 0.60. When considering the world's population as a whole, the Gini coefficient comes to 0.70—a figure so high that no country is known to have ever reached it.

Determining the Gini coefficient for global inequality requires making a number of simplifications and assumptions. Economists must accommodate gaps in domestic data—in Mexico, an extreme case, surveys of income and expenditures miss about half of all households. They need to come up with estimates for years in which national surveys are not available. They need to convert local incomes into a common currency, usually the U.S. dollar, and correct for differences in purchasing power. And they need to adjust for discrepancies in data collection among countries, such as those that arise when one state measures living standards by income and another by consumption per person or when a state does not collect data at all.

Such inexactitudes and the different ways of compensating for them explain why estimates of just how much global inequality has declined over the past two-plus decades tend to vary—from around two percentage points to up to five, depending on the study. No matter how steep this decline, however, economists generally agree that the end result has been a global Gini coefficient of around 0.70 in the years between 2008 and 2010.

The decline in global inequality is largely the product of the convergence of the economies of developing countries, particularly China and India, with those of the developed world. In the first decade of this century, booming economies in Latin America and sub-Saharan Africa also helped accelerate this trend. Remarkably, this decline followed a nearly uninterrupted rise in inequality from the advent of the Industrial Revolution in the early 19th century until the 1970s. What is more, the decline has been large enough to erase a substantial part of the inequality that built up over that century and a half.

Even as inequality among countries has decreased, however, inequality within individual countries has increased, gaining, on average, more than two percentage points in terms of the Gini coefficient between 1990 and 2010. The countries with the biggest economies are especially responsible for this trend—particularly the United States, where the Gini coefficient rose by five percentage points between 1990 and 2013, but also China and India and, to a lesser extent, most European countries, among them Germany and the Scandinavian states. Still, inequality within countries is not rising fast enough to offset the rapid decline in inequality among countries.

The good news is that the current decline in global inequality will probably persist. Despite the current global slowdown, China and India have such huge domestic markets that they retain an enormous amount of potential for growth. And even if their growth rates decline significantly in the next decade, so long as they remain higher than those of the advanced industrial economies, as is likely, global inequality will continue to fall. The prospects for growth are less favorable for the smaller economies in Latin America and sub-Saharan Africa that depend primarily on commodity exports, since world commodity prices may remain low for some time. All told, then, global inequality will likely keep falling in the coming decades—but probably at the slow pace seen during the 1990s rather than the rapid one enjoyed during the following decade.

The bad news, however, is that economists might have underestimated inequality within individual countries and the extent to which it has increased since the 1990s, because national surveys tend to underrepresent the wealthy and underreport income derived from property, which disproportionately accrues to the rich. Indeed, tax data from many developed states suggest that national surveys fail to account for a substantial portion of the incomes of the very highest earners.

According to the most drastic corrections for such underreporting, as calculated by the economists Sudhir Anand and Paul Segal, global inequality could have remained more or less constant between 1988 and 2005. Most likely, however, this conclusion is too extreme, and the increase in national inequality has been too small to cancel out the decline in inequality among countries. Yet it still points to a disheartening trend: increased inequality within countries has offset the drop in inequality among countries. In other words, the gap between average Americans and average Chinese is being partly replaced by larger gaps between rich and poor Americans and between rich and poor Chinese.

Interconnected and Unequal

The same factor that can be credited for the decline in inequality among countries can also be blamed for the increase in inequality within them: globalization. As firms from the developed world moved production overseas during the 1990s, emerging Asian economies, particularly China, started to converge with those of the developed world. The resulting boom triggered faster growth in Africa and Latin America as demand for commodities increased. In the developed world, meanwhile, as manufacturing firms outsourced some of their production, corporate profits rose but real wages for unskilled labor fell.

Economic liberalization also played an important role in this process. In China, the market reforms initiated by Deng Xiaoping in the 1980s contributed just as much to rapid growth as did the country's opening to foreign investment and trade, and the same is true of the reforms India undertook in the early 1990s. As with globalization, such reforms didn't just enable developing countries to get closer to the developed world; they also created a new elite within those countries while leaving many citizens behind, thus increasing domestic inequality.

The same drive toward economic liberalization has contributed to increasing inequality in the developed world. Reductions in income tax rates, cuts to welfare, and financial deregulation have also helped make the rich richer and, in some instances, the poor poorer. The increase in the international mobility of firms, wealth, and workers over the past two decades has compounded these problems by making it harder for governments to combat inequality: for example, companies and wealthy people have become increasingly able to shift capital to countries with low tax rates or to tax havens, allowing them to avoid paying more redistributive taxes in their home countries. And in both developed and developing countries, technological progress has exacerbated these trends by favoring skilled workers over unskilled ones and creating economies of scale that disproportionately favor corporate managers.

Maintaining Momentum

In the near future, the greatest potential for further reductions in global inequality will lie in Africa—the region that has arguably benefited the least from the past few decades of globalization, and the one where global poverty will likely concentrate in the coming decades as countries such as India leap ahead. Perhaps most important, the population of Africa is expected to double over the next 35 years, reaching some 25 percent of the world's population, and so the extent of global inequality will increasingly depend on the extent of African growth. Assuming that the economies of sub-Saharan Africa sustain the modest growth rates they have seen in recent years, then inequality among countries should keep declining, although not as fast as it did in the first decade of this century.

To maintain the momentum behind declining global inequality, all countries will need to work harder to reduce inequality within their borders, or at least prevent it from growing further. In the world's major economies, failing to do so could cause disenchanted citizens to misguidedly resist further attempts to integrate the world's economies—a process that, if properly managed, can in fact benefit everyone.

In practice, then, states should seek to equalize living standards among their populations by eliminating all types of ethnic, gender, and social discrimination; regulating the financial and labor markets; and implementing progressive taxation and welfare policies. Because the mobility of capital dulls the effectiveness of progressive taxation policies, governments also need to push for international measures that improve the transparency of the financial system, such as those the G20 and the Organization for Economic Cooperation and Development have endorsed to share information among states in order to clamp down on tax avoidance. Practical steps such as these should remind policymakers that even though global inequality and domestic inequality have moved in opposite directions for the past few decades, they need not do so forever.

Critical Thinking

1. What can the G-20 and the Organization for Economic Cooperation and Development do to eliminate global inequality?
2. Why has global inequality among states decreased?
3. Why has global inequality within states increased?

Internet References

The Group of 20
https://en.wikipedia.org/wiki'G20

The World Bank
www.worldbank.org

François Bourguignon is a professor of Economics at the Paris School of Economics, former Chief Economist of the World Bank, and the author of *The Globalization of Inequality.*

Article

Prepared by: Robert Weiner, *University of Massachusetts, Boston*

Inequality and Modernization

Why Equality Is Likely to Make a Comeback

RONALD INGLEHART

Learning Outcomes

After reading this article, you will be able to:

- Discuss the factors that have resulted in growing inequality in the developing countries.
- Explain the rise in postmaterial values in developed countries and the reaction it has provoked in the working class.

During the past century, economic inequality in the developed world has traced a massive U-shaped curve—starting high, curving downward, then curving sharply back up again. In 1915, the richest 1 percent of Americans earned roughly 18 percent of all national income. Their share plummeted in the 1930s and remained below 10 percent through the 1970s, but by 2007, it had risen to 24 percent. Looking at household wealth rather than income, the rise of inequality has been even greater, with the share owned by the top 0.1 percent increasing to 22 percent from 9 percent three decades ago. In 2011, the top 1 percent of U.S. households controlled 40 percent of the nation's entire wealth. And while the U.S. case may be extreme, it is far from unique: all but a few of the countries of the Organization for Economic Cooperation and Development for which data are available experienced rising income inequality (before taxes and transfers) during the period from 1980 to 2009.

The French economist Thomas Piketty has famously interpreted this data by arguing that a tendency toward economic inequality is an inherent feature of capitalism. He sees the middle decades of the 20th century, during which inequality declined, as an exception to the rule, produced by essentially random shocks—the two world wars and the Great Depression—that led governments to adopt policies that redistributed income. Now, that the influence of those shocks has receded, life is returning to normal, with economic and political power concentrated in the hands of an oligarchy.

Piketty's work has been corrected on some details, but his claim that economic inequality is rising rapidly in most developed countries is clearly accurate. What most analyses of the subject miss, however, is the extent to which both the initial fall and the subsequent rise of inequality over the past century have been related to shifts in the balance of power between elites and masses, driven by the ongoing process of modernization.

In hunting-and-gathering societies, virtually everyone possessed the skills needed for political participation. Communication was by word of mouth, referring to things one knew of firsthand, and decision-making often occurred in village councils that included every adult male. Societies were relatively egalitarian.

The invention of agriculture gave rise to sedentary communities producing enough food to support elites with specialized military and communication skills. Literate administrators made it possible to coordinate large empires governing millions of people. This much larger scale of politics required specialized skills, including the ability to read and write. Word-of-mouth communication was no longer sufficient for political participation: messages had to be sent across great distances. Human memory was incapable of recording the tax base or military manpower of large numbers of districts: written records were needed. And personal loyalties were inadequate to hold together large empires: legitimating myths had to be propagated by religious or ideological specialists. This opened up a wide gap between a relatively skilled ruling class and the population as a whole, which consisted mainly of scattered, illiterate peasants who lacked the skills needed to cope with politics at a distance. And along with that gap, economic inequality increased dramatically.

This inequality was sustained throughout history and into the early capitalist era. At first, industrialization led to the ruthless exploitation of workers, with low wages, long workdays, no labor laws, and the suppression of union organizing. Eventually, however, the continuation of the Industrial Revolution narrowed the gap between elites and masses by redressing the balance of political skills. Urbanization brought people into close proximity; workers were concentrated in factories, facilitating communication; and the spread of mass literacy put them in touch with national politics, all of which led to social mobilization. In the late 19th century and early 20th century, unions won the right to organize, enabling workers to bargain collectively. The expansion of the franchise gave ever more people the vote, and leftist political parties mobilized the working class to fight for its economic interests. The result was the election of governments that adopted various kinds of redistributive policies—progressive taxation, social insurance, and an expansive welfare state—that caused inequality to decline for most of the 20th century.

The emergence of a postindustrial society, however, changed the game once again. The success of the modern welfare state made further redistribution seems less urgent. Noneconomic issues emerged that cut across class lines, with identity politics and environmentalism drawing some wealthier voters to the left, while cultural issues pushed many in the working class to the right. Globalization and deindustrialization undermined the strength of unions. And the information revolution helped establish a winner-take-all economy. Together these eroded the political base for redistributive policies, and as those policies fell out of favor, economic inequality rose once more.

Today, large economic gains are still being made in developed countries, but they are going primarily to those at the very top of the income distribution, whereas those lower down have seen their real incomes stagnate or even diminish.

The rich, in turn, have used their privilege to shape policies that further increase the concentration of wealth, often against the wishes and interests of the middle and lower classes. The political scientist, Martin Gilens, for example, has shown that the U.S. government responds so attentively to the preferences of the most affluent ten percent of the country's citizens that "under most circumstances, the preferences of the vast majority of Americans appear to have essentially no impact on which policies the government does or doesn't adopt."

Because advantages tend to be cumulative, with those born into more prosperous families receiving better nutrition and health care, more intellectual stimulation and better education, and more social capital for use in later life, there is an enduring tendency for the rich to get richer and the poor to be left behind. The extent to which this tendency prevails, however, depends on a country's political leaders and political institutions, which in turn tend to reflect the political pressures emerging from mobilized popular forces in the political system at large.

The extent to which inequality increases or decreases, in other words, is ultimately a political question.

Today, the conflict is no longer between the working class and the middle class; it is between a tiny elite and the great majority of citizens. This means that the crucial questions for future politics in the developed world will be how and when that majority develops a sense of common interest. The more current trends continue, the more pressure will build up to tackle inequality once again. The signs of such a stirring are already visible, and in time, the practical consequences will be as well.

Not about the Money

For the first ⅔ of the 20th century, working-class voters in developed countries tended to support parties of the left, and middle- and upper-class voters tended to support parties of the right. With partisan affiliation roughly correlating with social class, scholars found, unsurprisingly, that governments tended to pursue policies that reflected the economic interests of their sociopolitical constituencies.

As the century continued, however, both the nature of the economy and the attitudes and behaviors of the public changed. An industrial society gave way to a postindustrial one, and generations raised with high levels of economic and physical security during their formative years displayed a "postmaterialist" mindset, putting greater emphasis on autonomy and self-expression. As postmaterialists became more numerous in the population, they brought new issues into politics, leading to a decline in class conflict and a rise in political polarization based on noneconomic issues (such as environmentalism, gender equality, abortion, and immigration).

This stimulated a reaction in which segments of the working class moved to the right, reaffirming traditional values that seemed to be under attack. Moreover, large immigration flows, especially from low-income countries with different languages, cultures, and religions, changed the ethnic makeup of advanced industrial societies. The rise of religious fundamentalism in the United States and xenophobic populist movements in western European countries represents a reaction against rapid cultural changes that seem to be eroding basic social values and customs—something particularly alarming to the less secure groups in those countries.

All of this has greatly stressed existing party systems, which were established in an era when economic issues were dominant and the working class was the main base of support for sociopolitical change. Today, the most heated issues tend to be noneconomic, and support for change comes increasingly from postmaterialists, largely of middle-class origin. Traditional political polarization centered on differing views about economic redistribution, with workers' parties on the left and conservative parties on the right. The emergence of changing values and new issues gave rise to a second dimension of partisan polarization, with postmaterialist parties at one pole and authoritarian and xenophobic parties at the opposite pole.

The classic economic issues did not disappear. But their relative prominence declined to such an extent that by the late 1980s, noneconomic issues had become more prominent than economic issues in Western political parties' campaign platforms. A long-standing truism of political sociology is that working-class voters tend to support the parties of the left and middle-class voters those of the right. This was an accurate description of reality around 1950, but the tendency has grown steadily weaker. The rise of postmaterialist issues tends to neutralize class-based political polarization. The social basis of support for the left has increasingly come from the middle class, even as a substantial share of the working class has shifted its support to the right.

In fact, by the 1990s, social-class voting in most democracies was less than half as strong as it was a generation earlier. In the United States, it had fallen so low that there was virtually no room for further decline. Income and education had become much weaker indicators of the American public's political preferences than religiosity or one's stand on abortion or the same-sex marriage: by wide margins, those who opposed abortion and the same-sex marriage supported the Republican presidential candidate over the Democratic candidate. The electorate had shifted from class-based polarization toward value-based polarization.

The Machine Age

In 1860, the majority of the U.S. work force was employed in agriculture. By 2014, less than 2 percent was employed there, with modern agricultural technology enabling a tiny share of the population to produce even more food than before. With the transition to an industrial society, jobs in the agricultural sector virtually disappeared, but this didn't result in widespread unemployment and poverty, because there was a massive rise in industrial employment. By the 21st century, automation and outsourcing had reduced the ranks of industrial workers to 15 percent of the work force—but this too did not result in widespread unemployment and poverty, because the loss of industrial jobs was offset by a dramatic rise in service-sector jobs, which now make up about 80 percent of the U.S. workforce.

Within the service sector, there are some jobs that are integrally related to what has been called "the knowledge

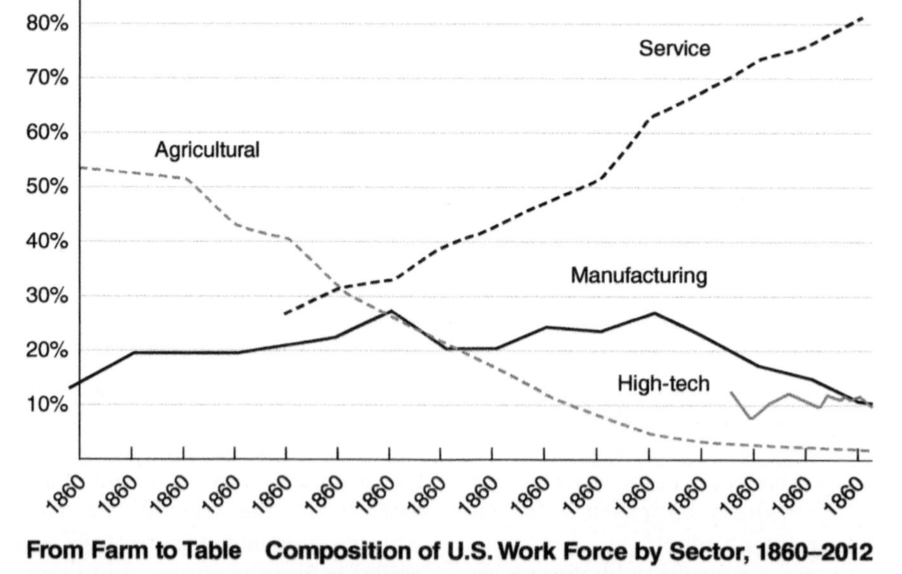

From Farm to Table Composition of U.S. Work Force by Sector, 1860–2012

SOURCES: Paul Hadlock, Daniel Hecker, and Joseph Gannon, "High Technology Employment: Another View," *Monthly Labor Review* (U.S. Bureau of Labor Statistics), July 1991; Daniel E. Hecker, "High Technology Employment: A NAICS-Based Update," *Monthly Labor Review* (U.S. Bureau of Labor Statistics), July 2005; Stanley Lebergott, "Labor Force and Employment, 1800–1960," in *Output, Employment, and Productivity in the United States After 1800*, ed. Dorothy S. Brady (National Bureau of Economic Research, 1966), 117–204; National Science Board, 2014; U.S. Bureau of Labor Statistics, 2014; U.S. Census Bureau, 1977.

NOTE: Data are not available for the service sector before 1900 or for the high-tech sector before 1986.

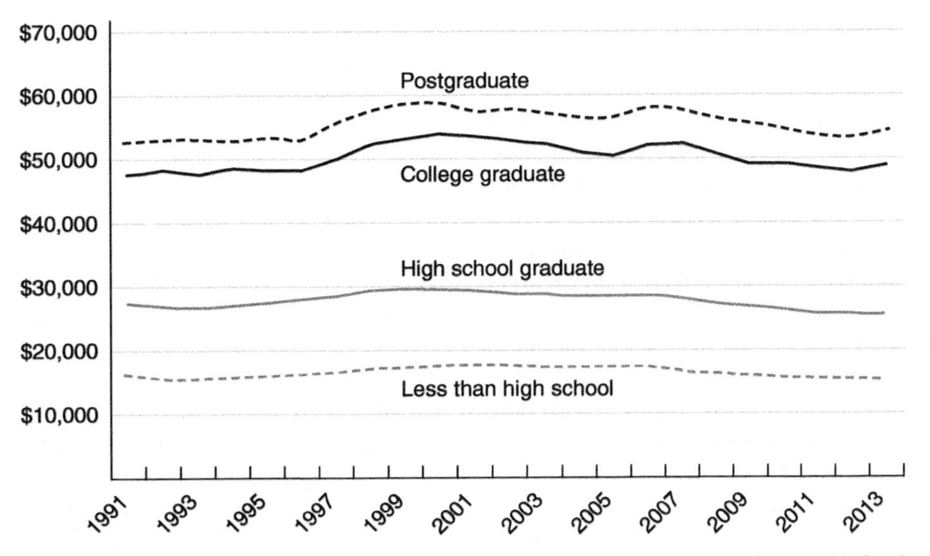

The World Is Flat Median Real Income by Educational Level in the United States, 1991–2013

SOURCE: U.S. Census Bureau, 2014.

NOTE: Incomes are in 2013 dollars.

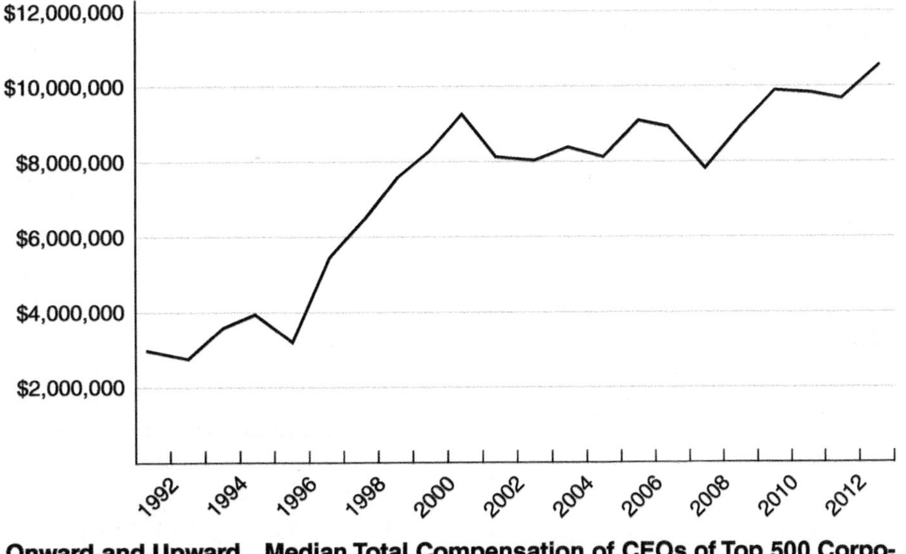

Onward and Upward Median Total Compensation of CEOs of Top 500 Corporations, 1992–2013

SOURCES: CEO salary data for 1998–2008 are from Carola Frydman and Dirk Jenter, "CEO Compensation," *Annual Review of Economics*, 2010; data for 2009–13 are from Joann S. Lublin, "CEO Pay in 2010 Jumped 11%," *Wall Street Journal*, May 9, 2011, and Ken Sweet, "Median CEO Pay Crosses $10 Million in 2013," Associated Press, May 27, 2014.

NOTE: Incomes are in 2013 dollars.

economy"—defined by the scholars Walter Powell and Kaisa Snellman as "production and services based on knowledge-intensive activities that contribute to the accelerated pace of technical and scientific advance." Because of its economic significance, the knowledge economy is worth breaking out as a separate category from the rest of the service sector; it is represented by what can be termed "the high-tech sector," which includes everyone employed in the information, finance, insurance, professional, scientific, and technical services categories of the economy.

Some assume that the high-tech sector will produce large numbers of high-paying jobs in the future. But employment in this area does not seem to be increasing; the sector's share of total employment has been essentially constant, since statistics became available about three decades ago. Unlike the transition from an agricultural to an industrial society, in other words, the rise of the knowledge society is not generating a lot of good new jobs.

Initially, only unskilled workers lost their jobs to automation. Today, even highly skilled occupations are being taken over by computers. Computer programs are replacing lawyers who used to do legal research. Expert systems are being developed that can make medical diagnoses better and faster than physicians. The fields of education and journalism are on their way to being automated. And increasingly, computer programs themselves may be written by computers.

As a result of such developments, even highly skilled jobs are being commodified, so that even many highly educated workers in the upper reaches of the income distribution are not moving ahead, with gains from the increases in GDP limited to those in a thin stratum of financiers, entrepreneurs, and managers at the very top. As expert systems replace people, market forces alone could conceivably produce a situation in which a tiny but extremely well-paid minority directs the economy, while the majority have precarious jobs, serving the minority as gardeners, waiters, nannies, and hairdressers—a future foreshadowed by the social structure of Silicon Valley today.

The rise of the postindustrial economy narrowed the life prospects of most unskilled workers, but until recently, it seemed that the rise of the knowledge society would keep the door open for those with sophisticated skills and a good education. Recent evidence, however, suggests that this is no longer true.

Between 1991 and 2013, real incomes in the United States stagnated across the educational spectrum. The highly educated still make substantially larger salaries than the less educated, but it is no longer just the unskilled workers who are being left behind.

The problem is not aggregate growth in the economy. During these years, U.S. GDP increased significantly. So where did the money go? To the elite of the elite, such as the CEOs of the country's largest corporations.

During a period in which the real incomes of even highly educated professionals, such as doctors, lawyers, professors, engineers, and scientists, were essentially flat, the real incomes of CEOs more than tripled. The pattern is even starker over a longer timeframe. In 1965, CEO pay at the largest 350 U.S. companies was 20 times as high as the pay of the average worker; in 1989, it was 58 times as high; and in 2012, it was 273 times as high.

Workers of the World, Unite?

Globalization is enabling half of the world's population to escape subsistence-level poverty but weakening the bargaining position of workers in developed countries. The rise of the knowledge society, meanwhile, is helping divide the economy into a small pool of elite winners and vast numbers of precariously employed workers. Market forces show no signs of reversing these trends on their own. But politics might do so, as growing insecurity and relative immiseration gradually reshape citizens' attitudes, creating greater support for government policies designed to alter the picture.

There are indications that the citizens of many countries are becoming sensitized to this problem. Concern over income inequality has increased dramatically during the past three decades. In surveys carried out from 1989 to 2014, respondents around the world were asked whether their views came closer to the statement "Incomes should be made more equal" or "Income differences should be larger to provide incentives for individual effort." In the earliest polls, majorities in ⅘ of the 65 countries surveyed believed that greater incentives for individual effort were needed. By the most recent surveys, however, that figure had dropped by half, with majorities in only two-fifths of the countries favoring that. Over a 25-year period in which income inequality increased dramatically, publics in 80 percent of the countries surveyed, including the United States, grew more supportive of actions to reduce inequality, and those beliefs are likely to intensify over time.

New political alignments, in short, might once again readjust the balance of power between elites and masses in the developed world, with the emerging struggle being between a tiny group at the top and a heterogeneous majority below. For the industrial society's working-class coalition to become effective, lengthy processes of social and cognitive mobilization had to be completed. In today's postindustrial society, however, a large share of the population is already highly educated, well informed, and in possession of political skills; all it needs to become politically effective is the development of an awareness of common interest.

Will enough of today's dispossessed develop what Marx might have called "class consciousness" to become a decisive political force? In the short run, probably not, because of the presence of various hot-button cultural issues cutting across economic lines. Over the long run, however, they probably will, as economic inequality and the resentment of it are likely to continue to intensify.

It was the rise of postmaterialist values, together with a backlash against the changes that the postmaterialists spearheaded, that helped topple economic issues from their central role in partisan political mobilization and install cultural issues in their place. But the continued spread of postmaterialist values is draining much of the passion from the cultural conflict, even as the continued rise of inequality is pushing economic issues back to the top of the political agenda.

During the 2004 U.S. presidential election, for example, the same-sex marriage was so unpopular in some quarters that Republican strategists deliberately put referendums banning it on the ballot in crucial swing states in the hope of increasing turnout among social conservatives in the middle and lower echelons of the income distribution. And they were smart to do so, for the measures passed in every case—as did virtually all others like them put forward from 1998 to 2008. In 2012, however, there were five new statewide referendums on the topic, and in four of them, the public voted in favor of legalization. Crosscutting cultural divisions still exist and can still divert attention from common economic interests, but the former no longer trump the latter as reliably as they used to. And the fact that not just all the Democrats but even several 2016 Republican presidential candidates have pledged to abolish the tax break on "carried interest" benefiting elite financiers might well be a portent of things to come.

The essence of modernization is the linkages among economic, social, ideational, and political trends. As changes ripple through the system, developments in one sphere can drive developments in the others. But the process doesn't work in just one direction, with economic trends driving everything else, for example. Social forces and ideas can drive political actions that reshape the economic landscape. Will that happen once again, with popular majorities mobilizing to reverse the trend toward economic inequality? In the long run, probably: publics around the world increasingly favor reducing inequality, and the societies that survive are the ones that successfully adapt to changing conditions and pressures. Despite current signs of paralysis, democracies still have the vitality to do so.

Critical Thinking

1. Why does the author think that economic issues are returning to the top of the political agenda?
2. What is meant by postmaterial values and why are they important?

Internet References

Group of 20
https://en.wikipedia.org/wiki/G20

The World Bank Group
www.worldbank.org

Ronald Inglehart is a professor of political science at the University of Michigan and Founding President of the World Values Survey.

Article

Prepared by: Robert Weiner, *University of Massachusetts, Boston*

As Objects Go Online: The Promise (and Pitfalls) of the Internet of Things

NEIL GERSHENFELD AND J. P. VASSEUR

Learning Outcomes

After reading this article, you will be able to:

- Understand what the Internet of Things is.
- Understand why the Internet of Things may represent a third industrial revolution.

Since 1969, when the first bit of data was transmitted over what would come to be known as the Internet, that global network has evolved from linking mainframe computers to connecting personal computers and now mobile devices. By 2010, the number of computers on the Internet had surpassed the number of people on earth.

Yet that impressive growth is about to be overshadowed as the things around us start going online as well, part of what is called "the Internet of Things." Thanks to advances in circuits and software, it is now possible to make a Web server that fits on (or in) a fingertip for $1. When embedded in everyday objects, these small computers can send and receive information via the Internet so that a coffeemaker can turn on when a person gets out of bed and turn off when a cup is loaded into a dishwasher, a stoplight can communicate with roads to route cars around traffic, a building can operate more efficiently by knowing where people are and what they're doing, and even the health of the whole planet can be monitored in real time by aggregating the data from all such devices.

Linking the digital and physical worlds in these ways will have profound implications for both. But this future won't be realized unless the Internet of Things learns from the history of the Internet. The open standards and decentralized design of the Internet won out over competing proprietary systems and centralized control by offering fewer obstacles to innovation and growth. This battle has resurfaced with the proliferation of conflicting visions of how devices should communicate. The challenge is primarily organizational, rather then technological, a contest between command-and-control technology and distributed solutions. The Internet of Things demands the latter, and openness will eventually triumph.

The Connected Life

The Internet of Things is not just science fiction; it has already arrived. Some of the things currently networked together send data over the public Internet, and some communicate over secure private networks, but all share common protocols that allow them to interoperate to help solve profound problems.

Take energy inefficiency. Buildings account for three-quarters of all electricity use in the United States, and of that, about one-third is wasted. Lights stay on when there is natural light available, and air is cooled even when the weather outside is more comfortable or a room is unoccupied. Sometimes fans move air in the wrong direction or heating and cooling systems are operated simultaneously. This enormous amount of waste persists because the behavior of thermostats and light bulbs are set when buildings are constructed; the wiring is fixed and the controllers are inaccessible. Only when the infrastructure itself becomes intelligent, with networked sensors and actuators, can the efficiency of a building be improved over the course of its lifetime.

Health care is another area of huge promise. The mismanagement of medication, for example, costs the health-care system billions of dollars per year. Shelves and pill bottles connected to the Internet can alert a forgetful patient when to take a pill, a pharmacist to make a refill, and a doctor when a dose is missed. Floors can call for help if a senior citizen has fallen, helping the elderly live independently. Wearable sensors could monitor one's activity throughout the day and serve as personal coaches, improving health and saving costs.

Countless futuristic "smart houses" have yet to generate much interest in living in them. But the Internet of Things succeeds to the extent that it is invisible. A refrigerator could communicate with a grocery store to reorder food, with a bathroom scale to monitor a diet, with a power utility to lower electricity consumption during peak demand, and with its manufacturer when maintenance is needed. Switches and lights in a house could adapt to how spaces are used and to the time of day. Thermostats with access to calendars, beds, and cars could plan heating and cooling based on the location of the house's occupants. Utilities today provide power and plumbing; these new services would provide safety, comfort, and convenience.

In cities, the Internet of Things will collect a wealth of new data. Understanding the flow of vehicles, utilities, and people is essential to maximizing the productivity of each, but traditionally, this has been measured poorly, if at all. If every street lamp, fire hydrant, bus, and crosswalk were connected to the Internet, then a city could generate real-time readouts of what's working and what's not. Rather than keeping this information internally, city hall could share open-source data sets with developers, as some cities are already doing.

Weather, agricultural inputs, and pollution levels all change with more local variation than can be captured by point measurements and remote sensing. But when the cost of an Internet connection falls far enough, these phenomena can all be measured precisely. Networking nature can help conserve animate, as well as inanimate, resources; an emerging "interspecies Internet" is linking elephants, dolphins, great apes, and other animals for the purposes of enrichment, research, and preservation.

The ultimate realization of the Internet of Things will be to transmit actual things through the Internet. Users can already send descriptions of objects that can be made with personal digital fabrication tools, such as 3D printers and laser cutters. As data turn into things and things into data, long manufacturing supply chains can be replaced by a process of shipping data over the Internet to local production facilities that would make objects on demand, where and when they were needed.

Back to the Future

To understand how the Internet of Things works, it is helpful to understand how the Internet itself works, and why. The first secret of the Internet's success is its architecture. At the time the Internet was being developed, in the 1960s and 1970s, telephones were wired to central office switchboards. That setup was analogous to a city in which every road goes through one traffic circle; it makes it easy to give directions but causes traffic jams at the central hub. To avoid such problems, the Internet's developers created a distributed network, analogous to the web of streets that vehicles navigate in a real city. This design lets data bypass traffic jams and lets managers add capacity where needed.

The second key insight in the Internet's development was the importance of breaking data down into individual chunks that could be reassembled after their online journey. "Packet switching," as this process is called, is like a railway system in which each railcar travels independently. Cars with different destinations share the same tracks, instead of having to wait for one long train to pass, and those going to the same place do not all have to take the same route. As long as each car has an address and each junction indicates where the tracks lead, the cars can be combined on arrival. By transmitting data in this way, packet switching has made the Internet more reliable, robust, and efficient.

The third crucial decision was to make it possible for data to flow over different types of networks, so that a message can travel through the wires in a building, into a fiber-optic cable that carries it across a city, and then to a satellite that sends it to another continent. To allow that, computer scientists developed the Internet Protocol, or IP, which standardized the way that packets of data were addressed. The equivalent development in railroads was the introduction of a standard track gauge, which allowed trains to cross international borders. The IP standard allows many different types of data to travel over a common protocol.

The fourth crucial choice was to have the functions of the Internet reside at the ends of the network, rather than at the intermediate nodes, which are reserved for routing traffic. Known as the "end-to-end principle," this design allows new applications to be invented and added without having to upgrade the whole network. The capabilities of a traditional telephone were only as advanced as the central office switch it was connected to, and those changed infrequently. But the layered architecture of the Internet avoids this problem. Online messaging, audio and video streaming, e-commerce, search engines, and social media were all developed on top of a system designed decades earlier, and new applications can be created from these components.

These principles may sound intuitive, but until recently, they were not shared by the systems that linked things other than computers. Instead, each industry, from heating and cooling to consumer electronics, created its own networking standards, which specified not only how their devices communicated with one another but also what they could communicate. This closed model may work within a fixed domain, but unlike the model used for the Internet, it limits future possibilities to what its creators originally anticipated. Moreover, each of these standards has struggled with the same problems the Internet has already solved: how to assign network names to devices, how to route messages between networks, how to manage the flow of traffic, and how to secure communications.

Although it might seem logical now to use the Internet to link things rather than reinvent the networking wheel for each industry, that has not been the norm so far. One reason is that manufacturers have wanted to establish proprietary control. The Internet does not have tollbooths, but if a vendor can control the communications standards used by the devices in a given industry, it can charge companies to use them.

Compounding this problem was the belief that special purpose solutions would perform better than the general-purpose Internet. In reality, these alternatives were less well developed and lacked the Internet's economies of scale and reliability. Their designers overvalued optimal functionality at the expense of interoperability. For any given purpose, the networking standards of the Internet are not ideal, but for almost anything, they are good enough. Not only do proprietary networks entail the high cost of maintaining multiple, incompatible standards; they have also been less secure. Decades of attacks on the Internet have led a large community of researchers and vendors to continually refine its defenses, which can now be applied to securing communications among things.

Finally, there was the problem of cost. The Internet relied at first on large computers that cost hundreds of thousands of dollars and then on $1,000 personal computers. The economics of the Internet were so far removed from the economics of light bulbs and doorknobs that developers never thought it would be commercially viable to put such objects online; the market for $1,000 light switches is limited. And so, for many decades, objects remained offline.

Big Things in Small Packages

But no longer do economic or technological barriers stand in the way of the Internet of Things. The unsung hero that has made this possible is the microcontroller, which consists of a simple processor packaged with a small amount of memory and peripheral parts. Microcontrollers measure just millimeters across, cost just pennies to manufacture, and use just milliwatts of electricity, so that they can run for years on a battery or a small solar cell. Unlike a personal computer, which now boasts billions of bytes of memory, a microcontroller may contain only thousands of bytes. That's not enough to run today's desktop programs, but it matches the capabilities of the computers used to develop the Internet.

Around 1995, we and our colleagues based at mit began using these parts to simplify Internet connections. That project grew into a collaboration with a group of the Internet's original architects, starting with the computer scientist Danny Cohen, to extend the Internet into things. Since "Internet2" had already been used to refer to the project for a higher-speed Internet, we chose to call this slower and simpler Internet "Internet 0."

The goal of Internet 0 was to bring IP to the smallest devices. By networking a smart light bulb and a smart light switch directly, we could enable these devices to turn themselves on and off rather than their having to communicate with a controller connected to the Internet. That way, new applications could be developed to communicate with the light and the switch, and without being limited by the capabilities of a controller.

Giving objects access to the Internet simplifies hard problems. Consider the Electronic Product Code (the successor to the familiar bar code), which retailers are starting to use in radio-frequency identification tags on their products. With great effort, the developers of the EPC have attempted to enumerate all possible products and track them centrally. Instead, the information in these tags could be replaced with packets of Internet data, so that objects could contain instructions that varied with the context: at the checkout counter in a store, a tag on a medicine bottle could communicate with a merchandise database; in a hospital, it could link to a patient's records.

Along with simplifying Internet connections, the Internet 0 project also simplified the networks that things link to. The quest for ever-faster networks has led to very different standards for each medium used to transmit data, with each requiring its own special precautions. But Morse code looks the same whether it is transmitted using flags or flashing lights, and in the same way, Internet 0 packages data in a way that is independent of the medium. Like IP, that's not optimal, but it trades speed for cheapness and simplicity. That makes sense, because high speed is not essential: light bulbs, after all, don't watch broadband movies.

Another innovation allowing the Internet to reach things is the ongoing transition from the previous version of IP to a new one. When the designers of the original standard, called IPv4, launched it in 1981, they used 32 bits (each either a zero or a one) to store each IP address, the unique identifiers assigned to every device connected to the Internet—allowing for over four billion IP addresses in total. That seemed like an enormous number at the time, but it is less than one address for every person on the planet. IPv4 has run out of addresses, and it is now being replaced with a new version, IPv6. The new standard uses 128-bit IP addresses, creating more possible identifiers than there are stars in the universe. With IPv6, everything can now get its own unique address.

But IPv6 still needs to cope with the unique requirements of the Internet of Things. Along with having limitations involving memory, speed, and power, devices can appear and disappear on the network intermittently, either to save energy or because they are on the move. And in big enough numbers, even simple sensors can quickly overwhelm existing network infrastructure; a city might contain millions of power meters and billions of electrical outlets. So in collaboration with our colleagues, we are developing extensions of the Internet protocols to handle these demands.

The Inevitable Internet

Although the Internet of Things is now technologically possible, its adoption is limited by a new version of an old conflict. During the 1980s, the Internet competed with a network called Bitnet, a centralized system that linked mainframe computers. Buying a mainframe was expensive, and so Bitnet's growth was limited; connecting personal computers to the Internet made more sense. The Internet won out, and by the early 1990s, Bitnet had fallen out of use. Today, a similar battle is emerging between the Internet of Things and what could be called the Bitnet of Things. The key distinction is where information resides:

in a smart device with its own IP address or in a dumb device wired to a proprietary controller with an Internet connection. Confusingly, the latter setup is itself frequently characterized as part of the Internet of Things. As with the Internet and bitnet, the difference between the two models is far from semantic. Extending IP to the ends of a network enables innovation at its edges; linking devices to the Internet indirectly erects barriers to their use.

The same conflicting meanings appear in use of the term "smart grid," which refers to networking everything that generates, controls, and consumes electricity. Smart grids promise to reduce the need for power plants by intelligently managing loads during peak demand, varying pricing dynamically to provide incentives for energy efficiency, and feeding power back into the grid from many small renewable sources. In the not-so-smart, utility-centric approach, these functions would all be centrally controlled. In the competing, Internet-centric approach, they would not, and its dispersed character would allow for a marketplace for developers to design power-saving applications.

Putting the power grid online raises obvious cybersecurity concerns, but centralized control would only magnify these problems. The history of the Internet has shown that security through obscurity doesn't work. Systems that have kept their inner workings a secret in the name of security have consistently proved more vulnerable than those that have allowed themselves to be examined—and challenged—by outsiders. The open protocols and programs used to protect Internet communications are the result of ongoing development and testing by a large expert community.

Another historical lesson is that people, not technology, are the most common weakness when it comes to security. No matter how secure a system is, someone who has access to it can always be corrupted, wittingly or otherwise. Centralized control introduces a point of vulnerability that is not present in a distributed system.

The flip side of security is privacy; eavesdropping takes on an entirely new meaning when actual eaves can do it. But privacy can be protected on the Internet of Things. Today, privacy on the rest of the Internet is safeguarded through cryptography, and it works: recent mass thefts of personal information have happened because firms failed to encrypt their customers' data, not because the hackers broke through strong protections. By extending cryptography down to the level of individual devices, the owners of those devices would gain a new kind of control over their personal information. Rather than maintaining secrecy as an absolute good, it could be priced based on the value of sharing. Users could set up a firewall to keep private the Internet traffic coming from the things in their homes—or they could share that data with, for example, a utility that gave a discount for their operating their dishwasher only during off-peak hours or a health insurance provider that offered lower rates in return for their making healthier lifestyle choices.

The size and speed of the Internet have grown by nine orders of magnitude since the time it was invented. This expansion vastly exceeds what its developers anticipated, but that the Internet could get so far is a testament to their insight and vision. The uses the Internet has been put to that have driven this growth are even more surprising; they were not part of any original plan. But they are the result of an open architecture that left room for the unexpected. Likewise, today's vision for the Internet of Things is sure to be eclipsed by the reality of how it is actually used. But the history of the Internet provides principles to guide this development in ways that are scalable, robust, secure, and encouraging of innovation.

The Internet's defining attribute is its interoperability; information can cross geographic and technological boundaries. With the Internet of Things, it can now leap out of the desktop and data center and merge with the rest of the world. As the technology becomes more finely integrated into daily life, it will become, paradoxically, less visible. The future of the Internet is to literally disappear into the woodwork.

Critical Thinking

1. What civil rights and security issues may be raised by the Internet of Things?
2. How may the Internet of Things transform the world?

Internet References

European Research Center on the Internet of Things
http://www.internet-of-things-research.eu/documents.htm

Goldman Sachs
http://www.goldmansachs.com/our-thinking/outlook/internet-ofthings/index.html?cid=PS_01_89_07_00_00_OIM

The Internet of Things Council
http://www.theinternetofthings.eu/

Pew Research Internet Project
http://www.pewinternet.org

Neil Gershenfeld is a Professor at the Massachusetts Institute of Technology and directs MIT's Center for Bits and Atoms. **JP Vasseur** is a Cisco Fellow and Chief Architect of the Internet of Things at Cisco Systems.

Article

Prepared by: Robert Weiner, *University of Massachusetts, Boston*

The Return of Europe's Nation-States: The Upside to the EU's Crisis

JAKUB GRYGIEL

Learning Outcomes

After reading this article, you will be able to:

- Discuss Europe's worst political crisis since World War II
- Explain why the United Kingdom voted to leave the European Union.

Europe currently finds itself in the throes of its worst political crisis since World War II. Across the continent, traditional political parties have lost their appeal as populist, Euroskeptical movements have attracted widespread support. Hopes for European unity seem to grow dimmer by the day. The euro crisis has exposed deep fault lines between Germany and debt-ridden southern European states, including Greece and Portugal. Germany and Italy have clashed on issues such as border controls and banking regulations. And on June 23, the United Kingdom became the first country in history to vote to leave the EU—a stunning blow to the bloc.

At the same time as its internal politics have gone off the rails, Europe now faces new external dangers. In the east, a revanchist Russia—having invaded Ukraine and annexed Crimea—looms ominously. To Europe's south, the collapse of numerous states has driven millions of migrants northward and created a breeding ground for Islamist terrorists. Recent attacks in Paris and Brussels have shown that these extremists can strike at the continent's heart.

Such mayhem has underscored the price of ignoring the geopolitical struggles that surround Europe. Yet the EU, crippled by the euro crisis and divisions over how to apportion refugees, no longer seems strong or united enough to address its domestic turmoil or the security threats on its borders. National leaders across the continent are already turning inward, concluding that the best way to protect their countries is through more sovereignty, not less. Many voters seem to agree.

As Europe's history makes painfully clear, a return to aggressive nationalism could be dangerous, not just for the continent but also for the world. Yet a Europe of newly assertive nation-states would be preferable to the disjointed, ineffectual, and unpopular EU of today. There's good reason to believe that European countries would do a better job of checking Russia, managing the migrant crisis, and combating terrorism on their own than they have done under the auspices of the EU.

Ever-Farther Union

In the years after World War II, numerous European leaders made a convincing argument that only through unity could the continent escape its bloody past and guarantee prosperity. Accordingly, in 1951, Belgium, France, Italy, Luxembourg, the Netherlands, and West Germany created the European Coal and Steel Community. Over the next several decades, that organization morphed into the European Economic Community and, eventually, the European Union, and its membership grew from six countries to 28. Along the way, as the fear of war receded, European leaders began to talk about integration not merely as a force for peace but also as a way to allow Europe to stand alongside China, Russia, and the United States as a great power.

The EU's boosters argued that the benefits of membership—an integrated market, shared borders, and a transnational legal system—were self-evident. By this logic, expanding the union eastward wouldn't require force or political coercion; it would simply take patience, since nonmember states would soon recognize the upsides of membership and join as soon as they could. And for many years, this logic held, as central and eastern European countries raced to join the union after the collapse of the Soviet Union. Eight countries—the Czech Republic, Estonia, Hungary, Latvia, Lithuania, Poland, Slovakia, and Slovenia—became members in 2004; Bulgaria and Romania followed in 2007.

Then came the Ukraine crisis. In 2014, the Ukrainian people took to the streets and overthrew their corrupt president, Viktor Yanukovych, after he abruptly canceled a new economic deal with the EU. Immediately afterward, Russia invaded and annexed Crimea, and it soon sent soldiers and artillery into eastern Ukraine, too. The EU's leaders had hoped that economic inducements would inevitably increase the union's membership and bring peace and prosperity to an ever-larger public. But that dream proved no match for Russia's tanks and so-called little green men.

Moscow's gambit was not, on its own, enough to cripple the EU. But soon, another crisis hit, and this one nearly pushed the union to its breaking point. In 2015, more than a million refugees—nearly half of them fleeing the civil war in Syria—entered Europe, and since then, many more have followed. Early on, several countries, especially Germany and Sweden, proved especially welcoming, and leaders in those states angrily criticized those of their neighbors that tried to keep the migrants out. Last year, after Hungary built a razor-wire fence along its border with Croatia, German Chancellor Angela Merkel condemned the move as reminiscent of the Cold War, and French Foreign Minister Laurent Fabius said it did "not respect Europe's common values." But early this year, many of these same leaders changed their tune and began pressuring Europe's border countries to increase their security measures. In January, several European governments warned Greece that if it did not find a way to stanch the flow of refugees, they would expel it from the Schengen area, a passport-free zone within the EU.

Consciously or not, the European politicians advocating open borders have failed to prioritize their own citizens over foreigners. These leaders' intentions may be noble, but if a state fails to limit its protection to a particular group of people—its nationals—its government risks losing legitimacy. Indeed, the main measure of a country's success is how well it can secure its people and borders from external threats, be they hostile neighbors, terrorism, or mass migration. On this score, the EU and its proponents are failing. And voters have noticed. The British people issued a strong rebuke to the bloc in June when they voted to leave the EU by a margin of 52 percent to 48 percent, ignoring warnings from the International Monetary Fund, the Bank of England, and the United Kingdom's Treasury that doing so would wreak economic disaster. In France, according to a recent Pew survey, 61 percent of the population holds unfavorable views of the EU; in Greece, 71 percent of the population shares these views.

Back when Europe faced no pressing security threats—as was the case for most of the last two decades—EU members could afford to pursue more high-minded objectives, such as dissolving borders within the union. Now that dangers have returned, however, and the EU has shown that it is incapable of dealing with them, Europe's national leaders must fulfill their most basic duty: defending their own.

Back to Basics

The EU's architects created a head without a body: they built a unified political and administrative bureaucracy but not a united European nation. The EU aspired to transcend nation-states, but its fatal flaw has been its consistent failure to recognize the persistence of national differences and the importance of addressing threats on its frontiers.

One consequence of this oversight has been the rise of political parties that aim to restore national autonomy, often by appealing to far-right, populist, and sometimes xenophobic sentiments. In 2014, the UK Independence Party won the popular vote in an election for the European Parliament—the first time since 1906 that any party in the United Kingdom had bested Labor and the Conservatives in a nationwide vote. Last December in France, Marine Le Pen's far-right National Front won the first round of the country's regional elections; then, in March in Germany, a right-wing Euroskeptical party, Alternative for Germany, won almost 25 percent of the vote in Saxony-Anhalt. And in May, Norbert Hofer, a candidate from the far-right Freedom Party, narrowly lost Austria's presidential election. (Austria's Constitutional Court later annulled that result, forcing a rerun of the election that will be held in October.)

Some of these parties have benefited from the enthusiastic support of Russia, as part of its campaign to buy influence in Europe. Until recently, Moscow could rely on European leaders who were friendly to Russia, including former German Chancellor Gerhard Schröder and former Italian Prime Minister Silvio Berlusconi. But now, as new parties take the place of established ones, the Kremlin needs fresh partners. It has given money to the National Front, and the U.S. Congress has asked James Clapper, the U.S. director of national intelligence, to investigate the Kremlin's ties to other fringe parties, including Greece's Golden Dawn and Hungary's Jobbik. Yet such parties would be surging even without Russian backing. Many Europeans are disenchanted with politicians who have supported EU integration, open borders, and the gradual dissolution of national sovereignty; they have a deep and lasting desire to reassert the supremacy of their nation-state.

Of course, most of Europe's Euroskeptical politicians don't seek to disband the union entirely; in fact, many of them continue to see its creation as a historic victory for the West.

They do, however, want greater national autonomy on social, economic, and foreign policy, especially in response to overreaching EU mandates on migration and the demand for controversial continent-wide laws on issues such as abortion and marriage. Many in the United Kingdom, for example, pushed for a British exit from the EU, or Brexit, out of frustration with the number of British laws that have come from Brussels rather than Westminster.

The bet against sovereignty has failed. But sovereignty's resurgence has conjured up many dark memories of the nationalism that twice brought the continent to the brink of annihilation. Many observers now worry that European politics are coming to resemble those of the 1930s, when populist leaders spewed hate to whip up support. Such fears are not wholly unfounded. The strident xenophobia of Austria's Freedom Party recalls the early days of fascism. Anti-Semitism has risen across Europe, sprouting up in parties that span the ideological spectrum, from the United Kingdom's Labour Party to Hungary's Jobbik. And in Greece, some members of the radical left-wing party Syriza have advocated Greek withdrawal from NATO, a prime example of a growing anti-Americanism that could undermine the foundation of European security.

Yet affirming national sovereignty does not require virulent nationalism. The support for Brexit in the United Kingdom, for instance, was less an expression of hostility toward other European countries than it was an assertion of the United Kingdom's right to self-govern. A return to nation-states entails not nationalism but patriotism, or what George Orwell called "devotion to a particular place and a particular way of life." It's also worth noting that one of the greatest threats Europe faced in the twentieth century was transnational in nature: communism, which divided the continent for 45 years and led to the deaths of millions.

Beyond the EU

A renationalization of Europe may be the continent's best hope for security. The EU's founders believed that the body would guarantee a stable and prosperous Europe—and for a while, it seemed to. But today, although the EU has generated wealth through its common market, it is increasingly a source of instability. The euro crisis has exposed the union's inability to resolve conflicts among its members: German leaders have had little incentive to address Greek concerns, and vice versa. The EU also suffers from what the German Federal Constitutional Court has called a "structural democratic deficit." Of its seven institutions, just one—the European Parliament—is directly elected by the people, and it cannot initiate legislation. Finally, the recent dominance of Germany within the EU has alienated smaller states, including Greece and Italy.

Meanwhile, the EU has failed to keep Europe safe. Since 1949, Europe has relied on NATO—and, in particular, the United States—to secure its borders. The anemic defense spending of most European countries has only increased their dependence on the United States' physical presence in Europe. The EU is unlikely to create its own army, at least in the near future, as its members have different strategic priorities and little desire to cede military sovereignty to Brussels.

Many of the EU's backers still insist that in its absence, anarchy will engulf the continent. In 2011, the French minister for European affairs, Jean Leonetti, warned that the failure of the euro could lead Europe to "unravel." In May, British Prime Minister David Cameron claimed that a British exit from the EU would raise the risk of war. But as the American theologian Reinhold Niebuhr wrote in the 1940s, "the fear of anarchy is less potent than the fear of a concrete foe." Today, the identifiable enemies that have arisen around Europe, from Russia to the self-proclaimed Islamic State (also known as ISIS), seem far more worrying to most people than the potential chaos arising from the dissolution of the EU. Their hope is that individual countries will provide the kind of safety that Brussels can't.

Special Relationships

From the United States' perspective, the fraying of the EU presents a serious challenge—but not an insurmountable one. In the decades after World War II, Washington sought to contain the Soviet Union not just through nuclear deterrence and a sizable military presence in Europe but also by promoting European integration. A united continent, the thinking went, would pacify Europe, strengthen the economies of U.S. allies, and encourage them to cooperate with Washington to ward off the Soviet menace. Today, however, the United States needs a new strategy. Because the EU no longer seems up to the task of protecting its borders or competing geopolitically, more American pressure for Europe to integrate will simply alienate the growing number of Europeans who have turned their backs on the EU.

Washington need not fear the dissolution of the EU. Fully sovereign European states may prove more adept than the union at warding off the various threats on its frontiers. When Russia invaded Ukraine, the EU had no answer besides sanctions and vague calls for more dialogue. The European states that border Russia have found little reassurance in the union, which explains why they have sought the help of NATO and U.S. forces. Yet where the EU has failed, individual countries may fare better. Only patriotism has the kind of powerful and popular appeal that can mobilize Europe's citizens to rearm against their threatening neighbors. People are far more willing to fight for their country—for their history, their soil, their common religious identity—than they are for an abstract regional body created by fiat. A 2015 Pew poll found that in the case of a Russian attack, more than half of French, Germans, and Italians would not want to come to the defense of a NATO—and thus likely an EU—ally.

The return of nation-states need not lead Europe to revert to an anarchic jumble of quarreling governments. Increased autonomy won't stop Europe's states from trading or negotiating with one another. Just as supranationalism does not guarantee harmony, sovereignty does not require hostility among nations.

In a Europe of revived nation-states, countries will continue to form alliances based on common interests and security concerns. Recognizing the weakness of the EU, some states have already done so. The Czech Republic, Hungary, Poland, and Slovakia, for example—normally a disjointed group—have joined forces to oppose EU plans that would force them to accept thousands of refugees.

The United States, for its part, needs a better partner in Europe than the EU. As the union dissolves, NATO's function in maintaining stability and deterring external threats will increase—strengthening Washington's role on the continent. Without the EU, many European countries, threatened by Russia and overwhelmed by mass migration, will likely invest more heavily in NATO, the only security alliance backed up by force and thus capable of protecting its members.

It's time for U.S. leaders and Europe's political class to recognize that a return to nation-states in Europe does not have to end in tragedy. On the contrary, Europe will be able to meet its most pressing security challenges only when it abandons the fantasy of continental unity and embraces its geopolitical pluralism.

Critical Thinking

1. Why would a Europe of nation-states be preferable to the European Union?
2. Why was the European Union created?
3. What explains the rise of populist nationalist parties on the right in Europe?

Internet References

European Parliament Information in the United Kingdom
http://www.europarl.org.uk/en/your-meps.html

UK Independent Party
http://www.ukip.org

Article Prepared by: Robert Weiner, *University of Massachusetts, Boston*

Is Africa's Land Up for Grabs?

Foreign Acquisitions: Some Opportunities, but Many See Threats.

ROY LAISHLEY

Learning Outcomes

After reading this article, you will be able to:

- Discuss a new form of neo-colonialism in Africa.
- Discuss what can be done to counter the negative effects of land grabs in Africa.

An apparent surge in the purchase of African land by foreign companies and governments to grow food and other crops for export has set alarm bells ringing on and off the continent. The headlines have been strident: "The Second Scramble for Africa Starts," "Quest for Food Security Breeds Neo-Colonists," "Food Security or Economic Slavery?"

2.5 mn hectares of African farmland allocated to foreign-owned entities between 2004 and 2009.

The outcries reflect the continuing impact of the continent's history, when as recently as the last century colonial powers and foreign settler populations arbitrarily seized African land and displaced those who lived on it, lending considerable emotion to the current volatile issue. Some agricultural experts have wondered whether such land deals could lead to a form of "neo-colonialism". But immediate, practical concerns are also prominent. "This is a worrisome trend," noted Akinwumi Adesina, the then vice president of the advocacy group Alliance for a Green Revolution in Africa (AGRA). Such foreign land acquisitions, he argued, have the potential to hurt domestic efforts to raise food production and could limit broad-based economic growth. Many deals have little oversight, transparency or regulation, have no environmental safeguards and fail to protect smallholder farmers from losing their customary rights to use land, added Mr. Adesina, now Nigeria's minister for agriculture.

The sheer size of some of the land agreements has added to the alarm. A deal to allow South Korea's Daewoo Corporation to lease 1.3 million hectares was a key factor in building support for the ouster of Madagascar's President Marc Ravalomanana in March 2009. In Kenya the government struggled to overcome local opposition to a proposal to give Qatar and others rights over some 40,000 hectares in the Tana River Valley in return for building a deep-sea port.

A number of international organizations reacted to this development. The Food and Agriculture Organization (FAO) and the World Bank commissioned studies into so-called "land grabs." At the 2009 summit of the Group of Eight (G-8) industrialized countries in Italy, Japan pushed for a code of conduct to govern such schemes. Any code of conduct is going to be difficult to negotiate, and it will be even more difficult for industrialized countries to apply to deals that are primarily worked out between countries in the South, the UN's Special Rapporteur on the Right to Food, Olivier De Schutter, told *Africa Renewal*.

In a report titled "Large-scale Land Acquisitions and Leases," Mr. De Schutter wrote that while such investments provide certain development opportunities, they also represent a threat to food security and other core human rights. "The stakes are huge," he told *Africa Renewal.* Unfortunately, "the deals as they have been concluded up to now are very meagre as far as the obligations of the investors are concerned." He also noted that agreements concerning thousands of hectares of farmland are sometimes just three or four pages long.

Yet for African countries agreeing to such deals, the possible advantages are also attractive. While African agriculture rarely attracts significant investments or external aid—and the current global economic downturn has made external financing even more scarce—leasing unused land to foreign governments and companies for large-scale cultivation can seem like a way to boost an underdeveloped sector and create new job opportunities.

A study by the International Institute for Environment and Development (IIED), a research group based in the UK, estimated that nearly 2.5 mn hectares of African farmland had been allocated to foreign-owned entities between 2004 and 2009 in just five countries (Ethiopia, Ghana, Madagascar, Mali and Sudan) it studied in depth. The sheer scale of many leases is unprecedented, said the IIED report, *Land Grab or Development Opportunity*?, which was prepared for the FAO and the UN's International Fund for Agricultural Development.

The surge in interest in African land has been driven by a number of factors. On the side of investors, those include a desire for food security back home and to a lesser extent rising demand for biofuels. Behind both is the expectation of rising costs of land and water as world demand for food and other crops continues to expand.

Many of the government-to-government deals are aimed at meeting food needs, especially in the states of the Arab Gulf and in South Korea. Indian companies, backed in part by their government, have invested millions of dollars in Ethiopia to meet rising domestic food and animal feed demand. Commercial enterprises, many of them European, as well as Chinese companies, have been in the lead in cultivating jatropha, sorghum and other biofuels in countries such as Madagascar, Mozambique and Tanzania.

Africa is a particular focus for this investment explosion because of the perception that there is plenty of cheap land and labour available, as well as a favourable climate, Mr. De Schutter points out. In Mozambique, Tanzania and Zambia, for example, only some 12% of arable land is actually cultivated.

Africa so far has been able to mobilize only limited financing to develop its arable land. Despite persistent calls for increased domestic investment, agriculture has lagged well behind other sectors. The African Union has urged governments to devote 10% of their spending to agriculture, but not many have actually met that target. Donor countries and institutions have also failed to play their part, with agriculture's share of aid tending to fall.

With land apparently in abundance, but money not, the offer by foreign investors to develop agricultural land appears very attractive. But with much of the land not as unused as it might seem and with actual returns on agricultural investment far lower than presented in initial feasibility studies, the political and economic reality for African governments can be very sobering.

"Governments are sitting on a box of dynamite," Namanga Ngongi, former president of AGRA, initiated by former UN Secretary-General Kofi Annan, told the media.

Towards a Strategic Approach

Recent assessments by IIED, FAO, the World Bank and the Washington-based International Food Policy Research Institute (IFPRI) all confirm the shortcomings and potential dangers. These include the risks of undermining domestic efforts to increase food production, the danger that agricultural projects aimed exclusively at foreign markets may do little to stimulate domestic economic activities, and the potential loss of land rights for local farmers.

Many of the studies also point to possible benefits for a sector strapped for cash. These include the creation of jobs, the introduction of new technologies, improvements in the quality of agricultural production and opportunities to develop higher-value agricultural processing activities. There might even be "an increase in food supplies for the domestic market and for export," the FAO says.

To reap the benefits of this new trend, says an IFPRI study, *"Land Grabbing" by Foreign Investors in Developing Countries: Risks and Opportunities*, governments need to develop the capacity to negotiate sound contracts and to exercise oversight. This can help create "a win-win scenario for both local communities and foreign investors." The studies advise African governments to be strategic in their approach. In his report, Mr. De Schutter puts forward a number of recommendations to guide such land deals. These include the free, prior and full participation and agreement of all local communities concerned—not just their leaders, the protection of the environment, based on thorough impact assessments that demonstrate a project's sustainability, full transparency, with clear and enforceable obligations for investors, backed by specified sanctions and legislation, as necessary and measures to protect human rights, labour rights, land rights and the right to food and development. Such comprehensive deals would be in the long-term interest of investors and local communities alike, IFPRI notes, pointing out that land disputes can become violent, and governments may quickly find themselves with no alternative but to change or rescind contractual arrangements.

Land Rights

Land ownership is a core issue. Only a relatively small portion of land in Africa is subject to individual titling. Much land is community-owned, and in some countries state-owned. Even land that is officially categorized as un- or under-utilized may in fact be subject to complex patterns of "customary" usage. "Better systems to recognize land rights are urgently needed," the FAO argues in a policy brief, From Land Grab to Win-Win.

The World Bank points to the importance of international bodies helping African governments develop land registry systems. The IIED study stresses that such schemes must allow for collective registration of community lands that protect "customary" land rights.

Mr. De Schutter argues that internationally agreed-upon human rights instruments can be used to protect such rights, including those of livestock herders and indigenous forest dwellers.

According to the IIED study, the bulk of recent large-scale land acquisitions in Africa have been based on the leasing of land to foreign entities with the intent of using labour to work the land. The study argues the need for governments to include clauses ensuring the use of local labour in contracts for such schemes. "Agreements to lease or cede large areas of land in no circumstance should be allowed to trump the human rights obligations of the states concerned," Mr. De Schutter argues.

Proposals for such ideal agreements, backed by necessary national legislation and enforcement principles, are being put forward. But, as the IIED study points out, there is already a large gulf between contractual provisions and their enforcement. The gap between the statute books and the reality on the ground may entail serious costs for local communities.

A code of conduct for host governments and foreign investors could help ensure that land deals are a "win-win" arrangement for investor and local communities alike, IFPRI suggests. It cites the Extractive Industries Transparency Initiative, which binds participating governments and companies to certain standards in mining and oil activities, as one possible model for large-scale land deals.

Mr. De Schutter is sceptical that such a code can be negotiated or enforced. He instead emphasizes the existing body of human rights laws, which can be applied to large-scale land acquisitions and used to get governments to meet their obligations to citizens.

Either way, experts agree that African governments must have the will and the ability to apply laws. "Strengthening the negotiation capacity is vital," Mr. De Schutter argues. And that capacity cannot be of governments alone, he says. Local communities must also be empowered and national parliaments must be involved. Achieving that, many fear, may be the most difficult gap to bridge.

Critical Thinking

1. Why are the "land grabs" in Africa considered harmful to African economic interests?
2. What advantages do African countries gain from land deals?
3. Why is Africa the focus of land investment by such countries as India and South Africa?

Internet References

African Development Bank
http://www.afdb.org

Food and Agriculture Organization
www.fao.org

UN Economic Commission for Africa
www.uneca.org

Laishley, Roy. "Is Africa's Land Up for Grabs?" *Africa Renewal Online*, 2014. United Nations Africa Renewal.

Article

Prepared by: Robert Weiner, *University of Massachusetts, Boston*

The Blood Cries Out

Burundi is about the size of Maryland but holds nearly twice as many people. Brothers are now killing brothers over mere acres of earth. Could africa's next civil war erupt over land?

JILLIAN KEENAN

Learning Outcomes

After reading this article, you will be able to:

- Understand why Burundi may be on the edge of civil conflict.
- Discuss the relationship between land and civil conflict.

When Pierre Gahungu thinks about the small farm in the Burundian hills where he grew up and started a family, he remembers the soil—rich and red, perfect for growing beans, sweet potatoes, and bananas. He used to bend over and scoop up a handful of the earth just to savor its moist feel. To Gahungu, now in his 70s, the farm was everything: his home, his livelihood, and his hope. After he was gone, he had always believed, the land would sustain his eventual heirs.

But then, in an instant, his dreams were thrown into jeopardy. On a dusky evening in 1984, Gahungu was walking home when he heard a noise behind him. He turned and found himself face to face with Alphonse, the son of a cousin. For months, Alphonse had been begging Gahungu, whom he called "uncle," for a portion of the farm. Alphonse's polygamous father had many sons—more than 20, Gahungu says—which meant each one would get just a tiny plot of his land. (In Burundi, generally only men may inherit property.) Alphonse wanted more space, a rapidly shrinking commodity, on which to build a house and a life. Gahungu had a much smaller family—ultimately, he and his wife would have three children, but only one boy, named Lionel—so he had plenty of land to share, Alphonse reasoned. Why shouldn't he get a piece of it? Gahungu, however, had refused repeatedly. When he saw Alphonse that night on the road, he assumed they were in for another round of the same exhausting refrain.

Alphonse, however, had not come to talk. Without saying a word, he raised a machete and brought it down onto his uncle's skull. Gahungu remembers feeling a flash of pain and hearing a bone crunch before everything went black.

"I was terrified," Gahungu says through an interpreter. He woke up wounded and later saw a doctor. He began recovering from his injury, but he feared that his farm would never be safe in his hands. Gahungu decided that if Alphonse couldn't kill him, the land's legal owner, his son's inheritance would be safe. So he left his family behind and moved alone to the nearby city of Muramvya, where he worked at a tailoring shop downtown.

Before long, more problems arose, but not with Alphonse (who, Gahungu says, died in a car accident). One of Alphonse's brothers built two houses on Gahungu's land without his uncle's permission. In 1991, on the eve of a brutal, 12-year civil war that pitted Burundi's two main ethnic groups, Hutu and Tutsi, against one another, Gahungu took the man to a local court, which ruled in the owner's favor. But winning the legal battle did nothing to change Gahungu's situation: To this day, he says, Alphonse's brother, the brother's family, and the two houses illegally occupy his farm.

Gahungu has tried to go back to Burundi's backlogged courts for help, but he doesn't have the money to pay for a case. Tragedy, too, has continued to follow him: Lionel died at just 19. As his own life draws toward an inevitable end, Gahungu lives alone in Muramvya. He now fears he will die before ever getting his beloved land back. "It was the perfect farm because it was my farm," Gahungu recalls. "It was my whole life."

Gahungu's experience mirrors other stories familiar to Burundi for decades—stories that are multiplying and worsening as the country copes with a veritable explosion of people. At 10,745 square miles, Burundi is slightly smaller than the U.S. state of Maryland, but it holds nearly twice as many people: about 10 million, according to the U.N. Development Programme, or roughly 40 percent more than a decade ago. The population growth rate is 2.5 percent per year, more than twice the average global pace, and the average Burundian woman has 6.3 children, nearly triple the international fertility rate. Moreover, roughly half a million refugees who fled the country's 1993–2005 civil war or previous ethnic violence had come back as of late 2014. Another 7,000 are expected to arrive this year.

The vast majority of Burundians rely on subsistence farming, but under the weight of a booming population and in the long-standing absence of coherent policies governing land ownership, many people barely have enough earth to sustain themselves. Steve McDonald, who has worked on a reconciliation project in Burundi with the Woodrow Wilson International Center for Scholars, estimates that in 1970 the average farm was probably between 9 and 12 acres. Today, that number has shrunk to just over one acre. The consequence is remarkable scarcity: In the 2013 Global Hunger Index, Burundi had the severest hunger and malnourishment rates of all 120 countries ranked. "As the land gets chopped into smaller and smaller pieces," McDonald says, "the pressure intensifies."

This pressure has led many people who want land, like Alphonse years ago, to take matters into their own hands—at times violently. The United Nations estimates that roughly 85 percent of disputes pending in Burundian courts pertain to land. Between 2013 and 2014, incidents of arson and attempted murder related to land conflict rose 19 percent and 36 percent, respectively. Violence sometimes occurs within families, but it also can play out between ethnic groups: Most returning refugees are Hutu, but the land they left behind has often been purchased by Tutsis. "The land issue comes into politics when parties say, 'I promise to return to you what is rightfully yours,'" says Thierry Uwamahoro, a Burundian political analyst based in the Washington, D.C. area.

Against this fragile backdrop, the Institute for Security Studies, a South African-based think tank, has warned that "attempts to politicise land management . . . risks reigniting ethnic tensions" before national elections scheduled for May and June. Many locals, however, fear that an even bigger disaster is looming. "The next civil war in Burundi will absolutely be over land," says a communications consultant in Bujumbura, the capital, who works for U.N. agencies and asked not to be named for security reasons. "If there is no new land policy, we won't last a decade."

There are no easy solutions to Burundi's mounting land crisis, but stories like Gahungu's offer a glimpse of what might happen if this ticking time bomb is not diffused. "In the past, this situation didn't exist," Gahungu says, standing outside the tailoring shop where he still works, cleaning and ironing clothes. "There was land for all, but not anymore. I fled because I feared that what happened to me before could happen again. It happens to someone every day now."

Before European colonizers arrived in Burundi, farmers cultivated the country's arable hilltops, while less desirable, low-lying swamplands went largely unclaimed. An aristocratic class, known as the Ganwa, technically owned the land, but farmers' access was administered at the local level by a network of "land chiefs," many of whom were Hutu. The chiefs also resolved land conflicts, according to Timothy Longman, director of Boston University's African Studies Center.

Under Belgian rule, which lasted from 1916 to 1962, this all began to change. The king, the head of the Ganwa, kept control of the highlands. (According to scholar Dominik Kohlhagen, the king was seen as the land's spiritual guardian.) But the state assumed ownership of the lowlands and began to encourage their cultivation. The colonial government also concentrated political power among the Tutsi minority, which comprised about 14 percent of the population, giving the group a near monopoly on Burundi's government, military, and economy. Among other actions, this consolidation involved gradually stripping the Hutu land chiefs of their authority. More broadly, too, it sowed the seeds of dangerous ethnic polarization.

Under colonialism, official land deeds and titles were few and far between, which meant that Burundians often could not prove that they owned acreage. In the early 1960s, as independence loomed, the government began offering land registration to parties that requested it, which, for the most part, were foreign businesses such as hotels. Families also had the option to register their land, but because the centralized system was inaccessible for most farmers and required a huge tax payment, few did. So land plots quickly fell into two categories: those with boundaries recognized by the state, and those with borders determined by custom—that is to say, residents understood trees, rocks, paths, creeks, and huts to mark de facto property lines.

The Catholic Church was among the institutions that benefited from the colonial approach to land. Missionaries, known as "White Fathers," began arriving in the late 19th century, and over several decades, the king gave them large tracts of land, which they used to establish churches, schools, hospitals, and farms. After colonialism ended, the self-sufficiency that land provided the church helped it retain influence, even as its relationship with the newly independent government grew fraught. Most notably, Jean-Baptiste Bagaza, a military leader who in the mid-1970s seized Burundi's presidency in a bloodless coup, saw the church as an extension of colonial power and a rival to his own, so he limited the hours in which congregations could gather, shut down a Catholic radio station, and used visa non-renewals and expulsions to decrease the number of missionaries in the country. Nevertheless, the church retained millions of

Burundian followers, along with plenty of land, though no one, it seems, knew exactly how much.

The Catholic Church was also complicit in nurturing Burundi's ethnic divisions; Catholic schools, for instance, were largely reserved for "elite" children, meaning Tutsis. Intensifying schisms led to various outbreaks of ethnic violence, and in 1972, the Tutsi-dominated military launched a series of pogroms targeting Hutus. More than 300,000 Hutus fled the country in under a year, leaving behind their land. Bujumbura took advantage of some of this newly vacated property and extended agricultural schemes called *paysannats* (derived from the French word for "peasantry"): The state leased the land to farmers, who would grow cotton, tobacco, and coffee and then sell these crops back to the government, the only legal buyer. Officials in Bujumbura hoped to boost Burundi's weak economy by reselling the crops on the international market.

But the paysannat system failed miserably due to corruption, inefficient government bureaucracy, and variations in global commodity prices. Seeking bigger profits than they were able to get in Burundi, farmers began to smuggle their harvests over the country's borders, and state-run agricultural buying programs floundered in the mid-1980s. Paysannats also ignored and often destroyed the physical markers that had defined traditional land boundaries. Along with the pervasive lack of legal documents showing land ownership, this made it impossible for most returning refugees to reclaim their lost acreage. (Today, the Hutus who left in 1972, some of whom have never come home or are only just doing so, are called "old-caseload" refugees.)

The government set up two commissions, in 1977 and 1991, to resolve land disputes, but they proved largely ineffective. Ethnic tensions continued to mount, coming to a head in 1993. That October, Tutsi extremists assassinated Burundi's first democratically elected Hutu president, Melchior Ndadaye, and civil war erupted as Hutu peasants responded by murdering Tutsis. In just the first year of conflict, tens of thousands of people were killed; by the time the war ended more than a decade later, some 300,000 Burundians, most of them civilians, had died. The war also produced a new wave of roughly 687,000 refugees.

When the dust settled, the effects of mass death and displacement were exacerbated by widespread poverty, food insecurity, and a host of other post-conflict challenges, all of which persist today. In 2014, the World Bank estimated Burundi's GDP growth rate at 4.0 percent, below the average of 4.5 percent for countries in sub-Saharan Africa. The bank forecasts that this gap will only widen in 2015, with Burundi's rate declining to 3.7 percent and the region's climbing slightly. The issue of land, meanwhile, has become a casualty in its own right, thrown into greater flux than ever before.

Today, there are dozens of scenarios under which people claim land, and the same plot, no matter how tiny, is often the subject of competing claims. Some families still say they own acreage because of paysannat leases; seeking to make a profit, tenants have even sold their land over the years, despite the fact that it is technically state-owned. According to the International Crisis Group, some 95 percent of Burundian land still falls under customary law: A family says it purchased its farm from neighbors before the war, but holds no formal deed, while another claims village elders approved the purchase of a few acres after the war, and so on. A centralized registration system does exist, and according to the country's land code (which was revised in 2011 for the first time since 1986), any person who owns property must hold a land certificate. The bureaucratic system, however, is complicated, and the government has done little in terms of enforcement. According to Kohlhagen, offices that issue certificates exist in only three cities, and as of 2008, only about 1 percent of the country's surface area was registered.

Complicating matters further is the continuous flow of refugees who return home to find their land occupied by new owners. In some cases, a Hutu farmer who fled the 1972 pogroms may come back to find two other people claiming his property: whoever lived on it up until 1993, and whoever claimed it after the civil war. The last resident may have purchased the land legally, even from the government itself, and may have been paying off mortgages for decades. The question then becomes, who should get to live on the land now—and how should the claimants who can't have it be compensated?

The government has no clear answer. "It's tricky to say what land policy is today, because there is not a uniform dispute-resolution strategy," says Mike Jobbins, a senior program manager at Search for Common Ground, a conflict-prevention NGO that works in Burundi. "Every case is decided on its own merits."

The Burundian courts and *bashingantahe,* or traditional panels comprising senior men in villages, are empowered to settle land disputes. But while courts issue legally binding rulings, cases are time-consuming and often prohibitively expensive. The bashingantahe, meanwhile, are free, yet operate according to customary law. A third body would seem to offer a more promising option: The National Commission for Land and Property, known by its French acronym, CNTB, was established in 2006 to resolve arguments over who owns land vacated by refugees. Its 50 members are required to be 60 percent Hutu and 40 percent Tutsi, and since its creation, the CNTB has processed nearly 40,000 cases.

But the CNTB has struggled to adopt a consistent approach to its judgments. At first, its preference was to divide land between valid claimants, both past and current owners. Since 2011, however, it has begun returning land to its original owners, usually Hutu refugees displaced in 1972. In some cases, it has even revised decisions on previously closed disputes. This has led to angry claims that the government of President Pierre Nkurunziza, a Hutu, is trying to curry support from Burundi's predominantly Hutu electorate.

The government did nothing to quiet these concerns when, in December 2013, it expanded the CNTB's mandate to review cases that predate even 1972, made it a criminal offense to obstruct the commission's actions, and allowed rulings to be appealed to a new "land court" that can issue binding decisions. This move to boost the CNTB's power created suspicion among critics of the government that the commission's biases would only become more firmly entrenched. "Far from uniting Burundians or reconciling them, this new law on the CNTB will divide them," Charles Nditije, then leader of an opposition political party, told the media at the time.

Other criticisms surround the government's failure to establish a compensation fund for people who do not win land disputes. The 2000 Arusha Peace and Reconciliation Agreement for Burundi, which outlined a plan for the country's postwar peace process, guaranteed compensation, but according to Thierry Vircoulon, the International Crisis Group's project director for central Africa, the state seems to have "completely ignored" this detail. Some government detractors say this is a deliberate, ethnically driven decision by Nkurunziza's administration, because many people eligible for compensation would be Tutsi.

Murky, controversial land policies have at times led to interethnic violence. In 2013, riots broke out in Bujumbura when the police tried to evict a Tutsi family from the house it had owned for 40 years in order to give the house back to its previous Hutu residents. "We are here to oppose injustice, to oppose the CNTB, which is undermining reconciliation in Burundi society," a protester was quoted by Agence France-Presse as shouting. Over a six-hour standoff, more than a dozen people were reportedly hurt, and 20 were arrested.

Ethnic tensions, however, are only part of the puzzle in Burundi's land crisis. Poor farming families are straining the country's limited ground space. About two-thirds of Burundians live in poverty, and families often have several male heirs who are forced to share plots of earth that barely fit a home and a few rows of crops. As a result, according to research conducted by land-rights consultant Kelsey Jones-Casey, "[T]he most destructive conflicts experienced by rural people in Burundi are intra-family disputes, most of which manifest over the issue of inheritance." Violence sometimes occurs within polygamous families, with sons born by different mothers fighting for finite land. In Muramvya, people speak in low voices about a woman who slit her husband's throat to accelerate a land inheritance for her son.

Violence over land is roiling a country that already clings to an uneasy peace. Nkurunziza's government has been accused of ordering convictions, murders, and disappearances of political adversaries, among other abuses. In January, the state claimed to have killed nearly 100 rebels who had crossed the border between Burundi and the Democratic Republic of the Congo with the intention of destabilizing the country and the upcoming national elections, in which Nkurunziza is expected to seek a third term despite a constitutional limit of two. Vital Nshimirimana of the Forum for the Strengthening of Civil Society in Burundi told Voice of America that officials had suggested the country's political opposition supported the rebels: "This is what leads us to think that it might be a fake explanation to actually take advantage of the same to arrest opposition leaders or some civil society [members]," he said. A few days later, youth leader Patrick Nkurunziza (no relation to the president) was arrested for his alleged connections to the rebels; at the same time, the government sentenced opposition leader and former Vice President Frédéric Bamvuginyumvira to five years in prison for bribery.

Violence over land is roiling a country that already clings to an uneasy peace.

Many Burundians fear that land could be the detail that pushes swelling political tension into something far worse. "Land is the blood and the flesh of any human being," says Placide Hakizimana, a judge in Muramvya, who notes that 80 percent of the cases he adjudicates pertain to property disputes. "Without land, we are condemned to death. No one will accept that. [Families] will fight. We prefer to die rather than live without land."

Policy reform may be a dead end or, at least, one that is too rife with corruption and partisan battles to ever solve the land crisis. This thinking is driving some people to focus on restricting population growth. "Family planning is the only exit point to the land problem," says Norbert Ndihokubwayo, a member of Burundi's parliament and president of a legislative commission on social and health issues. "No other solution is possible."

In 2011, the government approved a national development strategy called "Vision Burundi 2025" with ambitious demographic goals: to reduce national growth from its current rate, which would cause the population to double every 28 years, to 2 percent over the next decade, and to slash the birthrate in half. To hit these numbers, the government said it would partner with civil society to "stress . . . information and education on family planning and reproductive health." Ndihokubwayo says the government is also "absolutely" considering a law that would limit the number of children each family can have.

Many international donors are helping to expand access to family-planning services. The Netherlands chose Burundi in 2011 as one of 15 "partner countries" in which to emphasize programs that promote peace and stability, and according to Jolke Oppewal, the Dutch ambassador to Burundi, his country now donates 8 million euros annually to programs promoting sexual and reproductive health, among other human rights. In a 2005–2013 contract, the German government-owned development bank KfW dedicated more than 1.4 million euros to "strengthening and reorganizing [Burundi's] reproductive health and family planning services."

Some medical professionals are keenly aware of the role they are meant to play in keeping population growth in check. Christine Nimbona is a nurse at a secondary health clinic in Kayanza province, which, with nearly 1,500 inhabitants per square mile, is one of Burundi's most overpopulated regions. One day in August, several women waiting outside the clinic where Nimbona works nursed babies; dozens of children played nearby. Nimbona says that of the roughly 30 patients she sees each week, "almost all" cite fears about land resources and potential inheritance conflicts as their reasons for seeking family planning. "I know that by what I am doing, I am lighting the escalation of violence in my country," Nimbona says.

It's an uphill battle, littered with enormous, deep-seated obstacles. According to the United Nations, modern contraceptive use among females between the ages of 15 and 49 was just 18.9 percent in 2010. In Burundi's male-dominated society, women are often powerless to convince their husbands to use birth control. Then there is the Catholic Church: In addition to claiming an estimated 60 percent of Burundians as followers, the church has affiliations with roughly 30 percent of national health clinics, which are forbidden from distributing or discussing condoms, the pill, and other medical contraceptives. "Catholic teachings against birth control are very resonant with Burundian culture, which says that children are wealth," explains Longman, of Boston University. "Because the Catholic Church is so powerful and controls so much of the health sector, it creates a huge stumbling block for family-planning practice."

Bujumbura insists that the Catholic Church is a collaborative partner on land issues. The president even appointed a Catholic bishop, Sérapion Bambonanire, as head of the CNTB in 2011. But cracks do exist between church and state. In 2012, the Ministry of Public Health launched a series of "secondary health posts," which offer medical contraceptives; sometimes these clinics, including Nimbona's, are built right next door to existing Catholic ones.

There is also tension over a variable with unknown dimensions: how much of the land the Catholic Church held onto after colonialism it still owns today. "The Catholic Church can't keep owning all the land while Christians are starving," says a regional government employee in Kayanza, who spoke on condition of anonymity out of concern for his safety. According to him, in 2013 the government quietly launched a mapping program to determine, among other things, how much land the church controls. "National politics don't allow us to focus on the Catholic Church," he says, referring to the fact that the church's followers are also voters. "So the government thinks this indirect method is best."

Ndihokubwayo says a land-mapping program does exist, but won't confirm or deny whether it was created specifically to find out how much property the Catholic Church possesses. "This is a very delicate issue," he explains. "I'm not sure whether we'll ever find out how much land the church owns, but we'll keep trying." (Cara Jones, an assistant professor at Mary Baldwin College who studies Burundi, pointed out that the program would also give President Nkurunziza's ruling party information about how much land its political opponents own.)

Some religious leaders are on board with the push for family planning. Pastor Andre Florian, a priest in Burundi's Anglican Church, which has an estimated 900,000 followers, says he used to be part of the problem. From the pulpit of his small stone church in Kayanza, he once railed against the evils of contraception. Family planning, he told his congregation, was best left to God. Yet Florian watched with grave concern as members of his flock struggled to feed their babies. One day, he looked at a child with dull orange hair, a clear sign of advanced malnutrition, and asked himself: Was this really God's plan? Shaken, Florian isolated himself for three months, studying scripture and praying. "When I returned from my research, I realized that I had done wrong," Florian says. "If nothing happens, if we just keep doing what we're doing, tomorrow is not certain. We will see families killing each other. We will see chaos in the country. The day after tomorrow will disappear."

Other Burundians, however, fear that support for family planning is too little too late. Joaquim Sinzobatohana, a father of four in Ngozi province, says he first learned about vasectomies on a radio program. (Eighty percent of Burundians have a radio, and U.N.-funded songs and soap operas now dramatize stories of families that have suffered the burden of many pregnancies but are saved by family planning.) He decided to get the operation, a simple outpatient procedure, because he and his wife, Clautilde, are "very scared" that their small plot will provoke conflict among their children, and more offspring would only increase the chances of violence. But Sinzobatohana admits that even a demographic freeze might not save his family, or his neighbors. The numbers just don't add up: Already, too many people are squeezed onto too little land.

"It's unfortunate that these contraceptive programs came after we already had too many kids," Sinzobatohana says. "The damage has been done. Now we wait."

International experts say a comprehensive approach to Burundi's land crisis is necessary—one that combines policy reform, better dispute-resolution options, family planning, and new economic opportunities that will ensure fewer Burundians rely solely on the earth for survival. "People need to have economic opportunities besides agriculture, to incorporate people into other kinds of jobs and trades, so that not everyone is dependent on farming for their livelihoods," says Jobbins of Search for Common Ground. "Without some prospect for economic growth within the context of the region and the East African community, land scarcity will continue to be a stressor."

But the land problem is infinitely complex, with roots that run deep into Burundi's history. The resources and political will to deal with it are scarce. And whether in a new law or a family's decades-long story, there will always be critical details that go overlooked—details that could become matters of life and death.

In 1999, Emmanuel Hatungimana, an elderly farmer in northern Ngozi, could feel his body slipping away. Death was very close. So he gathered his family—two wives and 14 children—around his bedside. It was time to divide his farm.

At 37.5 acres, Hatungimana's lush plot of land was a decent size. An equitable division would have left his eight sons with roughly 4.7 acres each. The eldest sons of Hatungimana's two wives stepped up to represent his part of the family: Pascal Hatungimana for the four sons of the first wife, and Prudence Ndikuryayo for the four sons of the second wife. Hatungimana gave exactly half of his land to Pascal and his brothers, and the other half to Prudence and his brothers. The patriarch was satisfied, according to Pascal; his family's future was secure. A few days later, he died at peace.

But no one had considered the road.

One side of Hatungimana's land runs alongside a paved road, a very desirable quality because access to that lane makes it significantly easier to bring supplies in and out of the property. In dividing his land right down the middle, however, Hatungimana had ensured that only four of his sons could claim the road as a border.

Today, more than 15 years after Hatungimana's death, his family teeters on the brink of violence. Prudence says it's unfair that Pascal and his brothers have the better land, and that he is willing to fight to get what he deserves. But Pascal doesn't want to give up anything. So they've brought their case to a local bashingantahe. If the panel can't resolve the dispute, both brothers say they don't know what will happen.

Standing on the land outside his brother's house, Prudence looks left, over his shoulder, at Pascal, who listens nearby with his arms crossed. A dark expression falls over Prudence's face. "Around Burundi, brothers are killing brothers. Sons are killing fathers. And it's all for land," he says. "Hopefully our family won't reach that stage. But if something doesn't happen, everything will fall apart."

Critical Thinking

1. What is the cause of ethnic tension between the Hutus and the Tutsis in Burundi?
2. Why is less land available for farming in Burundi?
3. What is the legacy of colonial rule in Burundi?

Internet References

African Development Bank Burundi
www.afdb.org/en/countries/east-africa/Burundi/

Burundi Government
https://www.cia.gov/library/publications/the-world-factbook/geos/bu.html

IFAD's Strategy in Burundi
operations.ifad.org/web/ifad/operations/country/home/tags/burundi

---

Jillian Keenan (*@JillianKeenan*) is a writer based in New York. She is working on a book about Shakespeare and global sexuality. A grant from the United Nations Population Fund supported research for this article.

Article Prepared by: Robert Weiner, *University of Massachusetts, Boston*

Can a Post-Crisis Country Survive in the Time of Ebola?

Issues Arising with Liberia's Post-War Recovery.

JORDAN RYAN

Learning Outcomes

After reading this article, you will be able to:

- Learn about the responses to infectious diseases in a post-conflict country.
- Learn about the focus on development in post-Ebola Liberia.

We suspect that the first case of the current outbreak of Ebola Virus Disease (EVD) began with the illness of a two-year old child who died at the end of December 2013. This occurred in Guéckédou prefecture, Guinea, located in the sub-region adjoining Liberia and Sierra Leone. This area is well known for its porous borders and peoples who share ethnic and tribal identities, and it has been a cauldron for the brutal conflict that enveloped the area for well over 15 years.

We may never see the face or know the name of "Patient Si." That infant's death and the thousands of others since the EVD outbreak provoked the near collapse of the health systems in these three countries. It is a catastrophe that demands more than an emergency response from the world. Now at stake are health systems, their scope, quality, and impact—and more broadly, governance, policy choices, and progress.

At a more fundamental level, the EVD outbreak requires nothing less than a wholesale reordering of our priorities and the way in which we respond to crisis and to the emerging threats, including infectious diseases—especially in areas emerging from violent conflict.

This article will consider some of these issues in broad terms, drawing on my personal experience and association with Liberia's progress over the past nine years. Following a review of progress and challenges, it will provide a perspective on the lessons and priorities for doing development differently in post-Ebola Liberia and the neighboring countries. One point is clear: building on the creative energies of the Liberian people, the international system needs to learn to act in a proactive manner, rather than wait until a global crisis arises.

Some Starting Points

First, to be absolutely clear, there can be no doubt of the paramount need to support the all-out effort to stop the spread of EVD now. Although the initial response was unfortunately marked with far too much hesitation, there is now a robust Security Council approved mission, the UN Mission for Ebola Emergency Response (UNMEER). Now is the time for UNMEER, with all concerned national authorities, bilateral and other partners, including philanthropies and the private sector, to act in concert with one single goal. Donors need to provide immediate and generous support now, not next year when it will be too late.

Second, it will be important to look carefully at why the epidemic flourished and what factors allowed it to do so. This outbreak must serve as a wake-up call to the international community, for certainly this situation is and will not be an isolated incident. We are witnessing the dawn of a much more complicated world to come: one which is regularly challenged by upheavals that may at first sight appear local, but because of the nature of globalization, can have a dramatic impact upon populations living far away.

Finally, the unleashing of Ebola in the 21st century is not a case of science fiction coming true. It is instead the result of a series of failures: failures to invest in a timely manner in the right infrastructure; failures to build accountable systems; failures to concentrate on resilience; and failures to put an end to the taxing, time-consuming bureaucracy which saps the ability of people to focus on what matters most—making a difference in the lives of people, especially the poorest.

Liberia: A Story of Hope and Work

I arrived in Liberia in November 2005 to serve as the deputy special representative of the UN secretary-general (DSRSG) in the peacekeeping mission, UN Mission in Liberia (UNMIL), which had been established under a Security Council mandate to support a successful peace process.

Within the first days, I had the thrill of witnessing a former UN colleague, Ellen Johnson-Sirleaf, win the run-off presidential election to become Africa's first democratically elected female president. In the mid 1990s, she had held senior positions within the United Nations Development Programme (UNDP), and prior to that, the World Bank.

In her inaugural address on January 6, 2006, President Sirleaf called on her fellow compatriots to "break with the past," declaring: "The future belongs to us because we have taken charge of it. We have the resources. We have the resourcefulness. Now, we have the right government. And we have good friends who want to work with us."

Each year since then, the international community has invested over US$1.5 billion to support the government of Liberia's five-pillar strategy of security, economic revitalization, basic services, infrastructure, and good governance. The international community embraced this strategy and its aim to direct assistance to support tangible development gains for rural Liberians.

My primary task as the DSRSG for UNMIL was to coordinate the provision of life-saving humanitarian aid as well as the longer-range development assistance of the United Nations and international partners. Working closely with the Special Representative, I had the vast logistical, technical, and military resources of UNMIL at my disposal. I could also call on the UN agencies which were collectively known as the UN Country Team, a number of which—including UNDP, the UN Refugee Agency, UNICEF, and the World Food Programme—had been working in Liberia for many years.

My job was to harness and integrate the different strengths and activities of the peacekeeping mission and the UN Country Team in a "one UN" effort to provide relief assistance, strengthen the capacity of Liberian institutions to govern and deliver basic services (e.g., security, justice, policing, and social services like health and education), revive economic activity (particularly in rural areas), and, ultimately, to foster national healing and reconciliation.

As I realized during my first visit with the minister of health (whose office was unreliably lit by a single electric bulb dangling from a wire), the war had left the Liberian health sector (and many others) in shambles. The minister told me that the primary provider of health services, both during and after the war, had been international NGOs. The country's health facilities had been completely looted and vandalized during the war, and medical supplies were simply unavailable. A country of over 4 million people had only 26 practicing doctors. In most parts of rural Liberia, health services and referral systems (including any kind of maternal or reproductive health care services and information) simply did not exist. It was clear that in the health sector, as in several other sectors, the work of strengthening institutions would be a case of rebuilding them virtually from scratch.

The Liberian government and its international partners took several key steps to rebuild these institutions. We understood from the past that the concession economy and the politics of elite capture and bribery had mutually reinforced one another, creating the dynamics that led to civil war and the deprivation, suffering, and traumatization of the Liberian people.

This led to the initial institution-building focus on building up much needed capacities and systems within government institutions, with the aim of developing the systems for accountable and transparent financial management, budgeting, and procurement. Steps were taken to put in place accountability mechanisms to reduce corruption and increase transparency, introduce a cash management system, devise a new procurement commission, and establish a general auditing commission.

Another critical task was to restore trust and public confidence between the Liberian people and the government. This needed to begin early and with quick support from Sweden and the UNDP/UN Country Team, which simultaneously began the process of decentralizing governance by supporting local development initiatives in each of the 15 administrative regions. Steps were taken to foster citizen involvement to build peace and re-establish trust between the government and the general population.

Transparency International (TI) ranked the transparency of Liberia's government as the third best in Africa, citing the independence of the General Auditing Commission, support for the establishment of the Liberia Anti-Corruption Commission, the promotion of transparent financial management, public procurement and budget processes, and the establishment of a national law to ensure Liberia's compliance with the Extractive Industries Transparency Initiative. The economy was also growing at an impressive rate. Trade, production, commerce, and construction expanded rapidly.

A series of government plans were issued to outline the way that the Johnson Sirleaf Administration would capitalize upon this post-conflict economic boom, starting with her 150-day Action Plan to jump-start economic recovery. This was followed by an 18-month Interim Poverty Reduction Strategy, and in 2008, the government completed its first Poverty Reduction Strategy (PRS), whose formulation drew on countywide consultations and citizen engagement.

"Within months, the disease dislocated the institutional fabric of Liberia, disrupting not just the health system, but also the entire system of governance. Of the 10, 129 reported cases globally as of October 23, 2014, 4,665 are in Liberia and 2,705 have died."

Framed around the five pillars mentioned above, the "Lift Liberia" strategy was specifically designed to promote rapid, shared growth. Officials at all levels sought to assure the Liberian population that "growth without development," which prior to the war had generated extreme inequality and deprivation, was gone forever. Promoting shared growth entailed the provision of quality public services (especially education and health) and the revival of small-scale agriculture and rural livelihoods supported by the expansion of infrastructure (roads, bridges, water, and sanitation) throughout the whole country.

This was music to the international community's ears. Aid continued to pour in. As a result, investor confidence rose dramatically. Starting with the rubber plantations, the concession economy (iron ore, rubber, and timber) began attracting large-scale international investors. Private investment increased rapidly in residential and commercial property, telecommunications, and transport.

Growth Without Development: A Return of Despair?

When I left Liberia in 2009, the story was still one of hope, and there was still widespread confidence in the national leadership. In fact, the nation's narrative of peace, stability, and recovery was heralded as a prized example of post-conflict stability, reconstruction, and development. Up until the outbreak, Liberia had experienced a sustained peace, two successful democratic elections, improved access to justice and human rights, a restoration of public services, and a reemergence of private sector activity. In conjunction were unprecedented growth rates, showcasing Liberia's considerable strides since the August 2003 Comprehensive Peace Accord and the profound chaos and disorder the country found itself in at that time.

Since my departure, the robust transparency and accountability architecture that led to the country being ranked favorably by TI and various international watchdog groups have disintegrated as quickly as they rose. The 2010 TI Global Corruption Barometer graded Liberia as among the world's most corrupt countries, especially in the area of citizens who need to pay bribes to public servants.

The forward march and the bright future to which the president called her compatriots at her first inauguration appear to have stalled. Instead of building on a promising foundation of public hope that greeted Liberia's post-war government, its performance began to erode rather than continue to build trust.

It has become evident that the old, tired pursuit of "growth without development," as well as its perennial companion, the politics of greed, have indeed begun to settle in Liberia—as is far too often the case in many resource-rich countries. While the economy continued to grow, its impact on the lives of ordinary Liberians has been limited. The original promise of broad-based engagement with citizens in order to foster countrywide development faded as government's attention turned to the revival of the concession economy. A rail line from the port to the mines was rebuilt and the iron ore mines were reopened, while after a corruption-tainted start, large-scale timber concessions were granted and the rubber plantations were rehabilitated and expanded. With the discovery of oil along the Gulf of Guinea (especially in Ghana), Liberian officials began contemplating and preparing for the emergence of a petroleum industry.

The lofty mission to "Lift Liberia" as captured in the PRS, especially as it related to small-scale agriculture, rural infrastructure, and strengthening rural public services, has failed. Rural poverty has remained high, exacerbating the low levels of health, poor standards of education, and food insecurity. Sadly it appears that Liberia has firmly moved onto a trajectory that it has already been [on] before: once more growing but not developing.

Ebola in a Time of Crisis

It is against this background—one of governance failures—that the EVD outbreak began. Within months, the disease dislocated the institutional fabric of Liberia, disrupting not just the health system, but also the entire system of governance. Of the 10,129 reported cases globally as of October 23, 2014, 4,665 people are in Liberia and 2,705 have died. The estimated figures will be more alarming if the epidemic is not brought under control.

Many health centers have shut down as health workers abandoned their posts for fear of contracting the virus, leaving hundreds of Liberians without access to health services. These centers have done so due to poor conditions, and provide no protective equipment and incentive to perform the life-threatening work for which they were created. They have already seen over 95 of their fellow health workers die, and hundreds of others fighting for their lives.

Consequently, there are widespread reports that people with high blood pressure and diabetes are no longer cared for. Pregnant women have been turned away from hospitals. They are left to die or lose their babies before they are born. Desperate Liberians have abandoned neighbors or relatives suspected of having EVD to die slow and painful deaths. Both Liberia's society and culture are being challenged in many new and desperate ways. This is a different type of war now, not the civil war of the 1990s, but a war brought about in part by a health system unable to cope with the scale of the Ebola outbreak.

The Ebola epidemic is not just devastating the Liberian population. It is also severely crippling all sectors of the country's economy: notably health, trade and commerce, and education. The World Bank recently projected major reductions in the economy over the coming years—estimating that Liberia could see significant contractions of its growth. The impact of EVD has seen the original GDP growth projections revised downwards from the initial 8.7 percent, progressively to 5.9 percent, 2.5 percent, and most recently, to 1 percent. This will have a direct impact on the country's Human Development Index. With villages decimated by the disease and agricultural fields being abandoned, famine is becoming a reality. The prices of food have been rising due to shortages. Liberian professionals who hold foreign passports, many of whom returned with high hopes of contributing to the development of their motherland, are leaving the country. This will accentuate Liberia's deficiencies in human resources. Schools have closed, business has declined, and international connections (via air and sea transport) have been curtailed. In rural Liberia, communities shun many who contract the virus for bringing calamity upon their neighbors. This is further undermining the fragile social fabric that had been slowly rebuilding after the war.

Lessons for Re-engaging on Recovery in Liberia

There will be plenty of calls for lessons learned from future analysis of what happened in Guinea, Sierra Leone, and Liberia during the early months from the death of Si in December 2013. Looking even further back, the efforts and choices made by the government of Liberia, as well as other governments in the regional and international community, in strengthening institutions appear to be either inappropriate, misguided, or too superficial to support the country's development. We need to learn from this so that we can build new and robust local, national, regional, and global architectures that can effectively respond to current and future epidemics. For me, the following lessons are worth heeding:

(1) Effective and accountable institutions remain key in post-conflict recovery and transition out of fragility.

It is already expected that climate change will shift where infectious diseases break out. We should expect both an increasing number of epidemics and mega multi-hazards. The WHO has warned that climate change will see the rise of infectious diseases. Many of these would likely originate in the so-called conflict-affected fragile states, so we must learn quickly to engage these states in ways that increase their resilience as the first line of defense in our emerging complex new world.

In my role as director for the former Bureau for Crisis Prevention and Recovery in UNDP, we supported the work of the g7+ (a group of self-identified fragile states) that organized themselves under what is called the New Deal for Peacebuilding and Statebuilding. The specific goal was to determine how the countries could transition out of fragility to become more resilient. At the heart of the New Deal is the building of institutions to promote inclusive politics, security, justice, revenue and jobs, and basic services. These countries recognize that until they build more resilient and participatory governance systems, their prospects for peace and sustained development are limited.

We cannot continue this firefight since the resources and the know-how are simply not available to respond in an ad-hoc manner to all of these mega-hazards. Resilient institutions are essential. According to the World Bank, these are institutions that "can sustain and enhance results overtime, can adapt to changing circumstances, anticipate new challenges, and cope with exogenous shocks." Building such institutions requires that they be embedded in the societal, political, and geographical contexts from where they derive meaning and legitimacy.

Creating such an institutional context means investing in education so that these countries can have the critical mass through which a supportive institutional environment can develop. While institution-building is for the long-term, this is a great opportunity to experiment with the concept of the use of the country system, national ownership, and the rebuilding of trust between government and society as well as governments and international partners. These are the core principles of the New Deal for Peacebuilding and Statebuilding.

(2) Timely, targeted, coordinated, and coherent result-oriented response.

The Ebola crisis is not just a health emergency; it is a multi-dimensional social and humanitarian crisis. It requires a complex, multi-pronged response involving health, aid-coordination, personal security, food security, appropriate budgetary decision-making, and responsive governance, among others. It is a whole of government challenge. While this point cannot be ducked we in the international community regularly develop "whole of government" approaches in ways that overextend the agendas of already fragile countries well beyond their capacities to respond. We often call on the government to act in a coordinated manner, but as international partners we can be disorganized and consequently fail to act in a unified or coordinated manner ourselves.

Rolling back an epidemic is not the time for long complicated layers of bureaucracies and agency-driven interests. We need targeted and efficient responses that produce results rapidly. In their immediate response, these countries need enough ambulances to quickly collect the sick and the dead. They need health workers including infectious disease control doctors on the ground in all affected parts of the countries. They need funds to pay health workers adequately for undertaking such dangerous work. Most importantly, they need the international community to accompany them by nurturing the use of their respective country systems. This includes the training programs at the local and regional level that will continuously build capacity to stay abreast with medical science, technology, and innovation.

In the medium-term, the network of health workers across the countries must be strengthened to exchange experiences and build practices on a regular basis. But much more is needed if the countries are to rebound. They need considerable support to revitalize the productivity of their agricultural sectors, and they need innovative ways to open the schools. Early recovery activities should be prioritized, including cash transfers targeting not only the directly impacted, but also the affected households; enterprise recovery must be a key component as well.

These are concrete tasks and should be carried out without being subject to the typical bogs of bureaucracy and complexity. How can this be done differently? Where is the venture-capitalist mindset behind all the Silicon Valley startups for all of West Africa? We need to adopt modern methods of training rural health workers, the young women and men who are ready to stay in the provinces and counties and who are willing to provide real services to their fellow citizens, in exchange for being paid real salaries on time.

(3) Limit coordination layers.

There are multiple actors who are returning to these countries to help. They must be coordinated and the governments must be at the center of these coordination platforms, but these should not result in multiple and burdensome transaction costs for coordination. Coordination at the center of government is one of the core functions that needs to be strengthened, particularly in countries where such systems are still not fully consolidated. There is absolutely no time for competing layers of coordination.

As director of the former Bureau for Crisis Response and Recovery in UNDP, I saw firsthand how effective a network of actors across the government can be. In fact, it is critical. As the former UN resident coordinator in Vietnam, I witnessed firsthand that nation's response to SARS and the avian flu. The remarkable success in that country was primarily due to the cohesive response of the government, as well as its clarity of purpose and its decisiveness.

(4) From knee-jerk international mobilization to global solidarity to shared security.

We all hope the current epidemic will be brought under control as rapidly as possible. Soon we will need to face the next challenge: rebuilding the affected countries. Yes, there will be calls to build back better. But it will take much more than slogans this time. The world will be challenged to make the right investments. It cannot be business as usual nor can we allow ourselves to slip simply into old comfortable patterns of working. We should be measured as to whether we are doing the right things. Who are the best judges? The people on the ground are. Do they see an improvement in the education system, the delivery of health, and access to clean water, all of which make life livable?

It is no longer a cliche to say that our security and existence are intertwined even with that of remote villages and impoverished fragile states. It is no longer a world of them and us. Whatever support we give to affected countries is not an act of a good Samaritan. It is for our very own personal safety and well-being.

In our affluent and technologically sophisticated world, complacency is not an option. We cannot glibly dismiss seemingly faraway threats as problems of the poor and remote parts of the world. As the Ebola epidemic in West Africa has revealed in just a matter of months, it is in our personal interest to address those problems at their source before they escalate. This will reduce the tragic impact of the epidemic locally and avoid having it become a global crisis.

With 2015 approaching, and its world of conferences and goals, now would be [the] time to take decisive action that fundamentally reinvigorates the ability of the international system to work in a more effective and cohesive manner with human, physical, and financial resources upfront to support national response plans as well as those that transcend national boundaries in the way modern threats do. Whether this requires fine-tuning or a complete overhaul, now is the time for action, and hopefully for something a bit more ambitious than just making the United Nations "fit for purpose" which seems to mean simply "good enough to get the job done".

Critical Thinking

1. What strategy was followed by the UN team to promote the postwar economic development of Liberia?
2. What was the effect of the Ebola virus on the Liberian economy?
3. What lessons were learned from the Ebola epidemic in Liberia?

Internet References

Centers for Disease Control and Prevention
www.cdc.gov/

Liberia
https://www.cia.gov/library/publications/the-world-factbook/geos/li.html

UN Mission in Liberia
www.un.org/en/peacekeeping/missions/unmil/

WHO Ebola virus Disease outbreak
www.who.int/csr/disease/ebola/en/

Jordan Ryan headed the UN Development Programme's Bureau for Crisis Prevention and Recovery from 2009 until retiring this September. He has also served as the UN Secretary-General's Deputy Special Representative in Liberia from 2005 to 2009. Ryan has worked as a lawyer in Saudi Arabia and China and was a Visiting Fellow at the Harvard Kennedy School in 2001.

Article

Prepared by: Robert Weiner, *University of Massachusetts, Boston*

The Mobile-Finance Revolution: How Cell Phones Can Spur Development

JAKE KENDALL AND RODGER VOORHIES

Learning Outcomes

After reading this article, you will be able to:

- Understand how digital technology can contribute to economic development.
- Understand the relationship between microfinance and business in developing countries.

The roughly 2.5 billion people in the world who live on less than $2 a day are not destined to remain in a state of chronic poverty. Every few years, somewhere between 10 and 30 percent of the world's poorest households manage to escape poverty, typically by finding steady employment or through entrepreneurial activities such as growing a business or improving agricultural harvests. During that same period, however, roughly an equal number of households slip below the poverty line. Health-related emergencies are the most common cause, but there are many more: crop failures, livestock deaths, farming-equipment breakdowns, and even wedding expenses.

In many such situations, the most important buffers against crippling setbacks are financial tools such as personal savings, insurance, credit, or cash transfers from family and friends. Yet these are rarely available because most of the world's poor lack access to even the most basic banking services. Globally, 77 percent of them do not have a savings account; in sub-Saharan Africa, the figure is 85 percent. An even greater number of poor people lack access to formal credit or insurance products. The main problem is not that the poor have nothing to save—studies show that they do—but rather that they are not profitable customers, so banks and other service providers do not try to reach them. As a result, poor people usually struggle to stitch together a patchwork of informal, often precarious arrangements to manage their financial lives.

Over the last few decades, microcredit programs—through which lenders have granted millions of small loans to poor people—have worked to address the problem. Institutions such as the Grameen Bank, which won the Nobel Peace Prize in 2006, have demonstrated impressive results with new financial arrangements, such as group loans that require weekly payments. Today, the microfinance industry provides loans to roughly 200 million borrowers—an impressive number to be sure, but only enough to make a dent in the over two billion people who lack access to formal financial services.

Despite its success, the microfinance industry has faced major hurdles. Due to the high overhead costs of administering so many small loans, the interest rates and fees associated with microcredit can be steep, often reaching 100 percent annually. Moreover, a number of rigorous field studies have shown that even when lending programs successfully reach borrowers, there is only a limited increase in entrepreneurial activity—and no measurable decrease in poverty rates. For years, the development community has promoted a narrative that borrowing and entrepreneurship have lifted large numbers of people out of poverty. But that narrative has not held up.

Despite these challenges, two trends indicate great promise for the next generation of financial-inclusion efforts. First, mobile technology has found its way to the developing world and spread at an astonishing pace. According to the World Bank, mobile signals now cover some 90 percent of the world's poor, and there are, on average, more than 89 cell-phone accounts for every 100 people living in a developing country. That presents an extraordinary opportunity: mobile-based financial tools have the potential to dramatically lower the cost of delivering banking services to the poor.

Second, economists and other researchers have in recent years generated a much richer fact base from rigorous studies to inform future product offerings. Early on, both sides of the debate over the true value of microcredit programs for the poor relied mostly on anecdotal observations and gut instincts. But now, there are hundreds of studies to draw from. The flexible, low-cost models made possible by mobile technology and the evidence base to guide their design have thus created a major opportunity to deliver real value to the poor.

Show Them the Money

Mobile finance offers at least three major advantages over traditional financial models. First, digital transactions are essentially free. In-person services and cash transactions account for the majority of routine banking expenses. But mobile-finance clients keep their money in digital form, and so they can send and receive money often, even with distant counter-parties, without creating significant transaction costs for their banks or mobile service providers. Second, mobile communications generate copious amounts of data, which banks and other providers can use to develop more profitable services and even to substitute for traditional credit scores (which can be hard for those without formal records or financial histories to obtain). Third, mobile platforms link banks to clients in real time. This means that banks can instantly relay account information or send reminders and clients can sign up for services quickly on their own.

The potential, in other words, is enormous. The benefits of credit, savings, and insurance are clear, but for most poor households, the simple ability to transfer money can be equally important. For example, a recent Gallup poll conducted in 11 sub-Saharan African countries found that over 50 percent of adults surveyed had made at least one payment to someone far away within the preceding 30 days. Eighty-three percent of them had used cash. Whether they were paying utility bills or sending money to their families, most had sent the money with bus drivers, had asked friends to carry it, or had delivered the payments themselves. The costs were high; moving physical cash, particularly in sub-Saharan Africa, is risky, unreliable, and slow.

Imagine what would happen if the poor had a better option. A recent study in Kenya found that access to a mobile-money product called M-Pesa, which allows clients to store money on their cell phones and send it at the touch of a button, increased the size and efficiency of the networks within which they moved money. That came in handy when poorer participants endured economic shocks spurred by unexpected events, such as a hospitalization or a house fire. Households with access to M-Pesa received more financial support from larger and more distant networks of friends and family. As a result, they were better able to survive hard times, maintaining their regular diets and keeping their children in school.

To consumers, the benefits of M-Pesa are self-evident. Today, according to a study by Kenya's Financial Sector Deepening Trust, 62 percent of adults in the country have active accounts. And other countries have since launched their own versions of the product. In Tanzania, over 47 percent of households have a family member who has registered. In Uganda, 26 percent of adults are users. The rates of adoption have been extraordinary; by contrast, microlenders rarely get more than 10 percent participation in their program areas.

Mobile money is useful for more than just emergency transfers. Regular remittances from family members working in other parts of the country, for example, make up a large share of the incomes of many poor households. A Gallup study in South Asia recently found that 72 percent of remittance-receiving households indicated that the cash transfers were "very important" to their financial situations. Studies of small-business owners show that they make use of mobile payments to improve their efficiency and expand their customer bases.

These technologies could also transform the way people interact with large formal institutions, especially by improving people's access to government services. A study in Niger by a researcher from Tufts University found that during a drought, allowing people to request emergency government support through their cell phones resulted in better diets for those people, compared with the diets of those who received cash handouts. The researchers concluded that women were more likely than men to control digital transfers (as opposed to cash transfers) and that they were more likely to spend the money on high-quality food.

Governments, meanwhile, stand to gain as much as consumers do. A McKinsey study in India found that the government could save $22 billion each year from digitizing all of its payments. Another study, by the Better Than Cash Alliance, a nonprofit that helps countries adopt electronic payment systems, found that the Mexican government's shift to digital payments (which began in 1997) trimmed its spending on wages, pensions, and social welfare by 3.3 percent annually, or nearly $1.3 billion.

Savings and Phones

In the developed world, bankers have long known that relatively simple nudges can have a big impact on long-term behavior. Banks regularly encourage clients to sign off on automatic contributions to their 401(k) retirement plans, set up automatic deposits into savings accounts from their paychecks, and open special accounts to save for a particular purpose.

Studies in the developing world confirm that, if anything, the poor need such decision aids even more than the rich, owing to the constant pressure they are under to spend their money on immediate needs. And cell phones make nudging easy. For example, a series of studies have shown that when clients receive text messages urging them to make regular savings deposits, they improve their balances over time. More draconian features have also proved effective, such as so-called commitment accounts, which impose financial discipline with large penalty fees.

Many poor people have already demonstrated their interest in financial mechanisms that encourage savings. In Africa, women commonly join groups called rotating savings and credit associations, or ROSCAS, which require them to attend weekly meetings and meet rigid deposit and withdrawal schedules. Studies suggest that in such countries as Cameroon, Gambia, Nigeria, and Togo, roughly half of all adults are members of a ROSCA, and similar group savings schemes are widespread outside Africa, as well. Research shows that members are drawn to the discipline of required regular payments and the social pressure of group meetings.

Mobile-banking applications have the potential to encourage financial discipline in even more effective ways. Seemingly marginal features designed to incentivize financial discipline can do much to set people on the path to financial prosperity. In one experiment, researchers allowed some small-scale farmers in Malawi to have their harvest proceeds directly deposited into commitment accounts. The farmers who were offered this option and chose to participate ended up investing 30 percent more in farm inputs than those who weren't offered the option, leading to a 22 percent increase in revenues and a 17 percent increase in household consumption after the harvest.

Poor households, not unlike rich ones, are not well served by simple loans in isolation; they need a full suite of financial tools that work in concert to mitigate risk, fund investment, grow savings, and move money. Insurance, for example, can significantly affect how borrowers invest in their businesses. A recent field study in Ghana gave different groups of farmers cash grants to fund investments in farm inputs, crop insurance, or both. The farmers with crop insurance invested more in agricultural inputs, particularly in chemicals, land preparation, and hired labor. And they spent, on average, $266 more on cultivation than did the farmers without insurance. It was not the farmers' lack of credit, then, that was the greatest barrier to expanding their businesses; it was risk.

Mobile applications allow banks to offer such services to huge numbers of customers in very short order. In November 2012, the Commercial Bank of Africa and the telecommunications firm Safaricom launched a product called M-Shwari, which enables M-Pesa users to open interest-accruing savings accounts and apply for short-term loans through their cell phones. The demand for the product proved overwhelming. By effectively eliminating the time it would have taken for users to sign up or apply in person, M-Shwari added roughly one million accounts in its first three months.

By attracting so many customers and tracking their behavior in real time, mobile platforms generate reams of useful data. People's calling and transaction patterns can reveal valuable insights about the behavior of certain segments of the client population, demonstrating how variations in income levels, employment status, social connectedness, marital status, creditworthiness, or other attributes shape outcomes. Many studies have already shown how certain product features can affect some groups differently from others. In one Kenyan study, researchers gave clients ATM cards that permitted cash withdrawals at lowered costs and allowed the clients to access their savings accounts after hours and on weekends. The change ended up positively affecting married men and adversely affecting married women, whose husbands could more easily get their hands on the money saved in a joint account. Before the ATM cards, married women could cite the high withdrawal fees or the bank's limited hours to discourage withdrawals. With the cards, moreover, husbands could get cash from an ATM themselves, whereas withdrawals at the branch office had usually required the wives to go in person during the hours their husbands were at work.

Location, Location, Location

The high cost of basic banking infrastructure may be the biggest barrier to providing financial services to the poor. Banks place ATMs and branch offices almost exclusively in the wealthier, denser (and safer) areas of poor countries. The cost of such infrastructure often dwarfs the potential profits to be made in poorer, more rural areas. In contrast, mobile banking allows customers to carry out transactions in existing shops and even market stalls, creating denser networks of transaction points at a much lower cost.

For clients to fully benefit from mobile financial services, however, access to a physical office that deals in cash remains critical. When researchers studying the M-Pesa program in Kenya cross-referenced the locations of M-Pesa agents and the locations of households in the program, they found that the closer a household was to an M-Pesa kiosk, where cash and customer services were available, the more it benefited from the service. Beyond a certain distance, it becomes infeasible for clients to use a given financial service, no matter how much they need it.

Meanwhile, a number of studies have shown that increasing physical access points to the financial system can help lift local economies. Researchers in India have documented the effects of a regulation requiring banks to open rural branches in exchange for licenses to operate in more profitable urban areas. The data showed significant increases in lending and agricultural output in the areas that received branches due to the program, as well as 4–5 percent reductions in the number of people living in poverty. A similar study in Mexico found that in areas where bank branches were introduced, the number of people who owned informal businesses increased by 7.6 percent. There were also ripple effects: an uptick in employment and a 7 percent increase in incomes.

In the right hands, then, access to financial tools can stimulate underserved economies and, at critical times, determine whether a poor household is able to capture an opportunity to move out of poverty or weather an otherwise debilitating financial shock. Thanks to new research, much more is known about what types of features can do the most to improve consumers' lives. And due to the rapid proliferation of cell phones, it is now possible to deliver such services to more people than ever before. Both of these trends have set the stage for yet further innovations by banks, cell-phone companies, microlenders, and entrepreneurs—all of whom have a role to play in delivering life-changing financial services to those who need them most.

Critical Thinking

1. How does Wizzit bank contribute to microfinance?
2. What is the relationship between cell phones and banking?
3. What is the role of the International Finance Corporation in mobile banking?

Internet References

International Finance Corporation
http://www.ifc.org

Wizzit bank
http://www.wizzit.co.za

Jake Kendall is Senior Program Officer for the Financial Services for the Poor program at the Bill & Melinda Gates Foundation. **Rodger Voorhies** is Director of the Financial Services for the Poor program at the Bill & Melinda Gates Foundation.

Unit 4

UNIT

Prepared by: Robert Weiner, *University of Massachusetts, Boston*

Terrorism

Terrorism usually refers to non-state actors who engage in violent extremist action to draw attention to their goals and ideologies. Terrorists can consist of organizations or single, radicalized individuals, who may be motivated by ideology, religion, or ethno-nationalism. Terrorists believe that it is necessary to use extreme violence to draw attention to their cause. In the age of globalization in the twenty-first century, terrorism is a transnational phenomenon. Terrorists have taken advantage of modern technology to engage in violent acts, as when the United States found itself attacked by Al-Qaida on 9/11. Terrorism has evolved from bomb throwing and assassinations to the hijacking of aircraft to efforts to kill as many innocent and vulnerable victims as possible. According to experts, terrorist acts have increased fourfold since 9/11, especially marked by an increase in the number of suicide bombers. Even though Osama Bin Laden, the leader of Al-Qaida or the "base" in Arabic, was killed by US Seals in 2011, and many other members of the organization were decimated by drones, franchises of Al-Qaida and other terrorist organizations have morphed and proliferated throughout the Middle East, North Africa, Somalia, and Nigeria. These include organizations such as Al-Shabaab, Boko Haram, and Al-Nusra, among others.

The United States especially focused on the Islamic State in Iraq and the Levant (ISIL) also known as Daesh. By 2015, ISIL, which originally began with Baathist military officers from Saddam Hussein's regime, and was joined by disaffected Sunni tribes who felt repressed by the Shia government in Bagdad, had grown to 25,000–30,000 insurgents, whose recruits were drawn from around the world by ISIL's sophisticated use of the social media. New recruits flocked to ISIL from Europe and the United States. It is estimated that in 2015, about 250 recruits from the United States joined ISIL. Western governments worried about the dangers of blowback as foreign fighters returned to their home states. There was fear of more incidents like the murder of 12 staff members of the French satirical magazine *Charlie Hedro* in Paris by jihadists who had been trained in the Middle East, as indeed did occur in Belgium in 2016.

ISIL's conduct of the war in Iraq and Syria, which aimed at overthrowing the regimes in both countries, was marked by a great deal of brutality and cruelty, including beheadings, mass executions, persecution of Christians, and other non-believers and apostates, and was accused of committing genocide against a group known as Yazedis. ISIL also looted and destroyed Assyrian monuments, sculptures, and treasures in the ancient Assyrian city of Palmyra. ISIL has assumed the attributes of a state, as it attempts to reconstitute an Islamic Caliphate of the Middle Ages.

The United States led a counterterrorist coalition of about 60 countries to degrade and destroy ISIL. In 2015, despite repeated US airstrikes against ISIL, it still remained entrenched in parts of Iraq and Syria. However, by 2016, some progress had been made in pushing ISIL back from the territory it controlled in Iraq, as the coalition forces launched an assault on the city of Mosul which had been controlled by ISIL Russia also deployed combat aircraft and other military assets in Syria, to defend President Assad's regime, which reportedly controlled about 25 percent of the country. This added a new level of complexity to the conflict, as President Putin of Russia urged other countries to join Moscow in the fight against terrorism. Increasing Russian involvement in the Syrian civil war raised the danger of a potential confrontation between Russia and the US, since Russia supported the Assad regime and the US wanted Assad out of power. The Russian military intervention was also designed to establish a more permanent presence for Moscow in the Middle East, and restore its credibility as a Great Power. Much of the US foreign policy establishment criticized the Obama administration for not engaging in more robust military intervention in Syria. President Obama's successor would have to decide whether to continue a policy of disengagement in Syria and the Middle East. The question also remained as to whether Obama's policy of pursuing a détente with Iran would work?

Article

Prepared by: Robert Weiner, *University of Massachusetts, Boston*

ISIS Is Not a Terrorist Group: Why Counterterrorism Won't Stop the Latest Jihadist Threat

AUDREY KURTH CRONIN

Learning Outcomes

After reading this article, you will be able to:

- Define who and what ISIS is.
- Understand the relationship between ISIS and al Qaeda.

After 9/11, many within the U.S. national security establishment worried that, following decades of preparation for confronting conventional enemies, Washington was unready for the challenge posed by an unconventional adversary such as al Qaeda. So over the next decade, the United States built an elaborate bureaucratic structure to fight the jihadist organization, adapting its military and its intelligence and law enforcement agencies to the tasks of counterterrorism and counterinsurgency.

Now, however, a different group, the Islamic State of Iraq and alSham (ISIS), which also calls itself the Islamic State, has supplanted al Qaeda as the jihadist threat of greatest concern. ISIS' ideology, rhetoric, and long-term goals are similar to al Qaeda's, and the two groups were once formally allied. So many observers assume that the current challenge is simply to refocus Washington's now-formidable counterterrorism apparatus on a new target.

But ISIS is not al Qaeda. It is not an outgrowth or a part of the older radical Islamist organization, nor does it represent the next phase in its evolution. Although al Qaeda remains dangerous—especially its affiliates in North Africa and Yemen—ISIS is its successor. ISIS represents the post-al Qaeda jihadist threat.

In a nationally televised speech last September explaining his plan to "degrade and ultimately destroy" ISIS, U.S. President Barack Obama drew a straight line between the group and al Qaeda and claimed that ISIS is "a terrorist organization, pure and simple." This was mistaken; ISIS hardly fits that description, and indeed, although it uses terrorism as a tactic, it is not really a terrorist organization at all. Terrorist networks, such as al Qaeda, generally have only dozens or hundreds of members, attack civilians, do not hold territory, and cannot directly confront military forces. ISIS, on the other hand, boasts some 30,000 fighters, holds territory in both Iraq and Syria, maintains extensive military capabilities, controls lines of communication, commands infrastructure, funds itself, and engages in sophisticated military operations. If ISIS is purely and simply anything, it is a pseudo-state led by a conventional army. And that is why the counterterrorism and counterinsurgency strategies that greatly diminished the threat from al Qaeda will not work against ISIS.

Washington has been slow to adapt its policies in Iraq and Syria to the true nature of the threat from ISIS. In Syria, U.S. counterterrorism has mostly prioritized the bombing of al Qaeda affiliates, which has given an edge to ISIS and has also provided the Assad regime with the opportunity to crush U.S.-allied moderate Syrian rebels. In Iraq, Washington continues to rely on a form of counterinsurgency, depending on the central government in Baghdad to regain its lost legitimacy, unite the country, and build indigenous forces to defeat ISIS. These approaches were developed to meet a different threat, and they have been overtaken by events. What's needed now is a strategy of "offensive containment": a combination of limited military tactics and a broad diplomatic strategy to halt ISIS' expansion, isolate the group, and degrade its capabilities.

Different Strokes

The differences between al Qaeda and ISIS are partly rooted in their histories. Al Qaeda came into being in the aftermath of the 1979 Soviet invasion of Afghanistan. Its leaders' worldviews and strategic thinking were shaped by the ten-year war against Soviet occupation, when thousands of Muslim militants, including Osama bin Laden, converged on the country. As the organization coalesced, it took the form of a global network focused on carrying out spectacular attacks against Western or Western-allied targets, with the goal of rallying Muslims to join a global confrontation with secular powers near and far.

ISIS came into being thanks to the 2003 U.S. invasion of Iraq. In its earliest incarnation, it was just one of a number of Sunni extremist groups fighting U.S. forces and attacking Shiite civilians in an attempt to foment a sectarian civil war. At that time, it was called al Qaeda in Iraq (aqi), and its leader, Abu Musab al-Zarqawi, had pledged allegiance to bin Laden. Zarqawi was killed by a U.S. air strike in 2006, and soon after, aqi was nearly wiped out when Sunni tribes decided to partner with the Americans to confront the jihadists. But the defeat was temporary; aqi renewed itself inside U.S.-run prisons in Iraq, where insurgents and terrorist operatives connected and formed networks—and where the group's current chief and self-proclaimed caliph, Abu Bakr al-Baghdadi, first distinguished himself as a leader.

In 2011, as a revolt against the Assad regime in Syria expanded into a full-blown civil war, the group took advantage of the chaos, seizing territory in Syria's northeast, establishing a base of operations, and rebranding itself as ISIS. In Iraq, the group continued to capitalize on the weakness of the central state and to exploit the country's sectarian strife, which intensified after U.S. combat forces withdrew. With the Americans gone, Iraqi Prime Minister Nouri al-Maliki pursued a hard-line pro-Shiite agenda, further alienating Sunni Arabs throughout the country. ISIS now counts among its members Iraqi Sunni tribal leaders, former anti-U.S. insurgents, and even secular former Iraqi military officers who seek to regain the power and security they enjoyed during the Saddam Hussein era.

The group's territorial conquest in Iraq came as a shock. When ISIS captured Fallujah and Ramadi in January 2014, most analysts predicted that the U.S.-trained Iraqi security forces would contain the threat. But in June, amid mass desertions from the Iraqi army, ISIS moved toward Baghdad, capturing Mosul, Tikrit, al-Qaim, and numerous other Iraqi towns. By the end of the month, ISIS had renamed itself the Islamic State and had proclaimed the territory under its control to be a new caliphate. Meanwhile, according to U.S. intelligence estimates, some 15,000 foreign fighters from 80 countries flocked to the region to join ISIS, at the rate of around 1,000 per month. Although most of these recruits came from Muslim-majority countries, such as Tunisia and Saudi Arabia, some also hailed from Australia, China, Russia, and western European countries. ISIS has even managed to attract some American teenagers, boys and girls alike, from ordinary middle-class homes in Denver, Minneapolis, and the suburbs of Chicago.

As ISIS has grown, its goals and intentions have become clearer. Al Qaeda conceived of itself as the vanguard of a global insurgency mobilizing Muslim communities against secular rule. ISIS, in contrast, seeks to control territory and create a "pure" Sunni Islamist state governed by a brutal interpretation of sharia; to immediately obliterate the political borders of the Middle East that were created by Western powers in the twentieth century; and to position itself as the sole political, religious, and military authority over all of the world's Muslims.

Not the Usual Suspects

Since ISIS' origins and goals differ markedly from al Qaeda's, the two groups operate in completely different ways. That is why a U.S. counterterrorism strategy custom-made to fight al Qaeda does not fit the struggle against ISIS.

In the post-9/11 era, the United States has built up a trillion-dollar infrastructure of intelligence, law enforcement, and military operations aimed at al Qaeda and its affiliates. According to a 2010 investigation by *The Washington Post*, some 263 U.S. government organizations were created or reorganized in response to the 9/11 attacks, including the Department of Homeland Security, the National Counterterrorism Center, and the Transportation Security Administration. Each year, U.S. intelligence agencies produce some 50,000 reports on terrorism. Fifty-one U.S. federal organizations and military commands track the flow of money to and from terrorist networks. This structure has helped make terrorist attacks on U.S. soil exceedingly rare. In that sense, the system has worked. But it is not well suited for dealing with ISIS, which presents a different sort of challenge. Consider first the tremendous U.S. military and intelligence campaign to capture or kill al Qaeda's core leadership through drone strikes and Special Forces raids. Some 75 percent of the leaders of the core al Qaeda group have been killed by raids and armed drones, a technology well suited to the task of going after targets hiding in rural areas, where the risk of accidentally killing civilians is lower.

Such tactics, however, don't hold much promise for combating ISIS. The group's fighters and leaders cluster in urban areas, where they are well integrated into civilian populations and usually surrounded by buildings, making drone strikes and raids much harder to carry out. And simply killing ISIS' leaders would not cripple the organization. They govern a functioning pseudo-state with a complex administrative structure. At the top of the military command is the emirate, which consists of Baghdadi and two deputies, both of whom formerly served as generals in the Saddam-era Iraqi army: Abu Ali al-Anbari, who controls ISIS' operations in Syria, and Abu Muslim al-Turkmani, who controls operations in Iraq. ISIS' civilian bureaucracy is supervised by 12 administrators who govern territories in Iraq and Syria, overseeing councils that handle matters such as finances, media, and religious affairs. Although it is hardly the model government depicted in ISIS' propaganda videos, this pseudo-state would carry on quite ably without Baghdadi or his closest lieutenants.

ISIS also poses a daunting challenge to traditional U.S. counterterrorism tactics that take aim at jihadist financing, propaganda, and recruitment. Cutting off al Qaeda's funding has been one of U.S. counterterrorism's most impressive success stories. Soon after the 9/11 attacks, the FBI and the CIA began to coordinate closely on financial intelligence, and they were soon joined by the Department of Defense. FBI agents embedded with U.S. military units during the 2003 invasion of Iraq and debriefed suspected terrorists detained at the U.S. facility at Guantánamo Bay, Cuba. In 2004, the U.S. Treasury Department established the Office of Terrorism and Financial Intelligence, which has cut deeply into al Qaeda's ability to profit from money laundering and receive funds under the cover of charitable giving. A global network for countering terrorist financing has also emerged, backed by the UN, the EU, and hundreds of cooperating governments. The result has been a serious squeeze on al Qaeda's financing; by 2011, the Treasury Department reported that al Qaeda was "struggling to secure steady financing to plan and execute terrorist attacks."

But such tools contribute little to the fight against ISIS, because ISIS does not need outside funding. Holding territory has allowed the group to build a self-sustaining financial model unthinkable for most terrorist groups. Beginning in 2012, ISIS gradually took over key oil assets in eastern Syria; it now controls an estimated 60 percent of the country's oil production capacity. Meanwhile, during its push into Iraq last summer, ISIS also seized seven oil-producing operations in that country. The group manages to sell some of this oil on the black market in Iraq and Syria—including, according to some reports, to the Assad regime itself. ISIS also smuggles oil out of Iraq and Syria into Jordan and Turkey, where it finds plenty of buyers happy to pay below-market prices for illicit crude. All told, ISIS' revenue from oil is estimated to be between $1 million and $3 million per day. And oil is only one element in the group's financial portfolio. Last June, when ISIS seized control of the northern Iraqi city of Mosul, it looted the provincial central bank and other smaller banks and plundered antiquities to sell on the black market. It steals jewelry, cars, machinery, and livestock from conquered residents. The group also controls major transportation arteries in western Iraq, allowing it to tax the movement of goods and charge tolls. It even earns revenue from cotton and wheat grown in Raqqa, the breadbasket of Syria.

Of course, like terrorist groups, ISIS also takes hostages, demanding tens of millions of dollars in ransom payments. But more important to the group's finances is a wide-ranging extortion racket that targets owners and producers in ISIS territory, taxing everything from small family farms to large enterprises such as cell-phone service providers, water delivery companies, and electric utilities. The enterprise is so complex that the U.S. Treasury has declined to estimate ISIS' total assets and revenues, but ISIS is clearly a highly diversified enterprise whose wealth dwarfs that of any terrorist organization. And there is little evidence that Washington has succeeded in reducing the group's coffers.

Sex and the Single Jihadist

Another aspect of U.S. counterterrorism that has worked well against al Qaeda is the effort to delegitimize the group by publicizing its targeting errors and violent excesses—or by helping U.S. allies do so. Al Qaeda's attacks frequently kill Muslims, and the group's leaders are highly sensitive to the risk this poses to their image as the vanguard of a mass Muslim movement. Attacks in Morocco, Saudi Arabia, and Turkey in 2003; Spain in 2004; and Jordan and the United Kingdom in 2005 all resulted in Muslim casualties that outraged members of Islamic communities everywhere and reduced support for al Qaeda across the Muslim world. The group has steadily lost popular support since around 2007; today, al Qaeda is widely reviled in the Muslim world. The Pew Research Center surveyed nearly 9,000 Muslims in 11 countries in 2013 and found a high median level of disapproval of al Qaeda: 57 percent. In many countries, the number was far higher: 96 percent of Muslims polled in Lebanon, 81 percent in Jordan, 73 percent in Turkey, and 69 percent in Egypt held an unfavorable view of al Qaeda.

ISIS, however, seems impervious to the risk of a backlash. In proclaiming himself the caliph, Baghdadi made a bold (if absurd) claim to religious authority. But ISIS' core message is about raw power and revenge, not legitimacy. Its brutality—videotaped beheadings, mass executions—is designed to intimidate foes and suppress dissent. Revulsion among Muslims at such cruelty might eventually undermine ISIS. But for the time being, Washington's focus on ISIS' savagery only helps the group augment its aura of strength.

For similar reasons, it has proved difficult for the United States and its partners to combat the recruitment efforts that have attracted so many young Muslims to ISIS' ranks. The core al Qaeda group attracted followers with religious arguments and a pseudo-scholarly message of altruism for the sake of the ummah, the global Muslim community. Bin Laden and his long-time second-in-command and successor, Ayman alZawahiri, carefully constructed an image of religious legitimacy and piety. In their propaganda videos, the men appeared as ascetic warriors, sitting on the ground in caves, studying in libraries, or taking refuge in remote camps. Although some of al Qaedas affiliates have better recruiting pitches, the core group cast the establishment of a caliphate as a long-term, almost utopian goal: educating and mobilizing the ummah came first. In al Qaeda, there is no place for alcohol or women. In this sense, al Qaeda's image is deeply unsexy; indeed, for the young al Qaeda recruit, sex itself comes only after marriage—or martyrdom.

Even for the angriest young Muslim man, this might be a bit of a hard sell. Al Qaeda's leaders' attempts to depict themselves as moral—even moralistic—figures have limited their appeal. Successful deradicalization programs in places such as Indonesia and Singapore have zeroed in on the mismatch between what al Qaeda offers and what most young people are really interested in, encouraging militants to reintegrate into society, where their more prosaic hopes and desires might be fulfilled more readily.

ISIS, in contrast, offers a very different message for young men, and sometimes women. The group attracts followers yearning for not only religious righteousness but also adventure, personal power, and a sense of self and community. And, of course, some people just want to kill—and ISIS welcomes them, too. The group's brutal violence attracts attention, demonstrates dominance, and draws people to the action.

ISIS operates in urban settings and offers recruits immediate opportunities to fight. It advertises by distributing exhilarating podcasts produced by individual fighters on the frontlines. The group also procures sexual partners for its male recruits; some of these women volunteer for this role, but most of them are coerced or even enslaved. The group barely bothers to justify this behavior in religious terms; its sales pitch is conquest in all its forms, including the sexual kind. And it has already established a self-styled caliphate, with Baghdadi as the caliph, thus making present (if only in a limited way, for now) what al Qaeda generally held out as something more akin to a utopian future.

In short, ISIS offers short-term, primitive gratification. It does not radicalize people in ways that can be countered by appeals to logic. Teenagers are attracted to the group without even understanding what it is, and older fighters just want to be associated with ISIS' success. Compared with fighting al Qaeda's relatively austere message, Washington has found it much harder to counter ISIS' more visceral appeal, perhaps for a very simple reason: a desire for power, agency, and instant results also pervades American culture.

2015 ≠ 2006

Counterterrorism wasn't the only element of national security practice that Washington rediscovered and reinvigorated after 9/11; counterinsurgency also enjoyed a renaissance. As chaos erupted in Iraq in the aftermath of the U.S. invasion and occupation of 2003, the U.S. military grudgingly started thinking about counterinsurgency, a subject that had fallen out of favor in the national security establishment after the Vietnam War. The most successful application of U.S. counterinsurgency doctrine was the 2007 "surge" in Iraq, overseen by General David Petraeus. In 2006, as violence peaked in Sunni-dominated Anbar Province, U.S. officials concluded that the United States was losing the war. In response, President George W. Bush decided to send an additional 20,000 U.S. troops to Iraq. General John Allen, then serving as deputy commander of the multinational forces in Anbar, cultivated relationships with local Sunni tribes and nurtured the so-called Sunni Awakening, in which some 40 Sunni tribes or subtribes essentially switched sides and decided to fight with the newly augmented U.S. forces against aqi. By the summer of 2008, the number of insurgent attacks had fallen by more than 80 percent.

Looking at the extent of ISIS' recent gains in Sunni areas of Iraq, which have undone much of the progress made in the surge, some have argued that Washington should respond with a second application of the Iraq war's counterinsurgency strategy. And the White House seems at least partly persuaded by this line of thinking: last year, Obama asked Allen to act as a special envoy for building an anti-ISIS coalition in the region. There is a certain logic to this approach, since ISIS draws support from many of the same insurgent groups that the surge and the Sunni Awakening neutralized—groups that have reemerged as threats thanks to the vacuum created by the withdrawal of U.S. forces in 2011 and Maliki's sectarian rule in Baghdad.

But vast differences exist between the situation today and the one that Washington faced in 2006, and the logic of U.S. counterinsurgency does not suit the struggle against ISIS. The United States cannot win the hearts and minds of Iraq's Sunni Arabs, because the Maliki government has already lost them. The Shiite-dominated Iraqi government has so badly undercut its own political legitimacy that it might be impossible to restore it. Moreover, the United States no longer occupies Iraq. Washington can send in more troops, but it cannot lend legitimacy to a government it no longer controls. ISIS is less an insurgent group fighting against an established government than one party in a conventional civil war between a breakaway territory and a weak central state.

Divide and Conquer?

The United States has relied on counterinsurgency strategy not only to reverse Iraq's slide into state failure but also to serve as a model for how to combat the wider jihadist movement. Al Qaeda expanded by persuading Muslim militant groups all over the world to turn their more narrowly targeted nationalist campaigns into nodes in al Qaeda's global jihad—and, sometimes, to convert themselves into al Qaeda affiliates. But there was little commonality in the visions pursued by Chechen, Filipino, Indonesian, Kashmiri, Palestinian, and Uighur militants, all of whom bin Laden tried to draw into al Qaeda's tent, and al Qaeda often had trouble fully reconciling its own goals with the interests of its far-flung affiliates.

That created a vulnerability, and the United States and its allies sought to exploit it. Governments in Indonesia and the Philippines won dramatic victories against al Qaeda affiliates in their countries by combining counterterrorism operations with relationship building in local communities, instituting deradicalization programs, providing religious training in prisons, using rehabilitated former terrorist operatives as government spokespeople, and sometimes negotiating over local grievances.

Some observers have called for Washington to apply the same strategy to ISIS by attempting to expose the fault lines between the group's secular former Iraqi army officers, Sunni tribal leaders, and Sunni resistance fighters, on the one hand, and its veteran jihadists, on the other. But it's too late for that approach to work. ISIS is now led by well-trained, capable former Iraqi military leaders who know U.S. techniques and habits because Washington helped train them. And after routing Iraqi army units and taking their U.S.-supplied equipment, ISIS is now armed with American tanks, artillery, armored Humvees, and mine-resistant vehicles.

Perhaps ISIS' harsh religious fanaticism will eventually prove too much for their secular former Baathist allies. But for now, the Saddam-era officers are far from reluctant warriors for ISIS: rather, they are leading the charge. In their hands, ISIS has developed a sophisticated light infantry army, brandishing American weapons.

Of course, this opens up a third possible approach to ISIS, besides counterterrorism and counterinsurgency: a full-on conventional war against the group, waged with the goal of completely destroying it. Such a war would be folly. After experiencing more than a decade of continuous war, the American public simply would not support the long-term occupation and intense fighting that would be required to obliterate ISIS. The pursuit of a full-fledged military campaign would exhaust U.S. resources and offer little hope of obtaining the objective. Wars pursued at odds with political reality cannot be won.

Containing the Threat

The sobering fact is that the United States has no good military options in its fight against ISIS. Neither counterterrorism, nor counterinsurgency, nor conventional warfare is likely to afford Washington a clear-cut victory against the group. For the time being, at least, the policy that best matches ends and means and that has the best chance of securing U.S. interests is one of offensive containment: combining a limited military campaign with a major diplomatic and economic effort to weaken ISIS and align the interests of the many countries that are threatened by the group's advance.

ISIS is not merely an American problem. The wars in Iraq and Syria involve not only regional players but also major global actors, such as Russia, Turkey, Iran, Saudi Arabia, and other Gulf states. Washington must stop behaving as if it can fix the region's problems with military force and instead resurrect its role as a diplomatic superpower.

Of course, U.S. military force would be an important part of an offensive containment policy. Air strikes can pin ISIS down, and cutting off its supply of technology, weapons, and ammunition by choking off smuggling routes would further weaken the group. Meanwhile, the United States should continue to advise and support the Iraqi military, assist regional forces such as the Kurdish Pesh Merga, and provide humanitarian assistance to civilians fleeing ISIS' territory. Washington should also expand its assistance to neighboring countries such as Jordan and Lebanon, which are struggling to contend with the massive flow of refugees from Syria. But putting more U.S. troops on the ground would be counterproductive, entangling the United States in an unwinnable war that could go on for decades. The United States cannot rebuild the Iraqi state or determine the outcome of the Syrian civil war. Frustrating as it might be to some, when it comes to military action, Washington should stick to a realistic course that recognizes the limitations of U.S. military force as a long-term solution.

The Obama administration's recently convened "summit on countering violent extremism"—which brought world leaders to Washington to discuss how to combat radical jihadism—was a valuable exercise. But although it highlighted the existing threat posed by al Qaeda's regional affiliates, it also reinforced the idea that ISIS is primarily a counterterrorism challenge. In fact, ISIS poses a much greater risk: it seeks to challenge the current international order, and, unlike the greatly diminished core al Qaeda organization, it is coming closer to actually achieving that goal. The United States cannot single-handedly defend the region and the world from an aggressive revisionist theocratic state—nor should it. The major powers must develop a common diplomatic, economic, and military approach to ensure that this pseudo-state is tightly contained and treated as a global pariah. The good news is that no government supports ISIS; the group has managed to make itself an enemy of every state in the region—and, indeed, the world. To exploit that fact, Washington should pursue a more aggressive, top-level diplomatic agenda with major powers and regional players, including Iran, Saudi Arabia, France, Germany, the United Kingdom, Russia, and even China, as well as Iraq's and Syria's neighbors, to design a unified response to ISIS.

That response must go beyond making a mutual commitment to prevent the radicalization and recruitment of would-be jihadists and beyond the regional military coalition that the United States has built. The major powers and regional players must agree to stiffen the international arms embargo currently imposed on ISIS, enact more vigorous sanctions against the group, conduct joint border patrols, provide more aid for displaced persons and refugees, and strengthen UN peacekeeping missions in countries that border Iraq and Syria. Although some of these tools overlap with counterterrorism, they should be put in the service of a strategy for fighting an enemy more akin to a state actor: ISIS is not a nuclear power, but the group represents a threat to international stability equivalent to that posed by North Korea. It should be treated no less seriously.

Given that political posturing over U.S. foreign policy will only intensify as the 2016 U.S. presidential election approaches, the White House would likely face numerous attacks on a containment approach that would satisfy neither the hawkish nor the anti-interventionist camp within the U.S. national security establishment. In the face of such criticism, the United States must stay committed to fighting ISIS over the long term in a manner that matches ends with means, calibrating and improving U.S. efforts to contain the group by moving past outmoded forms of counterterrorism and counterinsurgency while also resisting pressure to cross the threshold into full-fledged war. Over time, the successful containment of ISIS might open up better policy options. But for the foreseeable future, containment is the best policy that the United States can pursue.

Critical Thinking

1. What is the difference between ISIS and Al Qaeda?
2. What is the best strategy for the United States to use to deal with ISIS and why?
3. What explains the success of ISIS so far?

Internet References

Defense Intelligence Agency
www.dni.gov/index.php/intelligence-community/members-of-the ic#dia

Department of Homeland Security
www.dhs.gov/

SITE Intelligence Group
https//:ent.siteintelgroup.com/

U.S. Department of State.Bureau of Counter-Terrorism
www.state.gov/j/ct/

Audrey Kurth Cronin is Distinguished Professor and Director of the International Security Program at George Mason University and the author of *How Terrorism Ends: Understanding the Decline and Demise of Terrorist Campaigns*. Follow her on Twitter @akcronin.

Article

Prepared by: Robert Weiner, *University of Massachusetts, Boston*

ISIS and the Third Wave of Jihadism

"There is no simple or quick solution to rid the Middle East of ISIS because it is a manifestation of the breakdown of state institutions and the spread of sectarian fires in the region."

FAWAZ A. GERGES

Learning Outcomes

After reading this article, you will be able to:

- Discuss what factors contributed to the emergence of ISIS.
- Explain the differences between ISIS and al-Qaeda.

In order to make sense of the so-called Islamic State (known as ISIS or ISIL, or by its Arabic acronym, Daesh) and its sudden territorial conquests in Iraq and Syria, it is important to place the organization within the broader global jihadist movement. By tracing ISIS's social origins and comparing it with the first two jihadist waves of the 1980s and 1990s, we can gauge the extent of continuity and change, and account for the group's notorious savagery.

Although ISIS is an extension of the global jihadist movement in its ideology and worldview, its social origins are rooted in a specific Iraqi context, and, to a lesser extent, in the Syrian war that has raged for almost four years. While al-Qaeda's central organization emerged from an alliance between ultraconservative Saudi Salafism and radical Egyptian Islamism, ISIS was born of an unholy union between an Iraq-based al-Qaeda offshoot and the defeated Iraqi Baathist regime of Saddam Hussein, which has proved a lethal combination.

Bitter Inheritance

The causes of ISIS's unrestrained extremism lie in its origins in al-Qaeda in Iraq (AQI), founded by Abu Musab al-Zarqawi, who was killed by the Americans in 2006. The US-led invasion and occupation of Iraq caused a rupture in an Iraqi society already fractured and bled by decades of war and economic sanctions. America's destruction of Iraqi institutions, particularly its dismantling of the Baath Party and the army, created a vacuum that unleashed a fierce power struggle and allowed non-state actors, including al-Qaeda, to infiltrate the fragile body politic.

ISIS's viciousness reflects the bitter inheritance of decades of Baathist rule that tore apart Iraq's social fabric and left deep wounds that are still festering. America's bloody vanquishing of Baathism and the invasion's aftermath of sectarian civil war plunged Iraq into a sustained crisis, inflaming Sunnis' grievances over their disempowerment under the new Shia ascendancy and preponderant Iranian influence.

Iraqi Sunnis have been protesting the marginalization and discrimination they face for some time, but their complaints fell on deaf ears in Baghdad and Washington. This created an opening for ISIS to step in and instrumentalize their grievances. A similar story of Sunni resentment unfolded in Syria, where the minority Alawite sect dominates the regime of President Bashar al-Assad. Thousands of embittered Iraqi and Syrian Sunnis fight under ISIS's banner, even though many do not subscribe to its extremist Islamist ideology. While its chief, Abu Bakr al-Baghdadi, has anointed himself as the new caliph, on a more practical level he blended his group with local armed insurgencies in Syria and Iraq, building a base of support among rebellious Sunnis.

ISIS is a symptom of the broken politics of the Middle East and the fraying and delegitimation of state institutions, as well as the spreading of civil wars in Syria and Iraq. The group has filled the resulting vacuum of legitimate authority. For almost two decades, "al-Qaeda Central" leaders Osama bin Laden and Ayman al-Zawahiri were unable to establish the kind of social movement that Baghdadi has created in less than five years.

Unlike its transnational, borderless parent organization, ISIS has found a haven in the heart of the Levant. It has done so by exploiting the chaos in war-torn Syria and the sectarian, exclusionary policies of former Iraqi Prime Minister Nuri Kamal al-Maliki. More like the Taliban in Afghanistan in the 1990s than al-Qaeda Central, ISIS is developing a rudimentary infrastructure of administration and governance in captured territories in Syria and Iraq. It now controls a landmass as large as the United Kingdom. ISIS's swift military expansion stems from its ability not only to terrorize enemies but also to co-opt local Sunni communities, using networks of patronage and privilege. It offers economic incentives such as protection of contraband trafficking activity and a share of the oil trade and smuggling in eastern Syria.

Sectarian War

Building a social base from scratch in Iraq, AQI exploited the Sunni-Shia divide that opened after the United States toppled Hussein's Sunni-dominated regime. The group carried out wave after wave of suicide bombings against the Shia. Zarqawi's goal was to trigger all-out sectarian war and to position AQI as the champion of the embattled Sunnis. He ignored repeated pleas from his mentors, bin Laden and Zawahiri, to stop the indiscriminate killing of Shia and to focus instead on attacking Western troops and citizens.

Although Salafi jihadists are nourished on an anti-Shia propaganda diet, al-Qaeda Central prioritized the fight against the "far enemy"—America and its European allies. In contrast, AQI and its successor, ISIS, have so far consistently focused on the Shia and the "near enemy" (the Iraqi and Syrian regimes, as well as all secular, pro-Western regimes in the Muslim world). Baghdadi, like Zarqawi before him, has a genocidal worldview, according to which Shias are infidels—a fifth column in the heart of Islam that must either convert or be exterminated. The struggle against America and Europe is a distant, secondary goal that must be deferred until liberation at home is achieved. At the height of the Israeli assault on Gaza during the summer of 2014, militants criticized ISIS on social media for killing Muslims while failing to help the Palestinians. ISIS retorted that the struggle against the Shia comes first.

Baghdadi has exploited the deepening Sunni-Shia rift across the Middle East, intensified by a new regional cold war between Sunni-dominated Saudi Arabia and Shia-dominated Iran. He depicts his group as the vanguard of persecuted Sunni Arabs in a revolt against sectarian-based regimes in Baghdad, Damascus, and beyond. He has amassed a Sunni army of more than 30,000 fighters (including some 18,000 core members, plus affiliated groups). By contrast, at the height of its power in the late 1990s, al-Qaeda Central mustered only 1,000 to 3,000 fighters, a fact that shows the limits of transnational jihadism and its small constituency compared with the "near enemy" or local jihadism of the ISIS variety.

The weakest link of ISIS as a social movement is its poverty of ideas.

Numbers alone do not explain ISIS's rapid military advances in Syria and Iraq. After Baghdadi took charge of AQI in 2010, when it was in precipitous decline, he restructured its military network and recruited experienced officers from Hussein's disbanded army, particularly the Republican Guards, who turned ISIS into a professional fighting force. It has been toughened by fighting in neighboring Syria since the civil war there began in 2011. According to knowledgeable Iraqi sources, Baghdadi relies on a military council made up of 8 to 13 officers who all served in Saddam Hussein's army.

Rational Savagery

In a formal sense, ISIS is an effective fighting force. But it has become synonymous with viciousness, carrying out massacres, beheadings, and other atrocities. It has engaged in religious and ethnic cleansing against Yazidis and Kurds as well as Shia. Such savagery might seem senseless, but for ISIS it appears to be a rational choice, intended to terrorize its enemies and to impress potential recruits. ISIS's brutality also stems from the ruralization of this third wave of jihadism. Whereas the two previous waves had leaders from the social elite and a rank and file mainly composed of lower-middle-class university graduates, ISIS's cadre is rural and lacking in both theological and intellectual accomplishment. This social profile helps ISIS thrive among poor, disenfranchised Sunni communities in Iraq, Syria, Lebanon, and elsewhere.

ISIS adheres to a doctrine of total war, with no constraints. It disdains arbitration or compromise, even with Sunni Islamist rivals. Unlike al-Qaeda Central, it does not rely on theology to justify its actions. "The only law I subscribe to is the law of the jungle," retorted Baghdadi's second-in-command and right-hand man, Abu Muhammed al-Adnani, to a request more than a year ago by rival militant Islamists in Syria who called for ISIS to submit to a Sharia court so that a dispute with other factions could be properly adjudicated. For the top ideologues of Salafi jihadism, such statements and actions are sacrilegious, "smearing the reputation" of the global jihadist movement, in the words of Abu Mohammed al-Maqdisi, a Jordan-based mentor to Zarqawi and many jihadists worldwide.

New Wave

The scale and intensity of ISIS's brutality, stemming from Iraq's blood-soaked modern history, far exceed either of the first two jihadist waves of recent decades. Disciples of Sayyid Qutb—a radical Egyptian Islamist known as the master theoretician of modern jihadism—led the first wave. Pro-Western, secular Arab regimes, which they called the "near enemy," would be the main targets. Their first major act was the assassination of Egyptian President Anwar Sadat in 1981.

This first wave included militant religious activists of Zawahiri's generation. They wrote manifestos in an effort to obtain theological legitimation for their attacks on "renegade" and "apostate" rulers, such as Sadat, and their security services. On balance, though, they showed restraint in the use of political violence. Conscious of the importance of Egyptian and wider Arab opinion, Zawahiri spent considerable energy over the years trying to explain the circumstances that led to the killing of two children in Egypt and Sudan, and repeatedly insisted that his group, Egyptian Islamic Jihad, did not target civilians.

The first wave had subsided by the end of the 1990s. During the 1980s, many militants had traveled to Afghanistan to fight the Soviet occupation, a cause that launched the second jihadist wave. After the withdrawal of the Soviets from Afghanistan, bin Laden emerged as the leader of the new wave. The focus shifted to the "far enemy" in the West—the United States and, to a lesser degree, Europe.

To win support, bin Laden justified his actions as a form of self-defense. He portrayed al-Qaeda's September 11, 2001, attack on the United States as an act of "defensive jihad," or a just retaliation for American domination of Muslim countries. Baghdadi, by contrast, cares little for world opinion. Indeed, ISIS makes a point of displaying its barbarity in its internet videos. Stressing violent action rather than theology, it has offered no ideas to sustain its followers. Baghdadi has not fleshed out his vision of a caliphate but merely declared it by fiat, which contradicts Islamic law and tradition.

Ironically, Baghdadi—who has a doctorate from the Islamic University of Baghdad, with a focus on Islamic culture, history, sharia, and jurisprudence—is more steeped in religious education than al-Qaeda's past and current leaders, bin Laden (an engineer) and Zawahiri (a medical doctor), who had no such credentials. Yet he surrounds himself with former Baathist army officers, rather than ideologues, and has not issued a single manifesto laying out his claim to either the caliphate or the leadership of the global jihadist movement. ISIS's brutality has alienated senior radical preachers who have publicly disowned it, though some have softened their criticism in the wake of US-led airstrikes against the group in Iraq and Syria, which one ideologue described as "the aggression of crusaders."

Bin Laden said, "When people see a strong horse and a weak horse, by nature they will like the strong horse." Baghdadi's slogan of "victory through fear and terrorism" signals to friends and foes alike that ISIS is a winning horse. Increasing evidence shows that over the past few months, hundreds, if not thousands, of die-hard former Islamist enemies of ISIS, including members of groups such as the Nusra Front and the Islamic Front, have declared allegiance to Baghdadi.

For now, ISIS has taken operational leadership of the global jihadist movement by default, eclipsing its parent organization, al-Qaeda Central. Baghdadi has won the first round against his former mentor, Zawahiri, who triggered an intra-jihadist civil war by unsuccessfully trying to elevate his own man, Abu Mohammed al-Golani, head of the Nusra Front, over Baghdadi in Syria.

Recruiting Tactics

However, the so-called Islamic State is much more fragile than Baghdadi would like us to believe. His call to arms has not found any takers among either top jihadist preachers or leaders of mainstream Islamist organizations, while Islamic scholars, including the most notable Salafi clerics, have dismissed his declaration of a caliphate as null and void. In fact, many of these same renowned Salafi scholars have equated ISIS with the extremist Kharijites of the Prophet's time. ISIS also threatens the vital interests of regional and international powers, a fact that explains the large coalition organized by the United States to combat the group.

Nevertheless, ISIS's sophisticated outreach campaign appeals to disaffected Sunni youth around the world by presenting the group as a powerful vanguard movement capable of delivering victory and salvation. It provides them with both a utopian worldview and a political project. Young recruits do not abhor its brutality; on the contrary, its shock-and-awe methods against the enemies of Islam are what attract them.

ISIS adheres to a doctrine of total war, with no constraints.

ISIS's exploits on the battlefield, its conquest of vast swaths of territory in Syria and Iraq, and its declaration of a caliphate have resonated widely, facilitating recruitment. Increasing evidence shows that the US-led airstrikes have not slowed down the flow of foreign recruits to Syria—far from it. The *Washington Post* reported that more than 1,000 foreign fighters are streaming into Syria each month. Efforts by other countries, especially Turkey, to stem the flow of recruits (many of them from European countries) have proved largely ineffective, according to US intelligence officials. ISIS fighters have also highlighted the important role of Chechen trainers in developing the group's military capabilities. Some reportedly have set up a Russian school in Raqqa for their children, to prepare them for jihad back home.

Muslims living in Western countries join ISIS and other extremist groups because they want to be part of a tight-knit community with a potent identity. ISIS's vision of resurrecting an idealized caliphate gives them the sense of serving a sacred mission. Corrupt Arab rulers and the crushing of the Arab Spring uprisings have provided further motivation for recruits. Many young men from Western Europe and elsewhere migrate to the lands of jihad because they feel a duty to defend persecuted coreligionists. Yet many of those who join the ranks of ISIS find themselves persecuting innocent civilians of other faiths and committing atrocities.

Hearts and Minds

Now that the United States and Europe have joined the fight against ISIS, the group might garner backing from quarters of the Middle Eastern public sphere that oppose Western intervention in internal Arab affairs, though there has been no such blowback so far. More than bin Laden and Zawahiri, Baghdadi has mastered the art of making enemies. He has failed to nourish a broad constituency beyond a narrow, radical sectarian base.

There is no simple or quick solution to rid the Middle East of ISIS because it is a manifestation of the breakdown of state institutions and the spread of sectarian fires in the region. ISIS is a creature of accumulated grievances, of ideological and social polarization and mobilization a decade in the making. As a non-state actor, it represents a transformative movement in the politics of the Middle East, one that is qualitatively different from al-Qaeda Central's.

The key to weakening ISIS lies in working closely with local Sunni communities that it has co-opted, a bottom-up approach that requires considerable material and ideological investment. The most effective means to degrade ISIS is to dismantle its social base by winning over the hearts and minds of local communities. This is easier said than done, given the gravity of the crisis in the heart of the Arab world. The jury is still out on whether the new Iraqi prime minister, Haider al-Abadi, will be able to appeal to mistrustful Sunnis and reconcile warring communities. Rebuilding trust takes hard work and time, both of which play to ISIS's advantage.

Equally important, there is an urgent need to find a diplomatic solution to the civil war in Syria, which has empowered ISIS, fueling its surge after its predecessor, AQI, was vanquished in Iraq. Syria is the nerve center of ISIS—the location of its de facto capital, the northern city of Raqqa, and of its major sources of income, including the oil trade, taxation, and criminal activities. More than two-thirds of its fighters are deployed in Syria, according to US intelligence officials.

In the short- to medium-term, it would take a political miracle to engineer a settlement in Syria, given the disintegration of the country and the fragmentation of power among rival warlords and fiefdoms, not to mention the regional and great power proxy wars playing out there. Until there is a regional and international agreement to end the Syrian civil war, ISIS will continue to entrench itself in the country's provinces and cities.

Yet even ISIS's dark cloud has a silver lining. Once Baghdadi's killing machine is dismantled, he will leave behind no ideas, no theories, and no intellectual legacy. The weakest link of ISIS as a social movement is its poverty of ideas. It can thrive and sustain itself only in an environment of despair, state breakdown, and war. If these social conditions can be reversed, its appeal and potency will wither away, though its bloodletting will likely leave deep scars on the consciousness of Arab and Muslim youth.

Critical Thinking

1. Why has ISIS been able to attract so many young recruits?
2. What is the difference between the concepts of the "Far Enemy" and the "Near Enemy"?
3. Is ISIS both a non-state actor and a state actor? why?

Internet References

Bureau of Counter-Terrorism,U.S. Department of State
www.state.gov/j/ct/

Department of Homeland Security
www.dhs.gov/

SITE Intelligence Group
https://ent.siteintelgroup.com

Fawaz A. Gerges is a professor of international relations and Middle Eastern politics at the London School of Economics and Political Science. His books include The Far Enemy: Why Jihad Went Global (Cambridge University Press, 2005) and, most recently, *The New Middle East: Protest and Revolution in the Arab World* (Cambridge, 2014).

Article Prepared by: Robert Weiner, *University of Massachusetts, Boston*

Strategic Amnesia and ISIS

DAVID V. GIOE

Learning Outcomes

After reading this article, you will be able to:

- Understand the relevance of the U.S. military history to the war against Islamic State of Iraq and Syria (ISIS).
- Discuss the problems associated with expeditionary forces on foreign soil.

Mark Twain observed, "history doesn't repeat itself, but it does rhyme." The study of military history teaches us valuable lessons that are applicable to today's most intractable strategic problems; yet, these lessons are underappreciated in current American strategy formulation. Throughout the history of American armed conflict, the United States has discerned, at great cost, four critical lessons applicable to containing and combating the Islamic State.

First, as war theorist Carl von Clausewitz noted, war is a continuation of politics by other means; but resorting to war rarely yields the ideal political solution envisioned at the start of hostilities. Second, the use of proxy forces to pursue American geopolitical goals is rarely an investment worth making because proxies tend to have goals misaligned with those of their American sponsors. True control is an illusion. The corollary to this axiom is that supporting inept and corrupt leaders with American power only invites further dependency, does not solve political problems and usually prolongs an inevitable defeat. Third, conflating the security of a foreign power with that of America leads to disproportionate resource allocation and an apparent inability at the political level to pursue policies of peace and successful war termination. Fourth, alliance formation through lofty rhetorical positions imperils rational analysis of geopolitical and military realities. Publicly staking out inviable political end states invites a strategic mismatch between military capabilities and political wishes, endangering the current enterprise as well as future national credibility. America has paid for these lessons in blood; our leaders ought to heed them.

The Obama administration's effort to again increase the number of American military advisors in Iraq, coupled with the reconstruction of a new base at Al-Taqaddum in Anbar Province, has given rise to accusations among both Democrats and Republicans about either mission creep (from doves and non-interventionists) or weak incrementalism (from hawks and liberal interventionists). Former defense secretary Robert Gates observed in May 2015 that there simply was no American strategy in the Middle East. Congressional hawks have used Gates's observation to criticize the Obama administration's cautious efforts in any ground campaign against the Islamic State, and some have called for thousands of American boots back on the ground in Iraq. However, during a 2011 visit to the U.S. Military Academy at West Point, Gates also told the cadets that any adviser who counseled deploying large land forces to the Middle East should "have his head examined," suggesting that a larger military footprint should not be confused with a robust strategy.

Using rhetoric reminiscent of George W. Bush's "War on Terror," in 2014 President Obama pledged to "ultimately destroy" the Islamic State, but over a year later the Obama administration itself admitted that its strategy is not yet "complete." Indeed, even complete strategies often do not survive first contact with the enemy. As military personnel often quip, "the enemy gets a vote," and the Islamic State seems to be visiting the ballot box early and often. A candid comment about strategy formulation makes for an interesting sound bite (or cudgel). But discretion may be the better part of valor when facing the slippery slope of another open-ended commitment in Iraq. Many observers took President Bush to task for suggesting that an ideology could be defeated by applying military force, but their critiques could apply just as well to Obama's turn at the helm.

Pledging victory implies an end state that is ultimately acceptable to one's adversary—whether it's forced upon it (like the unconditional surrender of Japan in 1945) or a negotiated political solution (as in Korea eight years later). The Islamic State seems to show little taste for negotiation, and why should it? Most wars are prolonged in the hope that each side will come to the negotiating table with a better hand to play. The Pentagon service chiefs appear reticent to get further involved in Iraq absent a political solution. But the Obama administration has been bullied into increasing troop numbers by Congressional hawks who have conflated the security of Iraqis with that of Americans and the cohesiveness of the Iraqi state with core U.S. national-security interests. America has been here before.

Unlike at the height of its post-Cold War military power, America no longer has the ability to dictate events globally—to the overplayed extent that it ever did. This is particularly true in North Africa, the Middle East, and South Asia. Those who argue that an American military campaign could defeat ISIS in a durable way place too much confidence in the ability of any American administration to control events abroad, especially in deeply rooted internecine conflicts. In fact, although what is happening in the Islamic State's Iraqi strongholds is both primitive and shocking, the state of affairs in Baghdad is what should be cause for even greater concern in Washington. Iraq, under its Shia-dominated government, has marginalized Sunnis and alienated Kurds, perhaps to the point of no return. Indeed, the billions of dollars invested in training Iraqi forces are for naught if the controlling political entity is a house divided against itself.

Iraq did not slide into its current state of affairs without outside help. The United States cannot escape some culpability for what Iraq (and Syria) have become, but expensive U.S. efforts to encourage good governance and interreligious and intertribal dialogue and cooperation have fallen short. Calling for a strategy to defeat an ideology or repair Iraq is tantamount to demanding that a physician devise a strategy to treat a patient admitted to the emergency room with a shotgun blast to the head. Even with the best of intentions, unlimited resources and the best expertise available, there isn't much that can be done to reach the status quo ante helium.

Since the end of World War II, the American military has struggled to translate tactical military success on the battlefield into durable political gains

Since the end of World War II, the American military has struggled to translate tactical military success on the battlefield into durable political gains. Although America has no peer when it comes to accumulating post-9/11 tactical victories, the record is not enviable at the political level. Witness the bin Laden raid of May 2011 and the daring May 2015 Army Special Forces raid into Raqqa, Syria. These were spectacular tactical successes, and perhaps necessary from a moral perspective, but they achieved little at the strategic level. To be sure, the world is a better place without Osama bin Laden and ISIS financier Abu Sayyaf, and they richly deserved their fates. The problem is what comes next. The United States has been eliminating the leadership of Al-Qaeda since the end of 2001 and has transferred those lessons to effectively remove the leadership of many Al-Qaeda franchises in Yemen and North Africa as well. No doubt the United States will further apply its lethal craft to ISIS in the near term. However, military history reveals that accumulating tactical successes does not equal strategic victory. The German military learned these lessons the hard way in both twentieth-century world wars. The Germans, although well equipped and tactically sound, were unable to realize their broader political desires through violence.

Soviet leader Joseph Stalin is said to have suggested that quantity has a quality all its own. If this is true, we must recognize American tactical successes for what they have achieved, even absent a broader strategy. The American homeland is arguably safer because those that seek to do it harm are impeded by those tactical successes. American military and intelligence operations have made enemy communications more difficult and secure staging bases hard to come by. The U.S. military killed or captured the top leadership of Al-Qaeda and it's like, retarding their operational planning and derailing their efforts to undertake spectacular attacks. If American strategy is threat mitigation through sustained special-operations raids and intelligence-driven covert action, it is working. Still, it is an open question how long this is sustainable, especially on the back of a shrinking, all-volunteer military force. American military and political leaders have spoken of a "generational war," yet they also shrink from serious discussion about national service or a draft.

Assume for a moment that the Obama administration were to pour troops into Iraq and loosen their rules of engagement, permitting direct American participation in the fighting. Could a couple of U.S. divisions retake ISIS strongholds? Absolutely. It would come at a bloody cost to young American soldiers, as when Fallujah fell to American forces in 2004 with 560 American casualties (and thousands more with psychological wounds), but the U.S. military could surely retake the large cities of Anbar province. Could they hold them? Not indefinitely. With the forces and resources available to the Pentagon, the United States could hold it for a time while building Iraqi capacity. This is a key pillar of the Army's counterinsurgency doctrine, but recent experiences in Iraq and Afghanistan lay bare the failure of this approach, at least on a timeline not measured in decades. The only dramatic success in Anbar province was a political one: the Sunni tribes "awakened" to turn against Al-Qaeda of Iraq and the monstrous tactics of AQI leader Abu Musab al-Zarqawi. Political settlement manifested on the battlefield—a much more promising proposition than the other way around.

Most war theorists conceive of war as a contest of wills. Put another way, the party who wants it most—and will thus sacrifice the most—usually wins in the long run. Consider, for instance, a group of mujahideen repelling the Soviets after a decade of fruitless bloodletting in Afghanistan. Or, for that matter, consider some of the same fighters showing NATO the exit a generation later. War's fundamental character has not changed over time, and the contest of wills remains a bedrock principle. To apply the concept, consider what the average ISIS fighter would do to secure the success of the Islamic State against what the average American would do to roll it back. As things stand, the ISIS fighter is considerably more committed to his cause—particularly in the absence of convincing proof that ISIS poses an existential threat to the American way of life—and only 1 percent of Americans are actually involved in the so-called war on terror. In both of America's greatest military successes in the twentieth century—the world wars—America came late, but with a total mobilization that called on the resources of a significantly larger proportion of the population. Moreover, that population was considerably more unified in purpose.

Additionally, expeditionary wars on foreign soil represent challenges on several fronts. Deploying and supporting troops, heavy machinery and the routine supplies of war, especially over great distances into landlocked countries with rugged terrain, complicates logistics. This is also expensive and relies on the continued support of the citizenry back home. The expeditionary force is most often at a disadvantage in that it must secure victory while fighting far from home. Its opponents, comfortable on their home soil, do not have to win—they just have to wait out the invading force and not suffer catastrophic battlefield defeat.

During the American Revolution, George Washington employed a Fabian strategy against the expeditionary British force. Washington avoided large engagements on unfavorable terms, as his goal was to preserve a true fighting force and wear down his enemy until Westminster decided to stop throwing men and material at the Continental Army. The militarily superior British pulled the plug on their colonial undertaking after six years of active fighting in North America. They had other strategic considerations and made a difficult choice to concede defeat in the North American theater of a larger war. Unfortunately, today Washington has more in common with Westminster: it holds a losing hand against a determined enemy pursuing a Fabian strategy. The British experience in America suggests that a professional military in an expeditionary capacity may come up short.

To use a parallel from the American Revolution, those under ISIS rule may from their own Iraqi committees of correspondence, Iraqi Sons of Liberty and Iraqi minutemen.

Recent media coverage of daily life inside of the Islamic State suggests that U.S. officials should not be so condescending as to think that those living under harsh ISIS rule are mere sheep awaiting rescue. The millions of Iraqis and Syrians now living under ISIS domination far outnumber their new masters. Those under ISIS rule have few good options, and the costs of rash action are high. The Iraqi army is apparently unable to retake any of ISIS core territory, at least not without the help of American airpower, advisors and (most problematically) Iranian-backed Shia militias. It is perhaps not a foregone conclusion that Sunnis living in the Islamic State would prefer militias backed by Iran's Quds Force as their liberators from ISIS. They may well view this sort of "liberation" as out of the frying pan and into the fire.

Although patience in a 24-hour, crisis-to-crisis news cycle is notable for its absence, given time some promising developments in ISIS territory could come to pass that undermine the Islamic State from within. Any lasting governing entity relies on some level of support or at least consent of the governed. The actions of ISIS toward its subjects suggest that over the long term they might not achieve this. Fear and brutality only go so far. Parents fed up with their children being indoctrinated with fundamentalist hate at school, women who cannot leave the house with their faces uncovered or without male relatives, men who are being extorted for ISIS taxes, citizens disgusted by summary executions and floggings, fathers who dread their sons becoming brainwashed to be martyrs and mothers who want their daughters to enjoy equal rights will begin to find common cause against the Islamic State. They may decide to cautiously provide tips to the Iraqi army's special forces on the locations of ISIS leaders or their weapons caches. They may themselves begin to hide weapons and supplies for when the popular mood shifts.

To use a parallel from the American Revolution, those under ISIS rule may form their own Iraqi committees of correspondence, Iraqi Sons of Liberty and Iraqi minutemen. They may seek their own outside allies and develop their own internal intelligence networks. In short, they will eventually resist, as the early sparks of the ill-fated Arab Spring will attest. From the Orange Revolution to the Prague Spring, from the Polish Solidarity movement to the Warsaw Ghetto uprising, the oppressed eventually resist. The desire for life, liberty and the pursuit of happiness is not exclusively American, but Americans cannot be the sole guarantors either. History has shown that peoples who perceive themselves to be oppressed usually organize into a creditable rebellious force, although this usually takes years for suitable levels of organized frustration to congeal into a counterforce.

As in the American Revolution example, defeating a superior force often requires powerful allies. The French contribution to the American war effort was a key development, but the French refrained from becoming openly involved until after the American victory at Saratoga in the fall of 1777. It was only after the Americans proved to be a formidable fighting force and fully devoted to their cause that the French arrived. Indeed, foreign military assistance and training for "internal defense" can be critical to success, but first the will and ability to win must be convincingly demonstrated. An increasingly vocal minority of Americans, whom we now refer to as "Patriots" and "Founding Fathers," spent the greater part of the early 1770s organizing themselves and secretly preparing for violence.

Consider a familiar scenario, so familiar that it reads like many recent headlines from Iraq: Enemy forces are gaining momentum and seizing territory at an alarming rate. They have a stronghold in the shape of a triangle just over an hour's drive from the capital. With the political dysfunction in the capital city and, even after American training, local troops underperforming, it seems that only an American-led search-and-destroy mission could root out the enemy, protect the capital and shift the battlefield momentum. After a period of airstrikes, American armored and helicopter-borne infantry forces duly arrive in the insurgent triangle and for three weeks attempt to clear the area of enemy forces, but they are unable to discern civilian from insurgent—it's possible they are one and the same, but absent uniforms or a recognizable chain of command, they are uncertain. These American troops are killed by booby traps and snipers, but never identify the enemy. Eventually declaring the area "cleared," American soldiers destroy some enemy weapons caches, and American senior officers brand it a successful operation.

No, the above scenario isn't Baghdad, the nearby Sunni Triangle, the crumbling Iraqi National Army, and the advancing Islamic State. The year was 1967 and the operation, called Cedar Falls, was to be the largest of the Vietnam War. Its purpose was to clear the "Iron Triangle" of Viet Cong irregular forces that were threatening Saigon—a nearly failed state propped up by American power. Frustratingly, the enemy would not stand and fight in the face of overwhelming American tactical and material superiority. The Viet Cong forces moved across the porous border into Cambodia and simply returned when the American forces departed the area. After the operation had ended, and at the cost of more than four hundred American casualties, many senior American officers counted Cedar Falls as a success, pointing to the numbers of weapons stockpiles that were destroyed and the fleeing enemy. In retrospect, the failure of Cedar Falls was emblematic of American efforts in Vietnam. In reality, the residents of the Iron Triangle, much like their countrymen throughout the rest of South Vietnam, found aspects of the Viet Cong message appealing, and surely no worse than the repressive and corrupt government in Saigon.

Confusing a messy, localized civil war with an existential threat to American national security is a strategic mistake.

In hindsight, the theory that Vietnam's fall to the Communists would make the rest of Asia topple like dominos into the Soviet sphere proved to be alarmist and false. Further, what appeared in 1965 to be in America's core national-security interests was identified by 1970 as irrelevant, a significant diversion of American resources from more pressing concerns and a major source of political and social tension on the home front. American political leaders declared that the Republic of Vietnam should be responsible for its own security and promptly began a period of "Vietnamization," in which American forces trained and equipped the South Vietnamese military, paving the way for an American withdrawal. Saigon fell two years after the American withdrawal, but that would have happened no matter what year the Americans finally decided to pull the plug. This is worth remembering when hawks seek to blame the Obama administration for the current state of affairs in Iraq because it pulled out American troops in 2011. Even if the United States had kept thousands of troops in Vietnam until 1983, Saigon would have fallen by 1985. Political problems can be papered over with military force for a long time, but in the end the result is the same.

Such parallels from the Vietnam War are haunting, and should not be tossed aside in current strategy formulation. These are the lessons learned at the cost of 58,220 American soldiers who gave the last full measure of devotion in Southeast Asia. In Vietnam, America paid a heavy price in lives and treasure to prop up a corrupt and unrepresentative government, which it hoped could function as a regional ally and bulwark against the seemingly prevailing ideology in the region. American military personnel attempted to both provide population-centric security and bring massive firepower to bear on the enemy. With the benefit of hindsight, many military historians have declared the Vietnam War "unwinnable," yet America was not less secure because of the loss.

In the same way that Viet Cong forces took advantage of the porous border with Cambodia during operation Cedar Falls, ISIS fighters in Iraq would just as easily slip across the border (which they control) into Syria and wait for the Americans to leave. And just like the residents of the Iron Triangle outside Saigon, not all Sunni residents of Anbar province view being governed by ISIS as particularly worse than a corrupt Iranian proxy government in Baghdad. The politics of the region, particularly the animus between Shia and Sunni Islam, are a jumble of tribalism, mistrust, anarchy and greed. The government in Baghdad is helplessly divided and, as history consistently reveals, American military efforts cannot fix a political problem.

As in Iraq and Afghanistan, not all wars are worthy of continued American involvement, and hardly any wars in the U.S. history could be considered existential. During the Korean War, President Harry S. Truman correctly elected not to expand the war into China, despite the vociferous urging of General Douglas MacArthur. Likewise, President Johnson did not permit an invasion of North Vietnam, despite the fact that in both cases the enemy center of gravity lay beyond the local battlefields of Korea and South Vietnam. Neither president opted to unleash the supposed guarantor of continued American existence—the nuclear triad. While it is true that these conflicts were limited wars without existential risks, it is proper that they were conducted as such by the U.S. administrations that oversaw them. Escalation to total war, or an existential fight for national survival, is only appropriate in the direst circumstances, in which a loss on the battlefield might mean national calamity. Despite the repulsive and brutal conduct of ISIS, the stakes for the United States are not that high. Confusing a messy, localized civil war with an existential threat to American national security is a strategic mistake.

American power toppled the Taliban and Saddam rapidly with modestly sized forces, but the maelstrom and "surges" that followed pulled hundreds of thousands of American and allied troops into its wake. Like the process of Vietnamization, in both Iraq and Afghanistan U.S. forces sought to train, advise and equip allies in the hopes that they could stand on their own and American troops could leave with some political gains realized. In Iraq, the United States spent nearly a decade and approximately $20.2 billion on a dubious mission to train the Iraqi army to secure the country. This army had years to develop under the tutelage of the finest American instructors and was the beneficiary of millions of dollars of U.S. military hardware. Yet it is the black flag of ISIS that waves atop U.S.-made Humvees, armor and heavy weapons. This suggests that motivation, loyalty and esprit de corps matter more than the latest technology, hardware and training cadre.

Again, war is a contest of wills, and the U.S. policy at present is to stiffen the spine of the locals who are expected to do the fighting. The U.S. Marine Corps tried this in South Vietnam with Combined Action Platoons, a small group of Marines and a Navy Corpsman residing in a rural hamlet, strengthening the local militia forces. The cap program is often judged as successful because it denied sanctuary to enemy forces, but successes at such a low level had little impact. The political dysfunction in Saigon overshadowed stability in rural hamlets.

Stiffening the spine of local forces, sometimes referred to as Foreign Internal Defense, can work if there is a baseline level of common mission already in existence among the host nation forces. American advisors can provide expertise and technology, but vision and commitment need to be homegrown. In addition to the lack of discipline and esprit de corps that accompanies good militaries, a major failing in the Iraqi army is the lack of a shared vision of the end state for Iraq. It isn't obvious that a Shia soldier in the Iraqi Army, from Basra for instance, considers it a good idea to fight ISIS in Anbar province. He may not view Anbar as his home or even part of his conception of Iraq. It is understandable, then, that he may want victory there less than an ISIS fighter does. Anbar just doesn't mean as much to him.

The fear of "losing face" has led many commanders to attempt to turn straw into gold with new strategies.

In the early sixteenth century, Machiavelli observed that troops who are not fighting for their own homeland are not inclined toward bravery because their "trifle of a stipend" is acceptable until war comes and then they "run from the foe." This begs the question whether members of the Iraqi army can be said to be fighting for the U.S. conception of a single federated Iraq, or for their own religious sect or tribe. American military and political leaders hoped that a reliably paid and equipped Iraqi army would fight like those defending their homeland. In fleeing before the ISIS advance, they proved to resemble the "mercenaries" and "auxiliaries" of whose dubious dedication Machiavelli warned.

What is to be done? Washington continues to substitute tactical action for strategy, and thus continues to throw good money and American lives at the chimera of a pluralistic and tolerant Iraq (and Afghanistan) while at the same time breeding dependency on America. American decision-makers would be well served to avoid ideologically guided wishful thinking as this often tempts the strategist to ignore history's warnings. Americans aren't the only ones who brush aside historical lessons with wishful thinking. Why would Adolf Hitler open himself up to a two-front war and invade the Soviet Union in June 1941? His extreme ideology compelled him to brush aside Napoleon's harsh lessons about invading Russia.

Another step in the right direction is to stop speaking in euphemisms when discussing the performance of the Iraqi (or any) army. Defense Secretary Ashton Carter noted that the Iraqi army "showed no will to fight" in Ramadi, but a White House spokesman characterized the dismal performance as a "setback." Investing rhetorically in an ally is a slippery slope, and almost always comes at the cost of sober and dispassionate analysis of battlefield performance. If unchecked, when "their" performance turns into "ours" and "they" starts to be "us," two unfortunate things usually follow. First, it conflates the security of the Iraqi state with that of American national security. And, more insidiously, cutting losses becomes harder. The fear of "losing face" has led many commanders to attempt to turn straw into gold with new strategies. It gets harder to withdraw absent a plausible "mission accomplished" narrative because of the inevitable argument that cutting losses is tantamount to forfeiting American military credibility. As Clausewitz reminded his readers, once the blood and treasure expended exceeds the value of an objective, that objective must be given up. Giving up an advise-and-assist mission for Iraqi allies will be politically impossible when the effort transforms from a military analysis of "them" into face-saving political measures involving "us." Moreover, despite some marginal but real tactical differentiation, publicly referring to forward operating bases (fobs) or combat outposts (cops), as "lily pads"—implying fleeting presence—is another misleading battlefield euphemism. With the Obama administration being bullied into dripping the U.S. Army back into Iraq a few hundred soldiers at time, it would be unsurprising if these "lily pads" remain into the next decade.

Military history warns that observers with skin in the game are unable to see strategy unfolding as it actually is. Despite evidence that the government in Saigon was increasingly corrupt and repressive, President Johnson observed in 1967, "Certainly there is a positive movement toward constitutional government." In June 2005, the Bush administration claimed that there were 160,000 Iraqi security forces who were trained, equipped and on the verge of independent operations. The results of this training were on full display in the May 2015 ISIS victory in Ramadi. The list of misstatements goes on when we view our allies as we wish to see them, not as they are. Even if intelligence assessments in private offer more accurate assessments, their own skin in the game, coupled with a guiding ideological approach, will always color the vision of political leaders. Not succumbing to the temptation to offer a continual drumbeat of rosy analysis for public consumption is a critical first step to avoiding foreign-policy missteps, or at least reversing those errors already committed.

In fairness, a few commentators are advocating a full-scale return to Iraq, but that's not how long-term commitments are usually undertaken. Again, the case of Vietnam is instructive. In the early 1960s, the Kennedy administration sent advisors to South Vietnam on a rather modest advise-and-assist mission. Once it became clear that this would be insufficient to accomplish the desired objective, "they" became "us" and "their task" became "our task." The inevitable mission creep set in; by the end of the Vietnam War, more than 2.5 million American troops had rotated through a country roughly half the size of New Mexico—and still lost. Some U.S. officials would reject that the new Iraq mission is anything like Operation Iraqi Freedom (2003–11), but it may look increasingly similar with the passage of time.

Eventually, the American people will tire of nonstop war. After the horror of the Islamic State's barbarism and the resulting surge of patriotism has subsided, the public will question the costs borne by so many troops, especially those who have done many tours without seeing any real progress. It is up to the Iraqis, and perhaps the greater Middle East, to decide their own fate. The international system seeks balance, and this often occurs through violence. We're seeing that now in the Middle East. Given the cast of players, things may even get worse before they get better as regional competitors become more involved and the stakes get higher. Yet the American people cannot want a pluralistic and tolerant Iraq more than the Iraqis do. Clausewitz noted that, "One country may support another's cause, but will never take it so seriously as it takes its own." The study of military history reveals an abundance of material for defense strategists, commanders and policymakers. It is accessible and directly applicable to contemporary strategic dilemmas. Ignoring these lessons would be a disservice to those who made the ultimate sacrifice to reveal them.

Critical Thinking

1. Why can't the U.S. translate tactical victories against ISIS into political gains?
2. Has the U.S. counterinsurgency strategy worked? Why or why not?
3. What is the relevance of Machiavelli to U.S. strategy and tactics in Iraq?

Internet References

Department of Homeland Security
www.dhs.gov/

U.S. Department of State, Bureau of Counter-Terrorism
www.state.gov/j/ct/

David V. Gioe is an assistant professor of Military History at the U.S. Military Academy at West Point. He previously served as a CIA operations officer. The opinions expressed here are his own and do not necessarily reflect the U.S. Army, the Department of Defense or the U.S. government.

Article

Prepared by: Robert Weiner, *University of Massachusetts, Boston*

Obama and Terrorism

Like It or Not, the War Goes On

JESSICA STERN

Learning Outcomes

After reading this article, you will be able to:

- Explain the pillars of Obama's counter-terrorism strategy.
- Discuss the causes of Middle Eastern terrorism.

U.S. President Barack Obama came into office determined to end a seemingly endless war on terrorism. Obama pledged to make his counterterrorism policies more nimble, more transparent, and more ethical than the ones pursued by the George W. Bush administration. Obama wanted to get away from the overreliance on force that characterized the Bush era, which led to the disastrous U.S. invasion of Iraq in 2003. That war, in turn, compromised the U.S. campaign against Al-Qaeda. During the past six-plus years, Obama has overseen an approach that relies on a combination of targeted killing, security assistance to military and intelligence forces in partner and allied countries, and intensive electronic surveillance. He has also initiated, although in a tentative way, a crucial effort to identify and address the underlying causes of terrorism. Overall, these steps amount to an improvement over the Bush years. But in many important ways, the relationship between Bush's and Obama's counterterrorism programs is marked by continuity as much as by change.

One important difference, however, is that whereas Bush's approach was sometimes marred by an overly aggressive posture, Obama has sometimes erred too far in the other direction, seeming prone to idealism and wishful thinking. This has hampered his administration's efforts to combat the terrorist threat: despite Obama's laudable attempts to calibrate Washington's response, the American people find themselves living in a world plagued with more terrorism than before Obama took office, not less. Civil war, sectarian tensions, and state failure in the Middle East and Africa ensure that Islamist terrorism will continue its spread in those regions—and most likely in the rest of the world as well. Most worrisome is the emergence in Iraq and Syria of the self-proclaimed Islamic State (also known as ISIS), a protean Salafi jihadist organization whose brutal violence, ability to capture and hold territory, significant financial resources, and impressive strategic acumen make it a threat unlike any other the United States has faced in the contemporary era. The rise of ISIS represents not only the failure of Bush era counterterrorism policies but also a consequence of Obama's determination to withdraw from Iraq with little regard for the potential consequences. Obama was right to seethe 2003 invasion of Iraq as a distraction from the war on Salafi jihadists. But his premature political disengagement from Iraq eight years later only made things worse.

The Obama years have put in stark relief the inescapable dilemma faced by any U.S. president trying to protect the United States and its allies from terrorism. Military responses, although frequently necessary in the immediate term, can end up serving terrorists' agendas; blowback is all but inevitable. Obama has talked up the potential of preventive strategies, such as civic engagement with communities where extremists recruit and the promotion of inclusive and effective governance. Such approaches are less risky than the use of force, but their effects take time to manifest and are difficult to measure. They also enjoy little support in Congress or among the American public.

Meanwhile, debates about U.S. counterterrorism policy remain mired in counterproductive partisan bickering and recriminations, with different Washington factions blaming one another for what went wrong. Whoever succeeds Obama as president will have to sort out the costs and benefits of his approach in a far more nuanced way. In counterterrorism—as in foreign policy more generally—it's easier to assess the limitations of the last president's approach than to develop a more effective new one, and it's easier to talk about transformative change than to carry it out.

Plus ÇA Change

Some of the changes Obama has made have been mostly rhetorical or have reflected a shift in emphasis rather than a truly substantive move. Ironically, the aspects of U.S. counterterrorism to which he has made the least significant changes are the very ones that he was initially most determined to alter. The Bush administration's "global war on terrorism" has been replaced by a campaign known as "countering violent extremism" to serve as the overarching U.S. strategy to combat transnational Salafi jihadist groups such as Al-Qaeda and ISIS. But the new phraseology masks many similarities. The "kinetic" fight—the use of deadly force by the U.S. military and intelligence agencies—has continued unabated, mostly in the form of drone strikes, since Obama took office. According to estimates collected by *The Long War Journal*, the United States has launched approximately 450 such attacks in Pakistan and Yemen during Obama's tenure, killing some 2,800 suspected terrorists and around 200 civilians.

And although Obama explicitly outlawed Bush's "enhanced interrogation techniques"—rightly classifying them as torture—and closed the so-called black sites where the CIA carried out the abuse, those changes were not as significant as they might appear. According to Jack Goldsmith, who headed the Office of Legal Counsel from October 2003 until June 2004, the Bush administration had halted the practice of waterboarding (without specifically declaring it illegal) by 2003, and the black sites had been largely emptied by 2007. And although Obama denounced abusive interrogations and extralegal detentions, he did so presumably knowing full well that a number of Washington's Middle Eastern allies in the struggle against Salafi jihadists would nonetheless continue to engage in such activities, and therefore, if those techniques happened to produce useful intelligence, the United States could still benefit from it.

Perhaps, the most surprising continuity between Bush's and Obama's counterterrorism records is the fact that the U.S. detention center in Guantánamo Bay, Cuba, remains open. One of Obama's first acts as president was to sign an executive order requiring that the Pentagon shut down the facility within a year. But in March 2011, after facing years of intense bipartisan congressional opposition to that plan, Obama ordered the resumption of military commissions at Guantánamo and officially sanctioned the indefinite detention of suspected terrorists held there without charge—two of the policies he had vowed to change. In this case, the president's idealistic goals became hard to sustain once the duty to protect American lives became his primary responsibility.

Another irony is that the most successful reversal of Bush's counterterrorism agenda that Obama managed to achieve is arguably the one that has brought him the most grief: the end of the U.S. war in Iraq. The Bush administration made many different arguments—often based on flawed or misleading intelligence—for why the United States had to invade Iraq. But all of them were rooted in an increased feeling of vulnerability produced by the 9/11 attacks; in that sense, although many factors contributed

2009 | **2010** | **2011** | **2012**

January 22, 2009
Obama signs an excecutive order calling for the closure of the U.S. detention facility in Guantánamo Bay, Cuba.

May 1, 2010
Faisal Shahzad attempts to detonate a car bomb in New York's Times Square.

May 2, 2011
U.S. forces kill Osama bin Laden at his compound in Abbottabad, Pakistan.

September 11, 2012
Militants attact the U.S. diplomatic mission in Benghazi, Libya, killing U.S. Ambassador Christopher Stevens and three other Americans.

to the invasion, it must be considered a centerpiece of Bush's "war on terror"—and it was the element of Bush's counterterrorism policy to which Obama most strongly objected.

Obama was elected with a mandate to end the war in Iraq and bring the troops home. During his campaign for the White House in 2008, Obama described Iraqi Prime Minister Nouri al-Maliki's request for a timetable for the withdrawal of U.S. troops from his country as "an enormous opportunity" that would enhance the prospects for "long-term success in Iraq and the security interests of the United States." In 2010, when he announced the end of U.S. combat operations in Iraq, Obama declared that "ending this war is not only in Iraq's interest—it's in our own."

But four years later, as Iraqi cities fell to ISIS, the administration and its defenders argued that the removal of U.S. troops had not really been Obama's decision to make. Maliki, they insisted, had refused to provide immunity for any U.S. troops who stayed in Iraq after the expiration of the status-of-forces agreement that Bush and Maliki had agreed to years earlier. There was some truth to that claim, but it was also true that Obama hadn't pressed Maliki very hard on the issue. And most damaging of all, Obama had abruptly reduced the level of diplomatic engagement between Iraq and the United States, leaving Sunnis feeling isolated and vulnerable to Maliki's overtly anti-Sunni sectarian regime.

Drones, Loans, and Phones

Although many of Obama's counterterrorism choices were framed as corrective responses to Bush's missteps, the administration also had its own vision of how to combat the threat, and it's worth considering the three main tools it has relied on.

First and foremost among these are armed drones. Unmanned aerial vehicles, as they are technically known, are significantly more discriminating than any other weapon fired from afar. That accuracy is one reason Obama has come to rely so heavily on them. But they are still imperfect. Their targeting is entirely dependent on the quality of the intelligence available to the pilots, and it is not possible to completely avoid civilian casualties. Still, according to figures collected from open sources and published by the think tank New America, among others, the accuracy of U.S. drones has improved over time; the amount of collateral damage they cause has decreased.

One legitimate concern raised by critics is that news coverage of drone attacks might help terrorists find new recruits. The use of drones to target suspected Al-Qaeda operatives in Yemen has been correlated with a rapid growth in membership in the group's Yemen-based affiliate. Some have argued that the drone attacks themselves have caused this rise; others, such as the political scientist Christopher Swift, suggest that the group has attracted "idle teenagers" not by stoking anger over drones but by offering relatively generous salaries, as well as cars, khat, and rifles.

It is certainly possible that drone strikes could inspire terrorist strikes on U.S. soil. Faisal Shahzad, who tried and failed to detonate a bomb in New York City's Times Square in 2010, reportedly claimed he acted to avenge a 2009drone strike that killed Baitullah Mehsud, the leader of the Pakistani Taliban. But I

June 5, 2013
The Guardian publishes the first of many revelations about the National Security Agency's surveillance.

August 8, 2014
Warplanes conduct the first U.S. air strikes against ISIS militants in northern Iraq.

February 17, 2015
The White House's Summit on Countering Violent Extremism begins.

2013 **2014** **2015**

June 29, 2014
ISIS, having captured a number of Iraqi and Syrian cities in the preceding months, declares a caliphate.

October 11, 2014
The United States Carries out its 400th drone strike in Pakistan, the 349th under Obama.

December 9, 2014
The U.S. Senate releases a summary of its investigation of the CIA's torture of detainees.

have interviewed terrorists for some 15 years, and I've found that rather than a single source of motivation, there are invariably a combination of factors—emotional, social, financial, ideological—that push people to engage in terrorist violence.

Drones are a terrifying instrument of war. They sometimes cause the deaths of innocents. There is something that feels not quite right about a weapon whose use entails no direct physical risk to the user. And although most Americans approve of the use of drones in counterterrorism operations, if drones were to someday target U.S. government officials or American citizens themselves, such opinions would quickly shift. But for now, drones are the least bad of a number of bad options for targeting high-level terrorists.

Obama has also relied extensively on other governments to supply ground forces to fight terrorist groups abroad; this represents a second major pillar in his strategy. The policy has obvious appeal: if the United States cares more about the threat than local authorities do, U.S. interventions are unlikely to succeed in the long run. But this policy, too, is fraught with risk and can lead to significant blowback. Critics argue that it is hard to identify potential enemies among the forces Washington trains: consider the many "green on blue" attacks that have taken place in Afghanistan in the past dozen years, in which Afghan soldiers or police officers have killed members of the coalition forces tasked with training them. In Syria, where the Obama administration is not partnering with the government in Damascus but instead hopes to train rebel forces to fight ISIS, U.S. officials have identified only 60 volunteers who have the "right mindset and ideology," according to U.S. Secretary of Defense Ashton Carter. Similar efforts in Iraq have also been slowed by a lack of acceptable recruits. Whatever the virtues of this policy, it will not work if Washington cannot identify suitable candidates.

The third and final main element of Obama's counterterrorism approach is a reliance on intensive electronic surveillance. Digital communication is far more widespread, and far more vulnerable to exploitation, than it was when Obama was elected, and government surveillance of communications has expanded dramatically under his watch, as the former National Security Agency contractor Edward Snowden revealed in 2013 by leaking enormous amounts of classified information about the NSA's operations. Opposition to these activities—especially the NSA's collection of metadata on all Americans' phone calls from the public, major Silicon Valley firms, and U.S. allies has resulted in the curtailment of some of the NSA's most aggressive techniques. But surveillance is an essential counterterrorism tool. It is less likely to result in the loss of innocent lives than most other counterterrorism tactics; indeed, it limits collateral damage by improving intelligence. And because it doesn't target Muslims in particular, it doesn't play into the jihadist narrative that the United States is engaged in a war against Islam. Looking forward, cyberterrorism and cyberwar will likely pose a more serious threat to Americans' well-being than the conventional terrorist violence, and government surveillance is and will remain an essential weapon against cyberattacks.

The Containment Store

The Obama administration's combination of drone strikes, security assistance to U.S. partners and allies, and aggressive surveillance has undoubtedly helped protect Americans. The core Al-Qaeda organization has been greatly degraded, and there have been no major attacks on U.S. soil. Obama also deserves credit for launching the risky 2011 raid in Pakistan that eliminated Osama bin Laden. But there is also no question that on Obama's watch, the global threat of jihadist terrorism has grown more acute, owing mostly to the rise of ISIS, a hybrid organization that combines elements of a proto-state, a millenarian cult, an organized crime ring, and an insurgent army led by highly skilled former Baathist military and intelligence personnel.

No Salafi jihadist organization, not even ISIS, poses an existential threat to the United States. Nor, in recent years, have Salafi jihadists posed the most direct terrorist threat to individual American citizens. Indeed, white supremacists and far-right extremists have committed nearly twice as many terrorist murders in the United States as have jihadists in the years since the 9/11attacks. But that narrow measure of the threat fails to capture the unique danger posed by Salafi jihadism: it is the only extremist ideology able to attract large numbers of committed fighters around the world, and it motivates ISIS, the only extremist organization able to threaten the stability of states and the regional order in the Middle East. In addition to the territory, the group now controls in Iraq and Syria, and its affiliates have established "provinces" in Egypt, Libya, and Yemen, among other places. ISIS is threatening many U.S. allies and inspiring or directing an unknown number of followers to act beyond the territory it controls. Its ultimate goal—a pipe dream, one hopes—is to destabilize and eventually take over Saudi Arabia, which would have profound consequences not only for the region but also for the world.

Until recently, Obama consistently underestimated the strength and international appeal of ISIS, which in early 2014 he infamously likened to a junior varsity basketball team in comparison to Al-Qaeda's professional squad. Even after ISIS had marched across Iraq and Syria and seized territory equal to the land area of the United Kingdom, Obama referred to it as "a terrorist organization, pure and simple" and promised to "degrade and ultimately destroy" the group—an impossible goal, especially given his claim that no ground forces would be required.

Given that Obama's preferred approach failed to prevent the rise of ISIS, it's fair to ask whether the updated strategy he put in place in reaction to the group's breathtaking advance will fare any better. ISIS is a totalitarian regime, and Washington's goal should be to contain it in much the same way the United States has other totalitarian regimes. And despite the White House' stalk of degrading, defeating, and destroying ISIS, Obama's strategy is really one of containment: air strikes, training and equipping some of ISIS' adversaries in Iraq and Syria, and bolstering efforts to stop the flow of fighters into and commodities out of the territory ISIS controls.

But even this more limited anti-ISIS strategy has been hard to execute. Money, goods, and personnel are still getting into and out of ISIS-controlled territory. A 2015 UN Security Council report concluded that 22,000 foreign fighters have made their way to Iraq and Syria to join jihadist groups. According to U.S. intelligence officials, approximately 3,400 of them have come from Europe and the United States.

And perhaps most troubling, ISIS ideology continues to spread, largely due to the group's impressive use of social media. Indeed, the most direct threat ISIS poses to the United States, at least for now, appears to come from people already in the United States who might become radicalized through their online contact with ISIS supporters or recruiters based throughout the world. Combating the spread of extremist ideologies and preventing recruitment at home and abroad have thus emerged as the most important elements of U.S. counterterrorism.

Winning the War of Ideas

Obama's effort to do just that represents perhaps the single biggest change the president has effected in U.S. counterterrorism—although it is still more an aspirational ideal than a fully implemented policy. The Bush administration framed the promotion of electoral democracy as the best way to defeat extremism. But that policy was destined to fail in the short term: nascent democracies often drift toward majoritarian rule, disenfranchising minority groups and creating fertile ground for extremist movements. In place of Bush's aggressive democracy promotion, the Obama administration has focused on addressing the underlying conditions that make certain individuals and communities ripe for recruitment. In February, the president convened what he called the White House Summit on Countering Violent Extremism and laid out what amounted to a three-part plan: discredit terrorist ideologies, address the political and economic grievances that terrorists exploit, and improve governance in the regions where groups such as ISIS recruit. The aim, he said, was to stop merely reacting to extremism and instead try to prevent it from spreading, by creating jobs for young people who might otherwise be susceptible to recruitment, fighting the corruption that impedes development, and promoting education, especially for girls.

Poverty and lack of education, in and of themselves, do not cause terrorism. But terrorist groups exploit failed governance in places where governments routinely violate human rights; when people don't feel safe, they sometimes conclude that a terrorist group is more likely to protect them than their government. "We can't keep on thinking about counterterrorism and security as entirely separate from diplomacy, development, education, all these things that are considered soft but in fact are vital to our national security—and we do not fund those," the president said in March.

The point is valid. But it's worth noting that, months later, it is still not clear how these preventive strategies will be funded or implemented. Nor is it clear just how such a program would break the vicious cycle in which autocratic rule encourages extremist violence, which in turn produces harsh government crackdowns, which leads to more extremism. An even deeper problem, the political scientist and terrorism expert Daniel Byman has pointed out, is that there is no single pathway to violent extremism. "It varies by country, by historical period and by person," By man has written.

Obama administration officials are hardly unaware of these complexities, and challenges and have engaged in a tug of war familiar from many previous administrations. On one side are those who say that the threat from extremists dictates that military cooperation with partners and allies take precedence over other policy options, such as promoting better governance. On the other side stand those who want U.S. policy to focus more squarely on addressing what they believe are the underlying causes of extremism's spread. As Tamara Cofman Wittes, who served as deputy assistant secretary of state for Near Eastern affairs from 2009 until 2012, put it to me: "Our policy rhetoric regularly acknowledges that extremists thrive on grievances and disorder driven by failures of governance, but our policy practice avoids addressing governance for fear of disrupting short-term security goals." And indeed, arguments in favor of more military action and aid tend to carry the day in the Obama White House. Wittes also pointed out that ever since the deadly jihadist assault on U.S. facilities in Benghazi, Libya, in 2012—which led the administration to prioritize the protection of diplomatic personnel—it has become even more difficult for diplomats to engage with local officials, politicians, and activists who are working to foster improved governance and the protection of minority rights.

In trying to erode the appeal of extremist ideology, the administration has sought to amplify the voices of people who can credibly counter jihadist ideas, including Islamic scholars and Muslim clerics from all over the world and "formers"—individuals who have abandoned jihadist organizations and can provide a more accurate picture of the jihadist way of life, which rarely lives up to the romantic image of heroic resistance that groups such as ISIS peddle. But governments—especially the U.S. government—are inherently limited in what they can achieve in this regard; they are hardly the most credible brokers for messages of this kind. And although leaders in Muslim communities have more standing to push back against extremism, boring speeches by learned and respected Islamic scholars are unlikely to change the minds of the young people attracted by ISIS and similar groups. What is needed is more involvement from the private sector: entertainment, Internet, and media companies know how to appeal to younger audiences and could play a much larger role in crafting counter narratives to fight ISIS, bringing to bear their considerable expertise in market research and messaging.

The Limits of Change

Overall, Obama's approach to counterterrorism has been a step in the right direction. The next U.S. president would do well to view the combination of targeted killing, security assistance, and intensive surveillance as a relatively effective, low-risk tool kit, and he or she should also continue to experiment with preventive policies, which potentially represent the best way to combat jihadism in the long term. Violent Islamist extremism cannot be defeated through force, but neither can it be addressed by soft power alone. The threat is constantly evolving, and it requires a constantly evolving response. If nothing else, one lesson the next president should learn from the Obama years is to resist the temptation to change counterterrorism policy solely for the sake of change, or to help differentiate himself or herself from the previous occupant in the White House. In the fight against terrorism, as Obama discovered, Washington's room to maneuver is constrained by the dynamics of terrorist violence, the persistent appeal of extremist ideas, and the limits of state power in confronting the complex social and political movements such ideas foster.

Critical Thinking

1. Why has the Obama administration underestimated ISIS?
2. How can the next U.S. president effectively combat ISIS?
3. What are the advantages and disadvantages of drone strikes?

Internet References

Department of Homeland Security
www.dhs.gov/

U.S. State Department, Bureau of Counter-Terrorism
www.state.gov/j/ct/

Jessica Stern is a lecturer in Government at Harvard University and a member of the Hoover Institution's Task Force on National Security and Law. She is a coauthor, with J. M. Berger, of *ISIS: The State of Terror*. Follow heron Twitter @JessicaEStern.

Article

Prepared by: Robert Weiner, *University of Massachusetts, Boston*

Fixing Fragile States

DENNIS BLAIR ET AL.

Learning Outcomes

After reading this article, you will be able to:

- Talk about the operations of Al Qaeda in Yemen.
- Discuss how terrorist organizations can establish themselves in fragile states.
- Understand the role of bureaucratic politics in implementing U.S. foreign policy.

Since the 9/11 attacks, the United States has waged major postwar reconstruction campaigns in Iraq and Afghanistan and similar but smaller programs in other countries that harbor Al Qaeda affiliates. Continued complex political, economic and military operations will be needed for many years to deal with the continuing threat from Al Qaeda and its associated organizations, much of it stemming from fragile states with weak institutions, high rates of poverty and deep ethnic, religious or tribal divisions. Despite 13 years of experience—and innumerable opportunities to learn lessons from both successes and mistakes—there have been few significant changes in our cumbersome, inefficient and ineffective approach to interagency operations in the field.

We believe the time has come to look to a new, more effective operational model. For fragile states in which Al Qaeda is present, the United States should develop, select and support with strong staff a new type of ambassador with more authority to plan and direct complex operations across department and agency lines, and who will be accountable for their success or failure. We need to develop the plans to protect American interests and strengthen these countries out in the field, where local realities are understood, before Washington agencies bring their inside-the-Beltway perspectives to bear. Congress and the executive branch need to authorize field leaders to shift resources across agency lines to meet new threats. It is, in short, a time for change—change that upends our complacent and antiquated approach toward foreign societies and cultures.

The 9/11 attacks offered us a painful reminder of an old verity, which is that fragile states unable to enforce their laws and control their territory are the progenitors of potent threats that can be carried out simply and effectively. Such states provide safe havens from which Al Qaeda and its affiliates plan and launch terror attacks against the United States and other countries. Al Qaeda in the Arabian Peninsula (AQAP) operates in Yemen; Al Qaeda in the Islamic Maghreb operates across Algeria, Mali and other neighboring countries; and Al Shabab operates in Somalia. Civil war in Syria, spreading violence in Iraq and continued turmoil in Africa will most likely open new havens for similar groups.

Until now, the American response to the threat from fragile states has had three major components. First, we have greatly strengthened the control of our own borders. Second, American intelligence and military forces, particularly the CIA and the U.S. Special Operations Command, have taken the fight to Al Qaeda. Third, the United States, along with other countries and international organizations, has increased economic and civil assistance to many fragile states using existing programs and authorities.

How much have these approaches achieved? The American-led reconstruction efforts in Iraq and Afghanistan have been prolonged and massive, but cannot be considered successful.

A dysfunctional system of authorities and procedures hampered effectiveness. Plans were made in Washington by committees of the representatives of different departments and agencies; individual departments and agencies sent instructions to their representatives in the field; and the allocations of resources to country programs were based in large part on individual departmental and agency priorities and available funding, not on overall national priorities. Each of the departments maintained direct authority over its field personnel and resources. Short-term staffing was endemic and cooperation in the field was voluntary, with neither the ambassador nor any official in Washington below the president authorized to resolve disputes or set overall priorities. Budget resources for a particular program could not be shifted smoothly to others when local conditions changed, and congressional oversight was split among committees that oversaw only individual aspects of the overall program in a country.

Even when the president, the National Security Council and an energetic interagency process in Washington were fully engaged—as they were in later years in Iraq and Afghanistan—the results have not matched the commitment of resources. Numerous accounts by journalists and memoirs of participants have documented the interdepartmental frictions, inefficient bureaucratic compromises and delayed decisions that have hampered progress. The authors of this article know personally most of those involved in leading the long wars in Iraq and Afghanistan. They are to a person—whether military officers or civilian officials—diligent and dedicated patriots. They have often worked across departmental lines to integrate security, governance and economic-assistance programs to achieve real successes. However, when officials and officers in the field did not get along, the deficiencies of the system allowed their disputes to bring in-country progress to a halt. What is needed is an overall system that will make cooperation and integration the norm, not the exception.

Yemen and Libya provide smaller-scale but more contemporary illustrations of the shortcomings of today's approaches. Although American officials have gained more experience, the authorities and procedures have not changed.

Yemen is the home base for AQAP, generally considered the most dangerous franchise of Al Qaeda. The speeches of American officials paint a picture of a comprehensive, balanced set of U.S. government programs not only to attack AQAP, but also to assist the current Yemeni government with both political and economic development. In congressional testimony in November 2013, for example, Deputy Assistant Secretary of State for the Arabian Peninsula Barbara Leaf emphasized American "support for Yemen's historic transition and continued bilateral security cooperation." She mentioned the $39 million that the United States had provided to support the national reconciliation process, U.S. encouragement of economic reform, its support for restructuring the Yemeni armed forces, and its participation in a weekly meeting among outside countries and international organizations to "compare notes, compare approaches, and coordinate tightly." She said nothing of the American military and intelligence attacks on AQAP fighters, yet these actions are the most costly U.S. programs dealing with Yemen, and they feature prominently in Yemeni popular opinion.

Even in this friendly hearing, however, the shortcomings of American and international programs were made clear. Congressman Ted Deutch noted, "U.S. assistance to Yemen totaled $256 million for Fiscal Year 2013, but these funds come from 17 different accounts, all with very different objectives." He asked a fundamental question that went unanswered: "What exactly is our long-term strategy for Yemen?"

The view on the ground in Yemen is considerably darker. Two weeks before Leaf's testimony, an op-ed in the Yemen Post under the headline "Law of the Jungle in Yemen" stated:

> People have lost hope in the National Dialogue. . . . Billions of US dollars are still looted in the poverty stricken Yemen with not one corrupt senior official prosecuted. . . . Laws are only practiced against the weak and helpless. . . . An internal war is ongoing in the north of Yemen. . . . Al-Qaeda is regrouping and seeking to become a power once again . . . Safety and security in Yemen is nowhere to be seen. Government authority and presence over many parts of the country is limited, and where they are present they are almost useless.

U.S. policy in Yemen has been cobbled together in Washington through the typical interagency process. Because congressional funding for counterterrorist programs, both military and intelligence, is still flowing relatively freely, they are the largest American programs in Yemen. According to press reporting, there are two independent task forces—one military, one CÍA—operating drones over Yemen, and U.S. security assistance to the Yemeni armed forces is focused on the creation of small, well-trained counterterrorist forces. The Saudis and other Gulf Cooperation Council (GCC) states have promised over $3 billion in economic support to Yemen. American economic assistance to Yemen is a small fraction of this amount; thus, the American plan must leverage the greater GCC contribution. The overall picture in Yemen, then, is one of unbalanced, uncoordinated and suboptimal U.S. and international programs based on no coherent plan.

Libya is another excellent illustration of an American assistance program that is not meeting the needs of the country. American commitment of financial resources to assist Libya has been modest: the State Department estimates about $240 million since the beginning of operations to oust Muammar el-Qaddafi in 2011. Far more money was spent by the United States on the NATO air operations that pushed Qaddafi out of power. American assistance to Libya has been spread across different government programs, depending on the other bills for those programs in the rest of the world. With few resources at their command, the country team needed an integrated plan to make the actions they could take effective, to set priorities and to leverage the actions of other countries. Yet the various American agencies working in Libya, as usual, cooperated as best they could, under no integrated plan, with little experienced leadership either in

Tripoli or Washington. As crises occurred and conditions deteriorated, responses were improvised.

Like Iraq in 2003, Libya was coming out of a long and brutal dictatorship. Rebuilding the country would require extraordinary actions by the Libyans themselves and by outside countries like the United States that had helped bring down Qaddafi and had a stake in a favorable outcome. International security-support programs, including those by the United States, have been notably weak. Any doubt about the conditions on the ground ended when Ambassador J. Christopher Stevens was murdered in Benghazi in September 2012, the first U.S. ambassador to die in the line of duty since 1979. Yet it was over another year later that the United States and several other European countries began belatedly to take actions to strengthen the army and police. NATO began a program to train about twenty thousand Libyan soldiers. The program is not scheduled to be completed for many more months, and the result will be trained soldiers who perform their duties with mostly inadequate medical, communications and logistical support.

Yet this belated program to strengthen basic security in Libya still is not part of an overall plan to help Libya become a competent, functioning state. According to two experts at the Atlantic Council and the European Council on Foreign Relations, respectively:

The current western agenda for Libya lacks a political strategy and is focused almost exclusively on the training of the Libyan army. If experience elsewhere is an indication, it will take between 5 and 13 years for that to conclude. The same experience tells us that "strengthening the central government" is an insufficient goal if the country is to become stable and under the rule of law.

Meanwhile, the official U.S. activities in Libya, as described by the current U.S. ambassador, Deborah K. Jones, are directed toward "a broad process encompassing a National Dialogue, constitutional development, and governance capacity-building to increase public confidence." In American pronouncements, there is little sense of priorities, combined programs, milestones or urgency.

The embassies of the United States and its international partners in these fragile countries must do more than just be supportive of individual areas needing improvement. Their approach has to be selective, hands-on, tailored, flexible and integrated.

Selective: Resources are limited, and the approach needs to be sustained over an extended period of time. It should be applied only to the handful of countries in which the threat is high and host government capacity is low.

Hands-on: The United States and other international partners cannot simply transfer money to government departments in fragile states, as it will likely be stolen or misused. Instead, they must take an active role in building competent local government organizations that can use increased resources effectively. American and other international operators cannot train local organizations to be replicas of their Western equivalents, or models of counterparts in other countries; conditions are too different. Likewise, they cannot simply fly in for a two-week stint and then head home; there will be no follow-through. Experienced, carefully selected and trained officials who can influence host officials and build local capacity without causing resentment are essential.

Tailored: Sometimes existing security or law-enforcement organizations or judicial systems can be strengthened; other times they must be created. Sometimes putting the national finances of a country in order will unleash economic growth; other times training and economic support in a particular region of the country are vital. Sometimes training and assisting central government officials is important; other times it is competent provincial officials that are essential for success. The key to a tailored approach is for the American representatives in a country to have the authority and responsibility both for planning and for carrying out the plan.

Flexible: Requirements are always dynamic in fragile states. Plans need to be revised quickly in response to events on the ground; money and personnel need to be shifted quickly to meet new problems and to take advantage of new opportunities.

Integrated: Integration depends on setting a common set of priorities across all programs. Once security forces stabilize a city or region, improved governance and economic opportunity must follow immediately, or the security gains will be wasted. Integration depends on realistic sequencing of different programs. Unless the judiciary and prison systems are improved along with police forces, criminals will be released or tortured after their arrest. Policemen can be trained or retrained in weeks and prison systems can be improved quickly, yet training a core of judges and lawyers takes years. There must be practical interim plans that will ensure progress.

Finally, current operations to capture or kill hardcore Al Qaeda members need to continue, without stirring up local resentment that will make it more difficult to make the necessary longer-term improvements. However, these operations need to be consolidated and integrated into an overall plan in each country.

There is duplication, overlap and sometimes competition between the traditional military operations of the Department of Defense and the covert paramilitary operations of the CIA against Al Qaeda. To fully understand the issue, it's important to be clear about the significant difference between clandestine and covert operations. A "clandestine" operation is one that is secret, and no government official is to talk about it. Clandestine operations are routinely conducted by the Department of Defense, and on occasion by other U.S. government agencies. A "covert" operation is one in which the involvement of the

U.S. government is to be kept secret, to the point of official denial. The CIA has generally conducted covert operations, and an executive order gives this preference, but the basic legislation authorizes them to be conducted by other departments or agencies as directed by the president.

Although geopolitical conditions have changed fundamentally since the Cold War, when covert operations were originally authorized, there has been no serious consideration of updating the authorities for covert action. The 9/11 Commission's recommendation to assign paramilitary operations to the Department of Defense was not adopted either by Congress or by two successive administrations. The result has been continued complicated, duplicative and costly operations against Al Qaeda. It has only been experienced, dedicated and mission-focused operators in the field that have permitted the current system to work, and their successes have obscured the need for clarity and simplification. This recommendation should be seriously revisited based on our experiences of the last decade.

Two types of armed operations against Al Qaeda are the most important: raids and armed drone strikes. For raids—the helicopter raid that killed Osama bin Laden in Abbottabad is the best known—all the operational skills are in the Department of Defense, mostly within components of the Special Operations Command. Yet, there are often questions and disputes about whether they should be conducted as clandestine traditional military operations commanded by the secretary of defense under Title 10, or covert intelligence operations controlled by the director of the CIA under Title 50.

In reality, the president has the legal authority to order these operations under either title, using either organization. In 2011, the president decided to authorize the Abbottabad raid, entirely conducted by Department of Defense personnel, as a Title 50 covert action, under the control of the director of the CIA. It was a "clandestine military operation" conducted under authorities that were designated for "covert action." There was never any reason or intention to deny the role of the U.S. government—the primary rationale for covert action—once the operation commenced and inevitably became public. To the contrary, government officials were running for the microphones as soon as the helicopters returned from Pakistani airspace. Fortunately, experienced military commanders made all the tactical decisions, and the raid was a success. Had anything gone wrong—the loss of a helicopter and the capture of its crew by Pakistan, or a dispute between CIA officers and special-operations officers during the raid, with each group appealing to its own chain of command—the results could have been quite different.

For armed attacks by drones, the CIA and the Department of Defense have set up duplicate organizations, each authorized by separate legislation. There are reasons for the current arrangements. The bottom line, however, is that it is the Department of Defense that is established, trained and authorized to kill enemy combatants. For reasons of competence, accountability and effectiveness, the armed drone campaign should be assigned to the secretary of defense, with the entire intelligence community, including the CIA, playing an essential role in identifying, prioritizing and tracking the targets.

A new model for interagency operations in fragile states would be strongest and longest lasting if it were established by legislation. However, much can be done by executive order, policy and practice.

The foundational process change should be to assign the task of developing a comprehensive plan for a fragile state to the team on the ground in that state, rather than to an interagency group in Washington. It is axiomatic in both business and military planning that a plan ought to be drafted by those responsible for carrying it out. Only in American interagency planning is it done by a committee at headquarters, then passed to the field for implementation. Washington's plans are subject to pressures that often make them unrealistic and unsuitable for conditions in the field. An in-country planning team is much more likely to deliver a plan that is balanced between the short and the long term, that includes the most effective applications of the capabilities of the different departments and that realistically matches the needs on the ground. During interagency review in Washington, there will be plenty of opportunity for adding other considerations and good ideas.

However, for an embassy to submit a good plan takes a uniquely qualified and experienced ambassador with a dedicated, competent supporting interagency staff, in addition to the usual country team, comprising the representatives from the various departments and agencies.

Foreign Service officers spend most of their careers in staff positions, responsible for observing, reporting, negotiating, and making policy recommendations that are heavily weighted toward the short term and tactical. Their career pattern develops a high level of expertise, observational and writing skills, and diplomatic abilities. The leader of American in-country operations in a fragile state needs high-order managerial and leadership skills for complex program execution as well as a deep knowledge of the capabilities and limitations of other American organizations, especially military and intelligence. Some Foreign Service officers who became ambassadors have developed these skills. James Jeffrey, Ryan Crocker and Anne Patterson are among several in the recent past. However, although such training has been recommended, the Foreign Service is not geared toward producing such skills broadly. A qualification-and-selection process is needed for ambassadors to places like Yemen, Libya, Pakistan, Mali, Somalia, Afghanistan and Iraq to identify candidates with the experience, knowledge and stature to direct an integrated, multiagency task force.

The current manning of embassies does not include a central staff to support an ambassador in designing and implementing a country plan. What is needed is a small, separate staff of perhaps a dozen experienced officers, drawn from different agencies, to help the ambassador formulate the plan, and then to monitor its execution to determine if it is achieving its objectives and recommend adjustments as circumstances on the ground change. While maintaining strong links back to their parent organizations for advice, support and guidance, these staff officers would primarily serve the ambassador in developing and coordinating his or her plans. Such help is beginning to be available from the State Department's Bureau of Conflict and Stabilization Operations, but this falls short of an integrated interagency effort.

Within the overall integrated plan in a fragile state, the ambassador should recommend the military and intelligence actions to be taken directly against Al Qaeda personnel and units. The country plan must establish the priority and scope of these activities within the overall mission of strengthening security. It needs to define the areas in which the raids and drone strikes will be conducted and the intensity of the campaign. The overall objective is to capture or kill more enemies than are created. The ambassador should recommend whether these actions be taken as military activities under Title 10 or intelligence activities under Title 50. With special-operations forces and CIA planners as part of his team, the ambassador is in the best position to recommend both the actions themselves and the most appropriate authorities under which to conduct them. During the course of the campaign, the ambassador needs the authority to approve direct actions—drone strikes as well as raids and conventional military strikes—to ensure that they are integrated into the overall plan.

When the ambassador has formulated an integrated plan for the country, incorporating diplomatic, economic, intelligence, military and other aspects, including milestones that the plan will achieve on specified dates, it should be sent to Washington for interagency comment and for the allocation of resources—people and money. The Office of Management and Budget should participate at all levels of interagency review to ensure that budget plans are realistic. Ultimately, a resourced, comprehensive plan for a fragile state should be approved by the president.

Virtually every fragile state both affects and is affected by its neighbors. Tribal, ethnic and religious influences cross national boundaries; borders are often porous; pressuring groups in one country pushes them into others. The cooperation of neighboring states is thus essential to success within fragile states. To ensure that these factors are considered to obtain regional buy-in, each country plan should be sent to neighboring embassies (in the case of the State Department and other agencies without regional organizations) and to regional and global combatant commands (in the case of the Department of Defense) for review and comment.

No plan survives first contact with the enemy; success in the implementation of a plan depends on flexibility and adaptation. Yet currently those carrying out military, economic, diplomatic and other programs in fragile states have very little authority or capability to react. A change in one aspect of a plan will always cause changes in other aspects, yet because authorities in the American national-security system pass directly from departments and agencies to representatives in the field, it is very difficult to gain approval for necessary adjustments. Congressional oversight, based on jurisdiction over appropriated budgets, further hinders flexibility. Economic-development programs depend on successful security operations, yet there is no authority in a country that can direct adjustments when setbacks in one area require changes in another. No financial or personnel reserves are available to cover unexpected problems or to take advantage of surprise opportunities—the budget incentive is "use it or lose it," whether or not a program is effective, or whether or not the money could be used more effectively elsewhere. Again, dedicated, hardworking officers and officials cooperate with each other as best they can, but the current system does not support flexibility.

The solution is to give the ambassador both the responsibility for overall progress on the plan and directive authority over the programs in country within the limits of the plan that was approved. Once budgets have been allocated to U.S. programs to strengthen a fragile state, an ambassador should be able to shift them, within realistic thresholds, as needs and opportunities develop.

These reforms will go a long way toward improving American support for fragile states and dealing with Al Qaeda groups that find refuge in them. However, additional improvements are needed.

It is only the Department of Defense that has either the authority or the tradition of assigning personnel to difficult overseas postings, whether they volunteer or not. All other agencies rely on volunteers. The result has been chronic shortchanging of the nonmilitary billets in fragile states—short assignments for officers, or the use of contractors. Authority must be granted to department and agency heads to assign their personnel as needed to support the national interest. Without this change, American campaigns in these countries will be unbalanced and heavily influenced by military considerations, since it is the military personnel who show up.

Although the Department of Defense has the authority to send personnel overseas as needed, some key skills for assisting fragile states have deteriorated within the military services in recent years. In the past, there were experienced civil engineers, utility company officials, local government administrators and transportation officials in the Army and Marine Corps Reserve. They had the skills and experience to help establish competent organizations to provide basic infrastructure in fragile states. The civil-affairs personnel in military reserve units are more junior and much less experienced now. The contractors and individually mobilized reservists that are now used to assist struggling government organizations in fragile states are inadequate for the importance and difficulty of the need. The Reserve Components of the Army and Marine Corps must reestablish strong civilian-affairs components.

Successive secretaries of state in recent administrations have made strong attempts to improve the numbers and qualifications of civilian officials sent to fragile states. Continued emphasis is needed, as the State Department still has difficulty filling even established billets in Afghanistan, language skills do not meet existing requirements in fragile states, and there are too many short-term assignments of personnel to jobs that require sustained interaction with local officials to build trust. The State Department must continue to develop a cadre of officers who can be effective in the tough tasks of strengthening fragile states.

In virtually every fragile state, some of the weakest institutions are the police, courts and prisons. The U.S. government has very little capacity to help strengthen them. The Department of Justice has only a limited training-and-advisory capacity similar to that in the Department of Defense or the Department of State, and generally requires outside funding to mount training programs. Other countries have some capacity, but retired state and local police officers, private contractors, or volunteer judges and prison officials man the American assistance programs. In Afghanistan, the mission of police training was assigned to the Department of Defense, despite the fundamental differences between the military and law-enforcement missions. We need to develop a cadre of advisers and trainers for police, courts and prisons, and a means to supplement the cadre with qualified and supervised private volunteers.

Finally, Congress will need to establish new oversight procedures for an integrated country strategy, rather than the disjointed current system in which generals testify in front of one committee, ambassadors in front of another, and no executive branch official below the president has the responsibility for overall success or failure to strengthen a fragile state in which the United States has important interests.

Countries with weak governments, high levels of poverty, and internal ethnic, religious and tribal tensions that provide sanctuaries for Al Qaeda or its affiliates will remain a perennial source of instability and threats for America and its allies. Most of the discussion of the challenges of dealing with fragile states has been dominated by abstract debates over vital American interests, fears of long-term commitments, sterile arguments over military versus civil components, disagreements over deadlines and often ill-informed applications of the perceived lessons of the U.S. experience in one country to another. This is unfortunate. What has been missing from the discussion is an understanding of the very segmented, rigid and inefficient system under which the United States attempts to help these countries stabilize their governments and societies and control outside terrorist groups. The current system guarantees that the resources—people and dollars—that are allocated to these countries do not produce the results they could and should. The United States should likely provide more funding for its programs in countries like Yemen, Libya and Mali. However, what is even more important is for the Obama administration and Congress to improve the basic system for organizing and conducting these programs. The improvements in authorities and procedures that we recommend here will go a long way toward making America safer with a very small expenditure of additional resources. It's time to replace decades of failure with a new approach that protects American security by transforming fragile states into genuinely secure ones.

Critical Thinking

1. Why has the United States failed to bring peace and stability to Yemen?
2. How can U.S. diplomacy stabilize Libya?
3. Should drones be used in armed attacks against Al Qaeda?

Internet References

Fragile States Index
fsi.fundforpeace.org

Libya
https://www.cia.gov/library/publications/the-world-factbook/geos/ly.html

Yemen
https://www.cia.gov/library/publications/the-world-factbook/geos/ym.html

Dennis Blair is the former Director of National Intelligence and former Commander in Chief of U.S. Pacific Command. Ronald Neumann is president of the American Academy of Diplomacy and former U.S. Ambassador to Algeria, Bahrain and Afghanistan. Eric Olson is the former Commander of U.S. Special Operations Command.

Unit 5

UNIT

Prepared by: Robert Weiner, *University of Massachusetts, Boston*

Conflict and Peace

There is no single cause of war, as wars in any event have multiple causes. For example, wars can be about scarce resources, such as water, especially because the world's water supply is not endless, and a small number of countries actually possess the bulk of the world's water supply. Political scientists argue that the causes of war can be found at different levels of analysis, ranging from the individual level, to the domestic level (regime type and system), to the international level (interstate relations), and to the global level (international communications such as cyberwar) and international terrorism. Recent empirical studies, conducted by credible research institutions, have concluded that there has been a decline in the amount of interstate warfare in the international system, but an increase in internal or civil conflicts (sometimes with significant external intervention) since the end of the Cold War.

In 2014, the 100th anniversary of World War I was observed, as the legacy of the "Great War" is still being felt around the globe. Historians and political scientists are still debating the factors that led to World War I. At the level of the international system, the emergence of two rival alliance systems—the Entente and the Central Powers—may have been an important factor that contributed to the eruption of the war. The outbreak of World War I shattered the "Long Peace" (1815–1914) that had prevailed in Europe since the end of the Napoleonic Wars. This century of peace has been seen as a period of "golden diplomacy" in which the maintenance of a finely calibrated balance of power preserved stability in Europe.

World War I was called the Great War because no one could ever imagine that another bout of such atavistic bloodletting would happen until the occurrence of World War II. Some historians have viewed the Second World War as a continuation of World War I, based on a 20-year interlude between the two wars. Moreover, some respected experts on international relations believe that another Great War will never take place. Like Britain before it, the United States is a great trading state, which finds its sea power position in Asia challenged by China, which is a rising power. Beijing seems to be pursuing a Grand Strategy that is based on establishing itself as a regional hegemon in the South China and East China Seas, as well as projecting its power into Eurasia. Beijing also sees a strategic opportunity to project its air and naval power as the United States declines. The U.S. response to China's rise has been to pursue a "pivot" or rebalancing to Asia by moving some of its military assets from Europe and the Middle East to the Pacific region, thereby raising the level of tension with China. Japan plays an important strategic role in the rebalancing, as Washington has supported a revision of Japan's self-defense forces to develop a more offensive capability. As part of the overall strategy of rebalancing, Japan has signed on to the US free trade arrangement known as the Trans-Pacific Partnership (TPP). The TPP consists of 12 Pacific nations, including Canada and Mexico.

Ethnic and religious sectarian differences are also major causes of civil conflicts. For example, separatists in Eastern Ukraine consolidated their position. The Ukrainian situation settled into the pattern of a "frozen conflict," somewhat similar to other frozen conflicts that had prevailed in other parts of the former Soviet Union, such as Transnistria. The inability of the West to persuade the Russians to withdraw their support from the Ukrainian separatists also resulted in a rise in the level of tension between the Baltic States and Russia. The Baltic States also contain significant Russian minorities, viewed by Moscow as kith and kin. NATO's reaction was to reconfigure its strategy, to reassure the security concerns of the Baltics, Poland, as well as Finland and Sweden.

President Obama had campaigned on the theme of ending the U.S. involvement in the wars in the Middle East and Afghanistan. The United States had claimed to withdraw its combat forces from Iraq in 2014, but reengaged in the conflict in Iraq and found itself drawn into the civil conflict in Syria, which had begun in 2011. According to media reports, the conflict in Syria had cost about 500,006 lives by 2016. The US strategy in Iraq and Syria was complicated by the victories that were scored by what the United States viewed as an extremist group of Sunni jihadists, who were responsible for the videotaped executions of American and British citizens. The group was known in Arabic as Daesh, and in English as the Thalamic State in Iraq and the Levant (ISIL). ISIL had been able to gain control of a significant amount of territory in Iraq and Syria, including the Iraqi city of Mosul, which was an important oil center. The Islamic State had been able to take advantage of the power vacuum that was created in Syria by the civil war, and used the country as a sanctuary and base from which to expand its control of Iraqi territory, thereby putting Iraq once again on the brink of disaster. However, by 2016, the coalition forces fighting ISIL had launched an assault on Mosul.

The Obama administration made the decision to launch air strikes against the Islamic State, even though the air strikes had the effect of helping the Assad regime maintain its power.

However, the policy of the Obama administration was Iraq-centric, and designed to degrade and destroy the Islamic State. The Islamic State received some support from dissatisfied Sunni tribes which had been excluded from key power-sharing arrangements by the Maliki administration. In 2015, Moscow also decided to engage in a military intervention in Syria to support its client, President Assad. With both Russian and US aircraft operating in Syria, Washington expressed its concern about the Russian move, since Russia was supporting Assad, and the U.S. was trying to remove him.

The Obama administration also pursued a policy of trying to extricate the United States from the long-running war in Afghanistan, but with a commitment to remain into 2017 to aid the new Afghan government that followed the Karzai regime. U.S. efforts to withdraw from Afghanistan illustrated the difficulties associated with terminating a war in which there was no clear-cut winner. President Obama was not able to achieve his objective of terminating what seemed to be an endless war in Afghanistan. Opportunities for peace talks with the Taliban had been missed, as another complicating factor was added with the involvement of the Islamic State in the war in Afghanistan.

Philosophers and political scientists have worked on the problem of creating a system of universal peace as the underpinning of world order for centuries. The idea is to realize Kant's age-old dream of instituting a global system of perpetual peace as opposed to the Hobbesian system of a cruel world order based on perpetual warfare, where life is mean, nasty, short, and brutish, and consists of the war of each against all. Advocates of the possibility of a world order based on peace and justice have a much more optimistic view of human nature than classical realists, who seem to believe that human beings are inherently evil. Liberal internationalists especially believe that human beings are rational creatures, who find it in their interest to cooperate with each other. Human beings also have the capacity to work with each other, and construct a peaceful world society. Liberals believe that international institutions like the League of Nations and the United Nations can make a difference in preventing conflicts from occurring in the first place. The central problem of international relations is the reconciliation of order with justice, and liberals argue that international law and morality can contribute significantly to the creation of a peaceful world order as well.

Another central tenet of liberalism is that the domestic political system of a state has an effect on its foreign policy in the international arena. Liberal democratic states, according to democratic peace theory, which some political scientists view as the closest thing to an iron law of political science, have less of a tendency to go to war with other liberal democratic states. The reasons for this may range from the system of checks and balances that function to mitigate the decision to go to war in a democratic state, the values and morality that are associated with democracy, and the fact that liberal democratic states may be connected by a set of economic and trade linkages that enmesh them in a web of cooperation. Liberals also stress that economic ties between states in general may reduce the likelihood of a war taking place between them, because the economic costs of a war may jeopardize the benefits of a peaceful relationship.

Finally, arms control and disarmament have been viewed as a means of reducing the possibility of conflict and war between states in the international system. The age-old dream has been for the creation of a system of general and complete disarmament where the biblical injunction of beating swords into plowshares will mean that humans will never make war on each other again. However, military technology and the trade in conventional weapons have made the task of establishing a system of general and complete disarmament extremely difficult. The international community, however, has made progress in dealing with the conventional weapons that are supplied by both governmental and private "merchants of death" with the conclusion of an international treaty regulating the global arms trade.

Furthermore, a network of treaties has been negotiated since the end of World War II to deal with reducing, and hopefully eliminating, weapons of mass destruction (WMDs), as the technology associated with these weapons has spread. The focus of WMD treaties has been to prevent the spread of nuclear weapons. Perhaps with lessons learned from the Cuban missile crisis, the United States recently focused its efforts on preventing Iran from developing the bomb. The United States and Iran reached an agreement in the summer of 2015, known as the Joint Comprehensive Plan of Action. The agreement had been approved by the Iranian Parliament. The nuclear deal between Iran and the United States, which was the result of months of intensive negotiations, represented an attempt to contain Iran's nuclear weapons program for 15 years. The Iranians agreed in essence to freeze their weapons program, by placing thousands of their centrifuges, which produced the enriched uranium needed to make the bomb, in storage. In return, US and international sanctions that Iran had been subjected to would be lifted, and the Iranians would also be able to export their oil and natural gas, as well as gain access to at least $100 billion in frozen assets.

Teheran also agreed to open up its nuclear program to inspection by the International Atomic Energy Agency (IAEA). However, this portion of the agreement was open to criticism by the opponents of the deal, because Iran would have a 24-day notice before the IAEA inspectors arrived at the sites. It was argued that the Iranians would have enough time to clean up any evidence that they were secretly working on the development of nuclear weapons in violation of the agreement. However, the agreement was seen by President Obama as a way to engage in a deeper dialogue with the Iranians. It was hoped that Iran would moderate its behavior in the region and restore a sense of equilibrium as the United States attempted to withdraw from the area.

Article

Prepared by: Robert Weiner, *University of Massachusetts, Boston*

The Growing Threat of Maritime Conflict

"What makes these disputes so dangerous . . . is the apparent willingness of many claimants to employ military means in demarking their offshore territories and demonstrating their resolve to keep them."

MICHAEL T. KLARE

Learning Outcomes

After reading this article, you will be able to:

- Describe why these conflicts are increasingly dangerous.
- Identify the role of oil and natural gas in these disputes.
- Identify specific zones of conflict.

For centuries, nations and empires have gone to war over disputed colonies, territories, and border regions. Although usually justified by dynastic, religious, or nationalistic claims, such contests have largely been driven by the pursuit of valuable resources and the taxes or other income derived from the inhabitants of the disputed lands. Many of the great international conflicts of recent centuries—the Seven Years War, the Franco-German War, and World Wars I and II, for example—were sparked in large part by territorial disputes of this type. By the end of the twentieth century, however, most international boundary disputes had been resolved, and few states possessed the will or the capacity to alter existing territorial arrangements through military force.

Yet, even as the prospects for conflict over disputed land boundaries seem to have dwindled, the risk of conflict over contested maritime boundaries is growing. From the East China Sea to the Eastern Mediterranean, from the South China Sea to the South Atlantic, littoral powers are displaying fresh resolve to retain control over contested offshore territories.

The most recent expression of this phenomenon, and one of the most dangerous, is the clash between China and Japan over a group of uninhabited islands in the East China Sea that are claimed by both. Friction over the islands—known as the Diaoyu in China and the Senkaku in Japan—has persisted for years, but it reached an especially high level of intensity in the summer of 2012 after Japanese authorities arrested 14 Chinese citizens who attempted to land on one of the islands to press China's claims, provoking widespread anti-Japanese protests across China and a series of naval show-of-force operations in nearby waters.

Senior Chinese and Japanese officials have met privately in an attempt to reduce tensions, but no solution to the dispute has yet been announced, and both sides continue to deploy armed vessels in the area—often in close proximity to one another. Although the Barack Obama administration would like to see a negotiated outcome to the dispute. China views Washington as too close to Japan, so Beijing has rebuffed US mediation efforts.

Risk of conflict has also arisen in another disputed maritime area, the South China Sea, where China is again one of the major offshore claimants. As in the East China Sea, the dispute centers on a collection of (largely) uninhabited islands: the Paracels in the northwest, the Spratlys in the southeast, and Macclesfield Bank in the northeast (known in China as the Xisha, Nansha, and Zhongsha islands, respectively). China and Taiwan claim all of the islands, while Brunei, Malaysia, the Philippines, and Vietnam claim some among them, notably those lying closest to their shorelines.

Friction over these contested claims led to a series of nasty naval encounters in 2012, some involving China and Vietnam, and some China and the Philippines. In one such incident, armed

Chinese marine surveillance ships blocked efforts by a Philippine Navy warship to inspect Chinese fishing boats believed to be engaged in illegal fishing activities, leading to a tense standoff that lasted weeks. Chinese officials announced recently that, beginning January 1, their patrol ships will be empowered to stop, search, and repel foreign ships that enter the 12-nautical-mile zone surrounding the South China Sea islands claimed by Beijing, setting the stage for further confrontations.

Maritime disputes of this sort, also involving the use or threatened use of military force, have surfaced in other parts of the world, including the Sea of Japan, the Celebes Sea, the South Atlantic, and the Eastern Mediterranean. In these and other such cases, adjacent states have announced claims to large swaths of ocean (and the seabed below) that are also claimed in whole or in part by other nearby countries. The countries involved cite various provisions of the United Nations Convention on the Law of the Sea (UNCLOS) to justify their claims—provisions that in some cases seem to contradict one another.

Because the legal machinery for adjudicating offshore boundary disputes remains underdeveloped, and because many states are reluctant to cede authority over these matters to as-yet untested international courts and agencies, most disputants have refused to abandon any of their claims. This makes resolution of the quarrels especially difficult.

What makes these disputes so dangerous, however, is the apparent willingness of many claimants to employ military means in demarking their offshore teritories and demonstrating their resolve to keep them. This is evident, for example, in both the East and South China Seas, where China has repeatedly deployed its naval vessels in an aggressive fashion to assert its claims to the contested islands and chase off ships from all the other claimants. In response, Japan, Vietnam, and the Philippines have also employed their navies in a muscular manner, clearly aiming to show that they will not be intimidated by Beijing. Although shots have rarely been fired in these encounters, the ships often sail very close to each other and engage in menacing maneuvers of one sort or another, compounding the risk of accidental escalation.

What accounts for this growing emphasis on offshore disputes at a time when few states appear willing to fight over more traditional causes of war?

For some governments, offshore disputes may be seen as a sort of release valve for nationalistic impulses that might prove more dangerous if applied to other issues, or as a distraction from domestic woes. China's conflict with Japan over the Diaoyu/Senkaku Islands, for example, has provoked strong nationalistic passions in both countries—passions that leaders on each side no doubt would prefer to keep separate from the more important realm of economic relations. Likewise, Argentina's renewed focus on the Falklands/Malvinas is widely considered to be a deliberate response to political and economic difficulties at home. But these considerations are only part of the picture; far more important, in most cases, is a desire to exploit the oil and natural gas potential of the disputed areas.

The Lure of Oil and Gas

The world needs more oil and gas than ever before, and an ever-increasing share of this energy is likely to be derived from offshore reservoirs. According to the Energy Information Administration (EIA) of the US Department of Energy, global petroleum use will rise by 31 percent over the next quarter-century, climbing from 85 million to 115 million barrels per day. Consumption of natural gas will grow by an even faster rate, jumping from 111 trillion to 169 trillion cubic feet per year. Older industrialized nations, led by the United States and European countries, are expected to generate some of this growth in consumption. But most of it is projected to come from the newer industrial powers, including China, India, Brazil, and South Korea. These four nations alone, predicts the EIA, will account for 57 percent of the total global increase in energy demand between now and 2035.

Until now, the world's ever-increasing thirst for oil has been satisfied with supplies obtained from fields on land or shallow coastal areas that can be exploited without specialized drilling rigs. But many of the world's major onshore fields have been producing oil for a long time, and are now yielding diminishing levels of output: likewise, production in shallow areas of the Gulf of Mexico and the North Sea has long since fallen from peak levels. Some of the loss from existing reservoirs will be offset through the accelerated extraction of petroleum from shale rock, made possible by new technologies like hydraulic fracturing. But any significant increase in global oil production will require the accelerated exploitation of offshore—especially deep-offshore—reserves.

According to analysts at Douglas-Westwood, a United Kingdom–based energy consultancy, the share of world oil production supplied by offshore fields will rise from 25 percent in 1990 to 34 percent in 2020. More important, the share of world oil provided by deep wells (over 1,000 feet in depth) and ultra-deep wells (over one mile) will grow from zero in 1990 to a projected 13 percent in 2020. Douglas-Westwood further projects that onshore and shallow-water fields will yield no additional production increases after 2015, so all additional growth subsequently will have to come from deep and ultra-deep reserves. Meanwhile, the world's reliance on natural gas is likely to exhibit a similar trajectory: Whereas in 2000 approximately 27 percent of the world's gas supply came from offshore fields, by the year 2020 that share is projected to reach 41 percent.

Driving this shift toward greater reliance on offshore oil and gas is not only the depletion of onshore fields but also advances in drilling technology. Until recently, it was considered impossible to extract oil or gas from reserves located in waters over a mile deep. Now drilling at such depths is becoming almost routine, and extraction at even greater depths—up to two miles—is about to commence. Specialized rigs have also been developed for operations in the Arctic Ocean, and in areas that pose unusual climatic and environmental challenges, such as the Caspian Sea and the Sea of Okhotsk, off Russia's Sakhalin Island. In the future, technology may allow the extraction of natural gas from so-called methane hydrates—dense nodules of frozen gas that are trapped in ice crystals lying at the bottom of some northerly oceans.

It follows from all this that the world's major energy consumers—led by China, the United States, Japan, and the European Union countries—will become increasingly reliant on oil and gas supplies derived offshore. Some of this energy can be acquired from fields in areas with no outstanding territorial disputes, such as the North Sea and the Gulf of Mexico. Other large reservoirs, such as Brazil's "pre-salt" fields in the deep Atlantic, lie far enough from other coastal states to eliminate the potential for boundary conflict. But many promising fields are located in bodies of water where maritime boundaries remain undefined. And, as the perceived value of these resources grows, the potential for discord to take a military form will increase as well. This risk is greatest in areas thought to harbor large reserves of oil and gas, where the contending parties have repeatedly rebuffed efforts to adopt precise, mutually acceptable offshore boundaries, and where one or more of the claimants have employed (or threatened the use of) military means.

Contested Seas

The risk is especially great in the East and South China Seas. Both regions are thought to sit atop substantial reserves of oil and gas, both lack mutually accepted offshore boundaries, and both have witnessed repeated military encounters. The East China Sea, bounded by China to the west, Taiwan to the south, Japan to the east, and Korea to the north, harbors several large natural gas fields in areas claimed by China, Japan, and Taiwan. The South China Sea, bounded by China and Taiwan to the north. Vietnam to the west, the island of Borneo (divided among Brunei, Indonesia, and Malaysia) to the south, and the Philippines to the east, is believed to possess both oil and gas deposits: China and Taiwan claim the entire region, while Brunei, Malaysia, Vietnam, and the Philippines claim large portions of it. All of these countries have engaged in negotiations aimed at resolving the various overlapping claims—without achieving notable success—and all have taken military steps of one sort or another to defend their offshore interests.

Considerable debate persists among industry professionals as to exactly how much oil and gas is buried beneath the East and South China Seas. Because limited drilling has been conducted in these areas (except on the margins), analysts possess little detailed information from which to derive estimates of recoverable reserves. Nevertheless, Chinese experts regularly offer highly optimistic assessments of the seas' potential. The East China Sea, they claim, contains between 175 trillion and 210 trillion cubic feet of natural gas—approximately equivalent to the proven reserves of Venezuela, the world's seventh largest gas power. Chinese estimates of the oil and gas lying beneath the South China Sea are even more exalted: These place the region's ultimate oil potential at over 213 billion barrels (an amount exceeded only by the proven reserves of Saudi Arabia and Venezuela), and that of gas at 900 trillion cubic feet (exceeded only by Russia and Iran). Western analysts, such as those employed by the EIA, are reluctant to embrace such lofty estimates in the absence of actual drilling results, but acknowledge the two areas' great potential.

Whatever the precise scale of the East and South China Seas' hydrocarbon reserves, the various littoral states clearly see them as promising sources of energy. China, Japan, Malaysia, Vietnam, and the Philippines have awarded contracts to different combinations of private and state-owned firms to exploit oil and gas reserves in the areas they claim, and more such awards are being announced all the time. The Chinese have been particularly active, drilling for natural gas in the East China Sea and for oil in the South China Sea. Their efforts took a big step forward in May 2012, when the China National Offshore Oil Corporation (CNOOC) deployed the country's first Chinese-made deep-sea drilling platform in the South China Sea, at a point some 200 miles southeast of Hong Kong.

The Vietnamese have long extracted oil and gas from their coastal waters, and are now seeking to operate in deeper waters of the South China Sea. Across the sea to the east, the Philippines' Philex Petroleum Corporation has been exploring a major natural gas find off Reed Bank—another uninhabited islet claimed by China as well as the Philippines and a site of recent clashes between Chinese and Filipino vessels. Although Chinese leaders say they want to promote cooperative development of the East and South China Seas, Beijing has often taken steps to deter efforts by its neighbors to explore for oil and gas in these areas. In May 2011, for example, Chinese patrol boats repeatedly harassed exploration ships operated by state-owned PetroVietnam in the South China Sea, in two instances slicing cables attached to underwater survey equipment.

Despite the high expectations for oil and gas extraction in the two seas, therefore, any significant progress will have to await the resolution of outstanding territorial disputes or some agreement allowing drilling to proceed without risk of interference. Yet none of the parties to these disputes appears willing to retreat from long-established positions or eschew the use of force. Efforts to seek negotiated outcomes have been frustrated, moreover, by contending historical narratives and a lack of clarity in international law regarding the demarcation of offshore boundaries.

Legal Confusion

In the East China Sea, both China and Japan draw on competing provisions of UNCLOS (which both have signed) to justify their maritime claims. Each set of provisions defines a state's outer maritime boundary in a different way: One set allows coastal states to establish an exclusive economic zone (EEZ) extending up to 200 nautical miles offshore, in which they possess the sole right to exploit marine life and undersea resources, such as oil and gas; the other allows coastal states to exert such control over the "natural prolongation" of their outer continental shelf, even if it exceeds 200 nautical miles.

China, citing the latter provision, says that its maritime boundary in the East China Sea is defined by its continental shelf, an underwater feature that extends nearly to the Japanese islands. Japan, citing the former provision, insists that the boundary should be drawn along a median line equidistant between the two countries, since the distance separating them is less than 400 nautical miles.

Lying between these two hypothetical boundary lines is a contested area of approximately 81,000 square miles (nearly the size of Kansas) that is thought to harbor large volumes of natural gas—a resource that each side claims is its alone to exploit. The contested Diaoyu/Senkaku Islands lie at the southern edge of this area, and so neither side is willing to relinquish control over them, each fearing that doing so would jeopardize its claim to the adjacent seabed. Negotiations to resolve the impasse have produced talk of joint development efforts in the contested area, but no willingness to compromise on the basic issues.

The dispute in the South China Sea is even more complex. Drawing on ancient maps and historical accounts, the Chinese and Taiwanese insist that the sea's two island chains, the Spratlys and the Paracels, were long occupied by Chinese fisherfolk, and so the entire region belongs to them. The Vietnamese also assert historical ties to the two chains based on long-term fishing activities, while the other littoral states each claim a 200-nautical mile EEZ stretching into the heart of the sea. When combined, these various claims produce multiple overlaps, in some instances with three or more states involved—but always including China and Taiwan as claimants. Efforts to devise a formula to resolve the disputes through negotiations sponsored by the Association of Southeast Asian Nations (ASEAN) have so far met with failure: While China has offered to negotiate one-on-one with individual states but not in a roundtable with all claimants, the other countries—mindful of China's greater wealth and power—prefer to negotiate en masse.

Again, the various claimants in these conflicts have, on a regular basis, employed military force to demonstrate their determination to retain control over the territories they have claimed and to deter economic activities in these areas by competing countries. Few such actions have resulted in bloodshed—one major exception was a 1988 clash between Chinese and Vietnamese warships near Johnson Reef in the Spratly Islands that resulted in the loss of more than 70 lives—but many have prompted countermoves by other countries, posing a significant risk of escalation. In September 2005, for example, Chinese warships patrolling along the median line claimed by Japan in the East China Sea aimed their guns at a Japanese Navy surveillance plane, nearly leading to a serious incident.

More such engagements have occurred in the South China Sea, where there are a larger number of claimants and greater uncertainty over the location of boundaries. In one such incident Vietnamese troops fired on a Philippine air force plane on a reconnaissance mission in the Spratlys; in another, Malaysian and Filipino aircraft came close to firing on each other while flying over a Malaysian-occupied reef in the Spratlys.

Recognizing the potential for escalation, leaders of the countries involved in such encounters have taken some steps to avert a serious clash. Chinese and Japanese officials have met on several occasions to discuss the boundary dispute in the East China Sea, pledging to avoid the use of force. Likewise, China and the 10 members of ASEAN signed a Joint Declaration on the Conduct of Parties in 2002, pledging to resolve their territorial disputes in the South China Sea by peaceful means. However, these measures have not prevented the major parties from continuing to employ military means to reinforce their bargaining positions. Worried that such activities could lead to more serious conflict, endangering vital US interests, the Obama administration has offered to act as a mediator—only to provoke a hostile response from Beijing, which sees this as an unwelcome form of American meddling in its backyard.

Girding for Conflict

If the East and South China Seas represent the most conspicuous cases of offshore territorial conflicts driven in large part by the competitive pursuit of energy resources, they are by no means the only ones with a potential to spark violence. Others that exhibit many of the same characteristics include quarrels over the Falklands/Malvinas, the eastern Mediterranean, and the Caribbean near Nicaragua and Colombia.

The dispute over the Falklands/Malvinas Islands and their surrounding waters, claimed both by Britain and Argentina, is well known from the 1982 war over the islands, in which the British defeated an invasion by Argentina. At the time, the primary impulses for conflict were thought to be national pride and the political fortunes of the key leaders involved: Margaret Thatcher in Britain, and an unpopular military junta in Argentina. Now, however, a new factor has emerged: competing claims to undersea energy reserves. Large reservoirs of oil are thought to lie beneath areas of the South Atlantic to the north and south of the islands, and both Argentina and Britain say the reserves belong exclusively to them. A number of companies have obtained permits from British and Falkland Islands authorities to sink test wells within a 200 mile EEZ surrounding the islands claimed by London after it ratified UNCLOS in 1997.

Until now, neither side has engaged in provocative military action of the sort seen in the other offshore disputes, but both sides appear to be girding for the possibility. The British have replaced older ships and aircraft in the Falklands with more modern equipment, including Typhoon combat aircraft of the type used during the 2011 Libyan campaign. The Argentines have responded by blocking access to Argentine ports for British cruise ships that first dock in the Falklands—a largely symbolic act, to be sure, but one that hints of stronger actions to come. How this will play out remains to be seen, but neither side has budged on any of the fundamental issues, and the prospect of significant oil production by British firms on what the Argentines consider to be their sovereign territory is bound to increase resentment in the years ahead.

The Eastern Mediterranean, like the Falklands/Malvinas, is also a site of earlier conflict. In addition to the recurring Arab-Israeli wars, there are the ongoing Greek-Turkish dispute over governance of Cyprus—the backdrop for a war in 1974—and a growing schism between Israel and Turkey. But now, again, the discovery of potentially vast energy reserves is aggravating traditional rivalries. The offshore Levant Basin, stretching from Cyprus in the north to Egypt in the south and bounded by Israel, Lebanon, and the Gaza Strip on the east, is thought to hold 120 trillion cubic feet of natural gas, and perhaps much more. Production of this gas could prove a boon to the nations involved—few of which have experienced any benefit from the oil boom in neighboring countries.

At this point, the most advanced projects are under way in Israeli-claimed territory. Noble Energy, a Houston-based firm, is developing a number of giant gas fields in waters off the northern port of Haifa. The largest of these, named Leviathan, lies astride the EEZ claimed by the Republic of Cyprus, where Noble has also found substantial gas reservoirs. Although significant hurdles remain, both Israel and Cyprus hope to extract natural gas from these fields by the middle of the decade and to ship considerable volumes to Europe via new pipelines to be installed on the Mediterranean sea-bed, or in the form of liquefied natural gas.

Seeing the potential for cooperation in exporting gas, Israel and Cyprus have discussed common transportation options and signed a maritime border agreement in December 2011. But both countries face significant challenges from other nations in the region. The Leviathan field and other gas reservoirs being developed by Noble are located at the northern edge of the EEZ staked out by Israel, in waters also claimed by Lebanon. Lebanese authorities, who refuse to negotiate with Israel, have urged the UN to pressure Israel to recognize Lebanon's sovereignty over the area, but to no avail. Far more worrisome are threats by Hezbollah, the Iranian-backed Shiite militia based in Lebanon, to attack Israeli drilling rigs in waters claimed by the Lebanese. These threats have prompted Israel's air force to deploy drones over the facilities, allowing for a prompt response to any potential terrorist attack. Meanwhile, Noble's operations in Cypriot-claimed waters have been challenged by Turkey, which does not recognize the Republic of Cyprus or its claim to an EEZ. The Turks have deployed air and naval craft off the Turkish Republic of Northern Cyprus, an ethnic separatist entity that only they recognize, in what is viewed as an implied threat to Noble and other companies operating in the Cypriot EEZ.

In the Western Hemisphere, a dispute has arisen between Colombia and Nicaragua over a swath of the Caribbean claimed by both of them. On November 19, 2012, the International Court of Justice in The Hague awarded control of some 35,000 square miles of the Caribbean—believed to harbor valuable undersea reserves of oil and gas—to Nicaragua. The decision infuriated the Colombians, who rejected the ruling and withdrew from a pact recognizing the court's jurisdiction over its territorial disputes. Leaders of both countries have pledged to seek a peaceful resolution, but the situation remains tense. "Of course no one wants a war," said Colombian President Juan Manuel Santos. "That is a last resort."

Options for Resolution

As should now be evident, the accelerated pursuit of oil and gas reserves in disputed offshore territories entails significant potential for international friction, crisis, and conflict. This is so because such efforts combine unusually high economic stakes with intense nationalism and the absence of clearly defined boundaries. Add to this the lack of clearly defined mechanisms for resolving boundary disputes of this sort, and the magnitude of the problem becomes apparent. Unless a concerted effort is made to resolve these and other such disputes, what is now latent or low-level conflict could erupt into full-scale violence.

The problem is not a lack of viable solutions. In several contested maritime regions, countries that were unable to agree on their offshore boundaries have been able to establish joint development areas (JDAs) in which drilling has proceeded while negotiations continue regarding the demarcation of final borders. The first of these special zones, the Malaysia-Thailand Joint Development Area, was created in 1979 and has been producing gas since 2005: Vietnam has also become a party to an additional slice of the JDA. A similar formula has been adopted by Nigeria and the island state of São Tomé and Principe to develop offshore fields in a contested stretch of the Gulf of Guinea. China and Japan once agreed to employ a solution of this sort to develop the contested area claimed by both in the East China Sea, but so far little has come of the effort.

Meanwhile, UNCLOS, as amended, incorporates various measures for resolving disputes over the location of offshore territories. Essentially, it mandates that such disputes be resolved peacefully, through negotiations among the affected parties. UNCLOS also includes provisions for arbitration by third parties and referral of disputes to the International Court of Justice, or to the newly established International Tribunal for the Law of the Sea (based in Hamburg). Also, to help determine the validity of a state's claim to offshore territories based on the natural prolongation of its continental shelf, the UN has established a Commission on the Limits of the Continental Shelf.

However, all of these measures have limitations. For one thing, they do not apply to countries that have failed to ratify UNCLOS, such as Turkey and the United States. They have little effect, moreover, when contending states refuse to negotiate, as is the case with Israel and Lebanon; or eschew arbitration and outside involvement, as China has done in the East and South China Seas. Clearly, something more is needed.

What appears most lacking in all of these situations is a perception by the larger world community that disputes like these pose a significant threat to international peace and stability. Were these disputes occurring on land, one suspects, world leaders would pay much closer attention to the risks involved and take urgent steps to avoid military action and escalation. But because they are taking place at sea, away from population centers and the media, they seem to have attracted less concern.

This is a dangerous misreading of the perils involved: Because the parties to these disputes appear more inclined to employ military force than they might elsewhere, and boundaries are harder to define, the risk of miscalculation is greater, and so is the potential for violent confrontation. The risks can only grow as the world becomes more reliant on offshore energy and coastal states become less willing to surrender maritime claims.

To prevent the outbreak of serious conflict, the international community must acknowledge the seriousness of these disputes and call on all parties involved to solve them through peaceful means, as quickly as possible. This could occur through resolutions by the UN Security Council, or statements by leaders meeting in such forums as the Group of 20 governments. Such declarations need not specify the precise nature of any particular outcome, but rather must articulate a consensus view that a resolution of some sort is essential for the common good. Arbitration by neutral, internationally respected "elders" can be provided as necessary. To facilitate this process, ambiguities in UNCLOS should be resolved and holdouts from the treaty—including the United States—should be encouraged to sign.

After Consensus

Assuming such a consensus can be forged, solutions to the various maritime disputes should be within reach. China and Japan should jointly develop the gas field in the disputed area of the East China Sea until a final boundary is adopted—an option already embraced in principle by the two countries. In the South China Sea, a JDA should be established on the model of the Malaysia-Thailand Joint Authority, consisting of representatives of all littoral states and empowered to award exploration contracts (and allocate revenues) on an equitable basis. A similar authority should oversee drilling in the waters surrounding the Falklands, the Israel-Lebanon offshore area, and the waters around Cyprus. At the same time, negotiations leading to a permanent border settlement in these areas should be undertaken under international auspices.

If the countries involved cannot agree to such measures, they should be pressured to submit their competing claims to an international tribunal with the authority to determine the final demarcation of boundaries, while international energy companies should be required to abide by the outcome of such decisions or face legal action and the possible loss of revenues.

Such measures are important for another reason: to help reduce the risk of environmental damage. As demonstrated by the Deepwater Horizon disaster of April 2010 in the Gulf of Mexico and more recent oil leakages from Brazil's pre-salt fields, deep offshore drilling poses a significant threat to the environment if not conducted under the most scrupulous production methods. Clearly, maritime areas that lack an accepted regulatory and jurisdictional regime, such as the South China Sea, are more likely to experience spills and other disasters than areas with well-established boundaries and effective supervision.

The establishment of clear maritime boundaries and the promotion of collaborative offshore enterprises rank among the most important tasks facing the international community as the global competition for resources moves from traditional areas of struggle, such as the Middle East, to seas where the rules of engagement are less defined. The exploitation of offshore oil and gas could help compensate for the decline of existing reserves on land, but will result in increased levels of friction and conflict unless accompanied by efforts to resolve maritime boundary disputes. Defining borders at sea may not be as easy as it is on land, where natural features provide obvious reference points, but it will become increasingly critical as more of the world's vital resources are extracted from the deep oceans.

Critical Thinking

1. How are these conflicts changing Japanese military policy?
2. How do these conflicts reflect the emergence of China as a military power?
3. What role is the United States likely to play in these maritime conflicts?

Internet References

Japanese Ministry of Defense
www.mod.go.jp/e

Institute for Defence Studies and Analysis
www.idsa.in

Chinese Ministry of National Defense
http://eng.mod.gov.cn

Association of Southeast Asian Nations
www.aseansec.org

U.S. Department of Defense: Quadrennial Defense Review
www.defense.gov/qdr

Michael T. Klare, a *Current History* contributing editor, is a professor at Hampshire College and the author, most recently, of *The Race for What's Left: The Global Scramble for the World's Last Resources* (Metropolitan Books, 2012)

Article Prepared by: Robert Weiner, *University of Massachusetts, Boston*

Afghanistan's Arduous Search for Stability

THOMAS BARFIELD

Learning Outcomes

After reading this article, you will be able to:

- Understand the relationship between the Obama administration and the Karzai regime.
- Explain the role of Pakistan in the Afghan conflict.
- Discuss the importance of factionalism in understanding Afghan politics.

Why, after the expenditure of so much blood and treasure in Afghanistan over the past decade and a half, does there seem to be so little to show for it? One key reason has been a lack of leadership and unity in Kabul.

For almost 15 years, Afghanistan appeared mired in a political rut while President Hamid Karzai dominated the scene until he finally left office in 2014. On the domestic front, he made use of a highly centralized administrative system to appoint all the country's provincial officials, block the formation of political parties, and reward his allies with patronage. Externally, he acted as the sole interlocutor with international backers (primarily the United States and its allies) that provided more revenue to the government than it got in taxes, financed and equipped its security forces, and funded development projects.

In 2002, Karzai was acclaimed at home and abroad as just the man to unite a fragmented country. Initially endorsed by a *loya jirga* (a grand assembly of tribal leaders) to serve a two-year interim term as president, he won Afghanistan's first presidential election in 2004, riding genuine popular enthusiasm. But domestic support for Karzai and the national government began to wane by 2006, as Afghans increasingly viewed the president as incompetent and his administration as corrupt. Karzai recognized that some of these criticisms were legitimate, but his dependence on personalized political deal making and patronage created a dilemma he could not easily resolve: Reforming the system risked bringing about its collapse. Indeed, it sometimes appeared that corruption was the glue that held the Karzai government together.

Since international aid provided the bulk of Karzai's patronage assets and paid the salaries of his officials, the donors in theory should have had leverage to push for reforms. In reality, international pressure rarely amounted to more than the distribution of critical but toothless reports. Laundry lists of specific reforms that the Afghan government agreed to implement were the centerpieces of every international donors' conference—along with complaints that the goals set by previous conferences had not been met.

Karzai, who did not lack shrewdness, could safely ignore such international criticism and threats to withhold aid. All he had to do was make the case that Afghanistan would become a renewed security problem if his government did not receive support. He rightly assumed that none of the big donors were willing to test that proposition.

The administration of US President George W. Bush was so preoccupied with its failing war in Iraq after 2003 that it put Afghanistan on the back burner in hopes of limiting American involvement. Changing the leadership of the Afghan government or building its capacity for action would have required more time, money, and military effort than Washington was willing to commit. Even after the reemergence of a Taliban insurgency in the mid-2000s, the Bush administration muted its criticism of Karzai and cultivated a close personal relationship with him. Afghans firmly believed that no rival political leader could hope to challenge Karzai as long as he was treated as indispensable by the Americans.

Obama's "Good War"

The belief that nothing could ever change in Afghanistan was eroded by elections in the United States in 2008 and in Afghanistan the next year. During the US presidential campaign, Senator Barack Obama criticized the Bush administration's unpopular Iraq policy in part by arguing that the war in Afghanistan was more central to US antiterrorism priorities. In making Afghanistan "the good war," Obama moved it back to the forefront of US policy debates; but in doing so, he laid considerable blame on Karzai for the persistent difficulties. Soon-to-be Vice President Joseph Biden made that clear on a visit to Kabul in February 2008, when he stormed out of a dinner at the presidential palace after Karzai responded to his questions about corruption by denying it existed.

A year later, Biden returned to Kabul with bad news for Karzai from the newly elected President Obama. The high level of personal attention that Bush had lavished on him would be ending. Biden bluntly told Karzai that he would "probably talk to [Obama] a couple of times a year," and certainly not every week. For an Afghan leader anticipating his reelection bid later that year, such a cold shoulder from the Americans had direct, negative political repercussions.

The split between Karzai and the Obama administration led many Afghans (including Karzai himself) to assume that the Americans would be seeking a replacement for him in the 2009 election. There was strong historical precedent. Both the British and the Soviets installed weak leaders when they occupied Afghanistan in the nineteenth and twentieth centuries, respectively, and they both replaced those leaders with stronger personalities when they decided to withdraw. It was assumed that Obama (whose endgame was to turn all security responsibilities over to the Afghan government) might also be in the market for a more competent and reliable leader.

Popular support for insurgencies based on driving foreigners out declines when the foreigners actually leave.

The election seemed to be an opportunity to engineer this outcome while simultaneously claiming a success for the country's new democratic process by letting the Afghan people make the choice in a free and fair vote. Karzai would likely lose to any challenger who could unite the domestic opposition by demonstrating clear American backing. The allies Karzai had attracted through patronage would have no trouble shifting their allegiance to someone else if it looked as if he was not in a position to deliver that largesse any longer. In Afghan politics, the perception of power was power itself.

Richard Holbrooke, Obama's special representative to Pakistan and Afghanistan, was determined to find a replacement for Karzai, and he made that well known. The difficulty was that the highly fragmented Afghan opposition was waiting for the Americans to anoint a challenger who could use that backing to build a domestic coalition, while the Americans were waiting for the opposition to make the first move by uniting behind a consensus candidate Washington could discreetly back while maintaining a stance of official impartiality.

When the opposition proved unable to agree on a single candidate, the Obama administration resigned itself to Karzai's reelection. Even without a unified opposition, Karzai ultimately needed to get a runoff election canceled and resort to massive vote fraud to win. He characteristically blamed all election irregularities on "foreign outsiders," but most of the fraud was committed by his own allies.

Karzai's Grievances

Karzai's relationship with the Obama administration remained hostile throughout his second term. Even as the United States and its NATO allies deployed over 100,000 additional troops to Afghanistan between 2010 and 2012 to ensure his regime's security, Karzai's sense of grievance never ceased, and he attacked his backers anytime an opportunity presented itself. In October 2011, he went so far as to declare, according to a Reuters report, "God forbid, if ever there is a war between Pakistan and America, Afghanistan will side with Pakistan."

Karzai's continued distrust of the United States was so deep that he adamantly refused to sign a Bilateral Security Agreement (BSA) that would have allowed some US troops to remain in Afghanistan after most were set to leave by the end of 2014, even after it was overwhelmingly approved by a national *loya jirga*. In this and other actions Karzai appeared to bear out the conclusions of US Ambassador Karl Eikenberry's leaked November 2009 assessment, which asserted that the Afghan leader was "not an adequate strategic partner" and "continues to shun responsibility for any sovereign burden."

Few Afghans believed Karzai when he announced that he would abide by the constitution's two-term limit and step down in 2014. No Afghan leader had ever relinquished power before death, except at the point of a gun. Conspiracy theories proliferated over his true intentions. Some said he planned to amend the constitution and run again; others said he would call on a *loya jirga* to declare him president for life without an election; still others insisted he would find a sycophant to serve as a figurehead president while he ruled from behind the scenes.

In the end, Karzai stood down and his chosen successor, Zalmai Rassoul, went down to a humiliating defeat in the first round of the presidential election in April 2014. Two months later, the electoral process came to a standstill when Abdullah

Abdullah, the winner of a large plurality in the first round, alleged that only massive fraud could explain the second-round victory of his rival, Ashraf Ghani. Once again, rumors abounded that Karzai would use the disputed outcome to declare himself interim ruler. But after personal pressure from the US Secretary of State John Kerry, the two finalists agreed in September 2014 to share power and form a national unity government, with Ghani as president and Abdullah in the newly created position of chief executive officer.

The Taliban have moved away from a purely religious identity to burnish their credentials as Afghan nationalists.

Mood Change

The establishment of the unity government was a pivotal and positive change from the Karzai era. The two new leaders, unlike Karzai, were eager to improve Afghanistan's international relations. Ghani, a former academic and World Bank official, had lived in the United States for 30 years before returning to Afghanistan in 2002 and serving in a number of cabinet posts. An able administrator popular with the foreign embassies in Kabul, he presented himself as a technocrat rather than a politician. Abdullah had never lived abroad and could claim a distinguished record of opposing both the Soviet occupation and the Taliban regime as part of the Northern Alliance. As a former foreign minister (and one of the best-dressed men in Kabul), he was also skilled at representing Afghan interests internationally.

While Ghani and Abdullah were domestic rivals, they were in general agreement on major foreign policy issues. As candidates, both had promised to sign the BSA with the United States. Immediately upon taking office, they did so together. With that accomplished, other members of the international coalition agreed to continue their own more limited military assistance.

The new government also sought to reposition itself regionally by cultivating closer ties with China, to which Ghani made his first official trip abroad in October 2014. Beijing had only recently begun to engage more proactively with Kabul, offering to facilitate peace talks with the Taliban and to invest in the economy. (Despite the immense amounts of money spent in Afghanistan by Western governments, few if any private businesses from Europe or North America appeared willing to risk their own capital there.)

On the public relations front, Ghani made an effort to thank Afghanistan's foreign allies for their sacrifice and support rather than accusing them, as Karzai had, of plotting against the country's interests. This change in mood was soon publicly reciprocated: Both Abdullah and Ghani came to Washington for a formal state visit in March 2015, and Ghani gave an address to Congress.

Cross Purposes

Although it appeared to be moving forward in harmony on the international front, the new government was hamstrung domestically by its internal divisions. The shotgun wedding that compelled Abdullah and Ghani to share power had also created dual staffs working at cross purposes within the executive branch—particularly when it came to the division of ministry appointments, long a source of patronage. Each faction felt shortchanged by the other, and there were not enough high-level positions to go around.

Even when the rival factions did agree, the parliament rejected a number of appointments for its own reasons. Vital positions in the 25-member cabinet, including national security offices, were still filled by interim appointees more than a year and a half after the government was formed. This deadlock extended down to the appointment of provincial governors. Ghani declared all the positions vacant but he was slow to fill them, leaving far too many provinces in the hands of lame-duck Karzai appointees.

Even seemingly noncontroversial reforms, such as requiring merit-based appointments, also ran into stiff opposition. It originated in the long-standing animosity between Afghan exiles who returned to the country with technical skills after decades abroad and the relatively uneducated resistance leaders who had stayed to fight. These mujahedeen commanders (sometimes pejoratively labeled warlords) considered a merit-based appointment system to be a power grab by Ghani and the technocratic class he represented. In late 2015, they joined former ministers and officials from the Karzai government to create an opposition group called the Protection and Stability Council, but it too split between those who wanted Karzai back and those who just wanted a stake in the national unity government.

To be fair, it is hard to find any government or insurgent group in Afghan history that has not been prone to factionalism. (In the 1980s, the regime of the People's Democratic Party of Afghanistan was composed of two autonomous and hostile communist factions, while the mujahedeen resistance based in Pakistan was divided into seven separate recognized parties.) The greater internal threat to the unity government's stability was not so much the factions themselves as the uncertain lines of authority.

Abdullah's position had been invented in the compromise that made Ghani president. It called for a *loya jirga* to be held in two years to approve a constitutional amendment to convert his CEO post to that of prime minister. Yet it was never clear just what authority such a position would have. The hyperactive and micromanaging Ghani acted more like the chief executive, leaving Abdullah marginalized and his supporters frustrated. And because Ghani had frequently expressed his opposition to a two-headed government, there was always considerable doubt about his commitment to the power-sharing deal. As the deadline to ratify the agreement approached, neither the wording of the prospective amendment nor the date of the *loya jirga* had been set.

Even with the best intentions, assembling a constitutional *loya jirga* was never going to be easy, given the requirement that nearly half of its 761 delegates must be elected district representatives. Although the constitution was adopted in 2004, the district elections it mandated were never held. A requirement for the participation of current members of parliament was also problematic. The legislature's five-year term had expired in June 2015, leaving it to operate under the questionable authority of a presidential decree.

Seeking to resolve these issues, the Independent Electoral Commission in January 2016 unexpectedly announced that elections for both a new parliament and district representatives would be held on October 15. Since this was the same commission that had badly botched the presidential balloting, Abdullah's representatives objected that the election should wait until voting procedures were reformed and the commission itself was restructured. The president's office was silent on the matter. The uncertainty threatened the political bargain that had created the unity government. In a land where the rule of law was more a distant goal than a current reality, there was no ready solution for a crisis of legitimacy pitting the government's most powerful players against each other. It was also the last thing a government facing an insurgency would care to contemplate.

Rush to the Exit

For over a decade, Mullah Omar, the reclusive leader of the Taliban, had encouraged his forces to keep fighting until they regained power. While the Taliban did make some progress in the mid-2000s, the surge of more than 100,000 US and NATO troops into Afghanistan from 2010 to 2012 put them on the defensive, particularly in the south. As long as they had sanctuaries in Pakistan, however, the Taliban could absorb the loss of territory and manpower while waiting for the announced date of the foreigners' withdrawal.

That timing was driven more by American domestic politics than by the security situation in Afghanistan. The first withdrawals coincided with the 2012 presidential election in which Obama won a second term. By early 2014, only 33,000 US troops remained (partially to ensure security during the Afghan presidential election). That number had dwindled to just under 10,000 when the BSA was signed in Kabul in September 2014. The agreement turned over all security responsibilities to the Afghan government and stated that "unless otherwise mutually agreed, United States forces shall not conduct combat operations in Afghanistan."

Washington's plan for the next two years was to restrict its remaining troops (to be reduced to 5,000 in 2015) to training and counterterrorism operations, not fighting the Taliban. By the end of 2016, all that remained of the US military presence in Afghanistan would be a Marine guard unit at the embassy. This would allow Obama to declare that he had wrapped up the war on his watch, as promised. But as critics noted, this schedule ignored the realities on the ground.

The rebuilding of the Afghan security forces did not begin in earnest until after the surge troops arrived in 2010, and they were not ready to take over the security responsibilities assigned to them. Nor was there an effective replacement for the air power, logistics systems, and medical support that the US and NATO supplied to the Afghan military. And Afghan leaders had a bigger worry: If all international troops left the country, would donor countries continue to provide the high level of funding Afghan security forces would need in the years ahead? More than boots on the ground, money in the pipeline was critical to the survival of any Afghan government. The lack of such support had led to the destructive civil war and the rise of the Taliban in the 1990s.

Fortunately for the government, the Taliban insurgency had its own problems that were not unlike those in Kabul. In 2015, the insurgent group was roiled by a succession crisis and confronted a changing international calculus on the danger posed by jihadist movements worldwide.

Taliban Trouble

The Taliban had been thoroughly defeated in 2001, but they were able to restart the insurgency in Afghanistan by 2005 from their sanctuary in Quetta, Pakistan. Although nominally an ally of the US coalition in Afghanistan, Pakistan had long played a double game of supporting the insurgency while also receiving large amounts of US military and financial aid. Pakistani strategists assumed that the United States would eventually pull out of Afghanistan, and that the Taliban would then be able to take power in Kabul, or at least allow Pakistan to use them as proxies. It was a replay of Islamabad's policy of supporting the mujahedeen against the Soviet-backed Afghan government in the 1980s. In both cases, the policy was based on two simple strategic assumptions: foreign troops were the only firewall preventing the fall of a Kabul government, and insurgents could expect a quick victory once the foreigners withdrew.

Almost counterintuitively, however, Afghan history demonstrated the opposite. Over the course of three wars and 150 years, Kabul governments that retained a world power's patronage and a continued flow of money and weapons proved difficult or impossible to dislodge. This was seen most dramatically after the Soviet withdrawal from Afghanistan in 1989, when the regime of Mohammad Najibullah continued to receive Soviet aid and kept the mujahedeen insurgency at bay. That regime did fall in 1992, but only after the Soviet Union itself collapsed.

Indeed, insurgent movements that had proved so successful at inducing foreign governments to withdraw their troops from Afghanistan had a dismal record of failure in attempting to seize control of the country. When rebels did succeed, it was never against foreign-backed regimes but during domestic civil wars when the national government had no world-power patron. This occurred in 1929 when the bandit rebel Habibullah Kalakani ousted King Amanullah, and again in the mid-1990s when the Taliban ousted the mujahedeen President Burhanuddin Rabbani. It is also worth noting that both of these insurgent regimes were driven from power when their enemies obtained the backing of a world power.

The Taliban ramped up hostilities as international troops withdrew in 2014, and went on the offensive in 2015. Casualties on both sides were much higher than in previous years. The Taliban made substantial gains in the southern province of Helmand and surprised the Kabul government by taking the northern provincial capital of Kunduz in October before being driven out by a counterattack. Kunduz was particularly significant because it was outside the normal areas of Taliban strength. However, the Kabul government did not collapse, nor did its troops in the region abandon the fight after a dispiriting defeat, as happened in Iraq during a 2014 offensive by Islamic State (ISIS) jihadists.

Much like the mujahedeen after 1989, the Taliban have been most successful in rural areas where Kabul's influence has always been weak, but they have proved unable to take over major populated areas where they could establish a rival government. The Kabul government still controlled 85 percent of the country's 398 districts at the end of 2015 while the Taliban controlled only about 7 percent, with the remaining districts being contested between the two. But as was the case historically in Afghanistan, government control in rural areas generally does not extend much beyond the district center whether insurgents are present or not.

Nor is the conflict just between the Taliban and the Kabul government. Worsening security has encouraged the formation (or revival) of autonomous local militias that defend their own interests against outsiders of all sorts. Ghani's attempt to shut down militias in Kunduz without providing additional national army troops to replace them was one reason the Taliban were able to take the city.

Despite their successes, the Taliban faced a classic Afghan insurgent political dilemma. Popular support for insurgencies based on driving foreigners out declines when the foreigners actually leave, and insurgents are then faced with fractured movements divided by local issues rather than united by national ones. This inherent problem was worsened by the July 2015 revelation that Mullah Omar had died more than two years earlier, in April 2013. As Commander of the Faithful and ruler of the Islamic Emirate of Afghanistan, he had been such a key figure of unity for the Taliban that his death was hidden from his followers. This allowed the Taliban leadership to continue issuing orders and opinions in his name, avoiding the contentious question of succession.

Confirmation of Omar's death came to light only when it was leaked to sabotage the opening of a second round of peace talks in Pakistan between the Taliban and the Kabul government. Mullah Akhtar Mansoor then declared himself leader of the Quetta *shura* (consultative council) and the new Commander of the Faithful. (Mansoor, a senior member of the *shura*, had been acting as Mullah Omar's deputy before he died.) This provoked violent opposition from some Taliban leaders inside Afghanistan, which Mansoor suppressed. Other dissident factions proclaimed their independence from Mansoor and the Taliban by swearing allegiance to Abu Bakr al-Baghdadi, the self-declared caliph of the Islamic State (ISIS).

More than boots on the ground, money in the pipeline was critical to the survival of any Afghan government.

ISIS Alarms

Although it was small and only loosely connected with the movement in Iraq and Syria, the branch of ISIS that appeared in Afghanistan changed the course of US policy. Disturbed by the rapid collapse of Iraqi forces that allowed ISIS to seize the country's second-largest city, Mosul, in the summer of 2014, and by the group's declaration of a transnational caliphate, the Obama administration had to reevaluate the risks of letting Afghanistan fend for itself.

While unwilling to formally scrap its policy of disengagement or send more troops, the administration quietly reversed course in 2015 and began to assist Afghan security forces directly in a number of other areas. Under the new arrangements, some Afghan forces were allowed to call in US air support, and US Special Operations units assigned as advisers to Afghan troops began fighting on their own as well, helping the army recover Kunduz and hold its ground in Helmand. Drone strikes that had been directed primarily at al-Qaeda leaders in Pakistan now targeted leaders of the ISIS faction in Afghanistan. Talk of reducing the US troop presence to an embassy guard by the end of Obama's presidency was shelved. This shift was facilitated by a cooperative relationship with the national unity government, which welcomed all the help it could get.

Despite the renewed focus on radical jihadists as a common transnational threat, ISIS and the Taliban differ in many significant ways. Mullah Omar never sought a worldwide caliphate, only an Islamic emirate in Afghanistan. Since 2001 the Taliban have also moved away from a purely religious identity to burnish their credentials as Afghan nationalists. Having to live down a reputation for intolerance and mismanagement from their time ruling the country, they have softened many of their positions on education and women's participation in public life. Unlike ISIS, the Taliban have not labeled all Afghans opposed to them as apostates and no longer attempt to enforce the rigid Salafist religious practices that alienated so many Afghans when they were in power. It should also go without saying that the Taliban share no common ethnic or linguistic ties with Sunni insurgents in Iraq or Syria.

The Sunni–Shia sectarian differences that drive conflicts in Iraq and Syria are largely absent in Afghanistan. Insurgents in Iraq can exploit Sunni Muslim discontent at being a minority in a Shia-majority country, while in Syria they can use their opposition to the rule of the Alawite minority over a Sunni majority to build support even among those who do not share their ideology. But in Afghanistan, 85 percent of the population is Sunni, as are almost all the leading figures of the government in Kabul.

Similarly, the Taliban can get only limited traction by exploiting ethnic divisions. They are mostly Pashtun and strongest in the Pashtun regions of the east and south. But the top national leaders in Kabul, including Karzai and Ghani, are also Pashtun. To the extent that the Taliban have sought to exploit Pashtun grievances over being made to share power with other ethnic groups, they have undercut their own ability to expand in non-Pashtun regions.

Those former Taliban factions that did ally themselves with ISIS after 2014 did so primarily as a way to maintain their autonomy and gain new resources. In this respect, they were similar to other non-Taliban insurgent groups like Gulbuddin Hekmatyar's Hizb-i-Islami, which had always refused to pledge allegiance to Mullah Omar, or the Haqqani network, which cast its lot with Omar but operated independently from the Taliban's Quetta command structure.

However, there is also a new player in Afghanistan's easternmost provinces: Pakistani Taliban factions that have been pushed across the border and have reorganized themselves under the ISIS banner. While the Pakistani government has long supported the Afghan Taliban, it is at war with their Pakistani namesakes, over which it has no control. With extreme violence, they have set up an insurgency within an insurgency. The Afghan Taliban and non-Taliban local militia work in parallel to protect their communities against the Pakistanis.

Diplomatic Hopes

When the national unity government took office in 2014, Ghani made a determined effort to cultivate closer ties with Pakistan. Since Islamabad was the main if unofficial sponsor of the Taliban, its cooperation wouldbe vital in negotiations to end the insurgency. In the past, Pakistan had prevented Taliban leaders from talking with the Afghan government, and until Islamabad changed its strategic policy of disrupting Afghanistan, the war would likely continue. (As in the 1990s, this policy assumed there was no reason to negotiate because the Taliban could win the war outright once the Americans left.) For many years, the cost of such disruption had been minimal. Afghanistan was at war but Pakistan was at peace. The United States needed Pakistan's cooperation to maintain a supply route to landlocked Afghanistan and so was unwilling to make it pay much of a price for its meddling.

That free ride began to end with the emergence of a domestic Pakistani Taliban whose focus was on Islamabad rather than Kabul. It was declared a terrorist group by the United States in 2010 and its leadership increasingly became the target of US drone strikes. In a series of ever more violent terrorist attacks on civilian targets, the Pakistani Taliban struck throughout the country, making many Pakistani cities more dangerous than Afghan ones.

The Pakistani army eventually responded by moving troops into the militants' northwestern strongholds, the Swat Valley and the so-called tribal areas of North and South Waziristan, pushing many of them across the border into Afghanistan. It was with no small sense of irony that Kabul responded coolly to complaints from Islamabad that insurgents were now using Afghan territory to stage attacks on Pakistan, which had long allowed militants to cross the border in the other direction.

Ghani was hopeful when Pakistan facilitated official talks with the Taliban in the summer resort of Murree in July 2015, a step that Pakistani Prime Minister Nawaz Sharif called "a major breakthrough." But the talks soon broke down; the Taliban were consumed with their succession struggle due to the death of Mullah Omar. Ghani's broader outreach to Pakistan, which included domestically unpopular security arrangements with the Pakistani military, fell apart after an enormous truck bomb leveled a poor neighborhood in Kabul and inflicted hundreds of casualties in early August. Blamed on the Pakistan-backed Haqqani network, the bombing forced Ghani into an admission that his policy of rapprochement had failed. Talks to strengthen ties with India, which had been put on hold in deference to Pakistan's concerns, now went forward.

China's new interest in Afghanistan constitutes the one remaining hope for peace talks between the government and the Taliban. Chinese President Xi Jinping made a state visit to Islamabad in April 2015 and announced a plan to build a $46 billion China–Pakistan Economic Corridor featuring a network of roads, railways, and pipelines linking Xinjiang in western China with Pakistan's Indian Ocean port of Gwadar in Baluchistan province. Such a huge investment could not be made if active insurgencies in both Afghanistan and the border regions of Pakistan persisted.

Also, China had long expressed its concerns about the danger Islamic radicals posed to its restive province of Xinjiang (which borders both Afghanistan and Pakistan), where the Muslim Uighur minority was increasingly viewed as potentially subversive. In May 2015, Beijing arranged for secret negotiations between the Taliban and the Kabul government in Urumqi, Xinjiang's capital, and in January 2016 it declared that as "a peaceful mediator of the Afghan issue, China supports the 'Afghan-led and Afghan-owned' reconciliation process."

Iran is another new player that might also help broker peace in Afghanistan. Released from an international sanctions and diplomatic isolation in January 2016, it has more in common with the United States than Pakistan on the question of stabilizing Afghanistan. Neither the United States nor Iran, which are fighting in parallel in Iraq and Syria, wants to see another state besieged by Sunni jihadists. That would bring an unwelcome wave of new Afghan refugees to Iran, which still hosts perhaps two million from previous wars.

Paying the Price

If international peace talks fail to gain traction, the more likely outcome is a continued stalemate in Afghanistan. But there is always the possibility of an internal deal among the Afghans themselves. What divides Afghans more than ethnicity or ideology is an unwillingness to share power and a winner-takes-all approach to politics. Historically this led to domestic stability after wars were settled because the losers either did not have the means to keep fighting or deemed it prudent to join the winners. Periods of intense violence were followed by long periods of peace. But this changed after 1978, when domestic factions began to draw on outside resources that encouraged continued fighting rather than compromise or surrender.

By playing on Cold War fears, the People's Democratic Party of Afghanistan gained Soviet support. The mujahedeen used the same Cold War fears to get money and weapons from the United States, and they played the Islamic brotherhood card to obtain funding from Saudi Arabia. With the fall of the Soviet Union, the Afghan state collapsed, but the subsequent civil war was funded by new flows of international support. Private donors in the Arab world such as Osama bin Laden backed the Taliban's Islamic emirate. Pakistan also gave money and arms to the Taliban in hopes of gaining "strategic depth" and an ally against India.

When Al-Qaeda, based in Afghanistan, attacked the United States in 2001, US forces and their Afghan allies toppled the Taliban in 10 weeks but failed to follow up on the victory. Washington did nothing to midwife a structure of peace at a time when all factions were open to reconciliation. Instead, Afghanistan drifted back into a vortex of conflict that fed on itself. Today, the United States and its allies fund the Kabul government while Pakistan funnels support to its opponents.

Throughout this period of international intervention, Afghan factional leaders (in the government or among the insurgents) have been more intransigent than their followers. Supported by outside money and arms, they have assumed there is no need to compromise since victory could be just over the horizon. Yet over the course of four decades, it never was, and it never will be. No people are more practical than the Afghans, but until they can agree among themselves that dying in conflicts over issues mostly irrelevant to their own lives is a price not worth paying, the war will continue.

Critical Thinking

1. How can China play a role in mediating the conflict in Afghanistan?
2. How has the Islamic State played a role in changing the conflict in Afghanistan?
3. Why was the relationship between President Karzai and President Obama strained?

Internet References

Afghanistan Research and Evaluation Unit
http://areu.org/af/default.aspx?

The Afghanistan Analysts Network
www.Afghanistan-analysts.org/

The Taliban in Afghanistan
http://www.cfr.org/afghanistan/taliban-afghanistan/p/0551

Thomas Barfield is a professor of anthropology at Boston University. His books include *Afghanistan: A Political and Cultural History* (Princeton University Press, 2010).

Article

Prepared by: Robert Weiner, *University of Massachusetts, Boston*

Water Wars

A Surprisingly Rare Source of Conflict

GREGORY DUNN

Learning Outcomes

After reading this article, you will be able to:

- Understand the problems caused by the growing scarcity of freshwater.
- Understand what has contributed to the growing scarcity of freshwater.

Water seems an unlikely cause of war, but many commentators believe it could define 21st century conflict. A February 2013 article in U.S. News and World Report warns that "the water-war surprises will come," and laments that "traditional statesmanship will only take us so far in heading off water wars." A 2012 article in Al Jazeera notes that "strategists from Israel to Central Asia" are preparing for strife caused by water conflict. Even the United States National Intelligence Estimate predicts wars over water within 10 years. Their concern is understandable—humanity needs fresh water to live, but a rise in population coupled with a fall in available resources would seem to be a perfect catalyst for conflict. This thinking, although intuitively appealing, has little basis in reality—humans have contested water supplies for ages, but disputes over water tend to be resolved via cooperation, rather than conflict. Water conflict, rather than being a disturbing future source of conflict, is instead a study in the prevention of conflict through negotiation and agreement.

To understand the problems with arguments about the importance of water wars, it is first important to understand the arguments themselves. Drinking water is fundamentally necessary for humans to survive, and thus every human needs a reliable source of water to survive. If people are denied access to water they face death, and thus are more likely to go to war—even a war with only a small chance of resulting in access to water is preferable to certain death through dehydration. In ancient times, this sort of calculus was not necessary, since migration allowed humans to travel to areas that had water if water supplies were exhausted or inaccessible. However, the development of nations, cities, and governments has restricted the extent to which humans can migrate in pursuit of clean water. Additionally, in some areas—notably, the deserts of the Middle East and Africa—water may be so scarce that migration is futile. Additionally, industrial growth has exacerbated water scarcity in some areas. Dammed rivers, water diversion for irrigation, the extraction of water from underground aquifers, and the pollution of water supplies has made water even scarcer for some, and, critically, climate change threatens to dry up many people's sources of water. As water becomes scarcer, people without access to water resources face the choice of fighting or dying of dehydration, and water wars erupt. These wars are not necessarily world-encompassing conflagrations, but they are deadly conflicts between armed parties spurred by water scarcity. This logic of calamity driven by resource scarcity is in many ways simply an updated version of resource scarcity-based apocalypse that have been around since Malthus.

However, a casual look at dryer areas of the world suggests that Malthusian resource scarcity might finally be occurring. In East Africa, diplomatic rows between nations along the Nile grow increasingly heated, and lack of access to water fuels Somalia's conflict and division. Many of the governments in this region have been or are currently being threatened by insurgencies, waging war against the government and thus the current system of resource allocation. Southern Asia, Pakistan, India, and Bangladesh all face issues with regards to water, and the Southern Asian region remains a source of conflict and instability. Even in the developed United States of America, drug wars rage in the Southwest of the country, a desert region supplied by rivers whose water is increasingly diverted for agricultural purposes.

Given these seemingly disturbing conditions, it is not surprising that the United States National Intelligence Estimate on Water, one of the most useful documents for understanding how nations think about water issues, predicts that beyond the year 2022 upstream nations are likely to use their ability to control water supplies coercively, and water scarcity "will likely increase the risk of instability and state failure, exacerbate regional tensions and distract countries from working with the United States on important policy objectives." There is little doubt that climate change will deny people access to the water they need to survive, which seems a convincing argument that future conflict will occur to secure this valuable resource.

A Familiar Concern

However, this analysis does not take into account the economic, geopolitical, and governmental contexts that such changes will occur in. Economic growth, international organizations, and political leaders are powerful forces that dampen the tendency for water scarcity to cause conflict. The most powerful reason why the future does not hold water wars is the reason typically used to refute Malthusian arguments—technological and economic growth. Malthus correctly predicted the explosion in human population, and the amount of humans on earth would increase by five billion by the year 2000. However, the collapse of society Malthus envisioned failed to occur. The failure of human society to collapse was largely due to the economic and technological developments that occurred around the world. Economic growth allowed more access to resources, thus enabling people to invest in technology to increase their productivity. This investment in technology enabled incredible leaps in the productivity of farmers, thanks to devices like tractors, new practices in irrigation and crop rotation, and improvements in crops due to breeding and genetic modification. Although the data is somewhat inconclusive, estimates in literature reviewing the increase in farming productivity agree that farm productivity has increased many times over since the publication of the Essay on the Principle of Population, thus averting the collapse Malthus predicted.

A similar line of thought can he applied to water. Currently, many people access water from wells or rivers, sources that are susceptible to environmental changes. However, technological and economic growth allows for the development of aqueducts to service areas with little water, and the adoption of more efficient methods of using water (notably, watering plants with drip irrigation results in substantially less water loss), resulting in greater water availability. As evidenced by the development of the arid West of the United States, a lack of water does not necessarily mean that humans cannot survive, it merely means that technology and capital is required for survival. As nations continue to grow economically, they can acquire more resources and develop new technologies, such as water sanitation and treatment or desalination, to give their people better access to water, thus decreasing water scarcity over time. In fact, the University of California, San Diego's Erik Gartzke notes that global warming is associated with a reduction, rather than increase, in interstate conflict. He goes on to note that while resource depletion associated with global warming may contribute to instability, the economic growth that is associated with it results in an overall reduction of crime. Gartzke concludes that the only way climate-induced conflict might come about is if efforts to stem global warming at the expense of economic growth lead to a loss of wealth, and thus conflict. Although water scarcity may be a factor that can cause conflict, the economic development associated with modem water scarcity results in more peace, not more war. As nations develop, they gain the technology by which they can mitigate the effects of climate change, and the capital with which to implement these technological advances.

Modern times are associated with increasing rates of water depletion, but also with a rise of international institutions, diplomacy, and conflict mediation. History has shown that these forces are not always powerful enough to overcome wars fought for political or strategic reasons (notably, the Iraq war was launched to destroy the military threat of Weapons of Mass Destruction). However, water scarcity is a problem related to economic development. Thus, wars associated with water scarcity are not based in the wishes of leaders, but rather a failure of environment or leadership. International organizations are able to respond to a nation's failures, and leaders are generally willing to receive aid to complete tasks they have been unable to accomplish. Failures in water supply and distribution can be remedied with aid, which can install wells, aqueducts, and water purification facilities to improve access to clean water. Additionally, educational aid can help develop better practices for water use and conservation in an area of water scarcity. A large proportion of drinkable water is wasted or contaminated before it is available to those who need it to survive, a problem that can be solved through proper education and infrastructure development.

Examples of the power of aid to solve water issues are plentiful. In the United States, the state of California used federal assistance to construct an aqueduct from the wet North of the state to the arid South, allowing the city of Los Angeles to prosper as well as providing water to farmers along the fertile Central Valley of California. The international Non-Governmental Organization WaterAid approached the city of Takkas, in Nigeria. They installed wells, latrines, and instructed locals in best practices with regards to sanitation, resulting in a decrease in waterborne disease and an increase in water availability and thus quality of life. However, doubts about the long-term sustainability of water development projects remain since many nations do not have the capability to perform maintenance on the facilities provided to them. Thus, in terms of development, aid serves as a stopgap measure, providing critical water resources until economic growth allows nations to develop the infrastructure to indigenously refine and maintain water infrastructure. However, with regards to war, water aid is extremely effective, since temporary aid can be used to reduce tempers in the short term. Although a series of stop-gap measures is not substitute for indigenous production and maintenance of water supplies, stop-gap measures can prevent the humanitarian issue of water scarcity from causing international conflicts.

Although international aid and involvement are effective tools in development assistance, international aid is perhaps even more effective in aiding negotiations regarding the provision of water. Conflict over water is relatively easy to detect, since water scarcity builds over time. International tensions regarding water trigger a series of escalating diplomatic incidents and concerns that are easy to identify and thus resolve. Since the potential conflict is over a future where one or more parties lack access to water, rather than a nation's immediate needs, international organizations can foster negotiations to solve the problem before it gets out of hand.

Not Water Wars, Water Deals

Perhaps the best example of international organization facilitating water resource allocation is the Indus Waters Treaty. The Indus River, a key source of water for Pakistan, has headwaters and tributaries in both Pakistan and India. When the partition between India and Pakistan occurred, there was great animosity between the two nations, which eventually led to a series of wars. One future source of conflict was the Indus River, a river whose resources were contested by two bitter rivals. While in the late 1950s Pakistan and India were not at war, there was great potential for water to play a role in future hostilities between the nations, perhaps exacerbating conflict. At the time, the World Bank was playing an active role in the region, seeking to aid the development of the new countries. They held substantial sway in the region thanks to their ability to provide loans to the new nations, and were therefore able to bring both India and Pakistan to the negotiating table to determine use of the river. Pakistan was concerned that India could use water as a weapon in future conflict, while India was concerned that Indians (especially those in the north of the country) would be unable to access water resources that had historically been theirs. Over a period of 6 years from 1954 to 1960, the World Bank helped orchestrate talks which determined which river systems were under control of India, which systems were under control of Pakistan, and how infrastructure necessary for the control of water in the river system was to be developed and funded. In 1960, thanks in part to development assistance provided by the United States and the United Kingdom, an agreement was found and the treaty was signed. After the signing of the treaty, three wars occurred, but the treaty was not broken, a testament to the power of the international agreement. Water allocation difficulties are a problem of developing nations, since developed nations can make up for scarcity with infrastructure. Thus, developing nations are most prone to water conflict, but they are also in the most need of staying in the good graces of the international community. Therefore, these countries are quick to negotiate with international organizations, making treaties and negotiation a powerful tool in addressing water conflict.

". . .the government has little incentive to start a war over water shortages impacting those the state is already failing—their protests are inevitable, and the shortages do not impact the government."

Furthermore, the involvement of international organizations can redirect anger, turning potential conflicts into political matters. In 2000, the World Bank compelled Bolivia to privatize the water provider in Cochabamba, a large Bolivian city, to fund the construction of a dam. This move proved massively unpopular, sparking widespread riots. This anger over the provision of water was not directed at the Bolivian government, but instead the anger was directed at the World Bank, an international organization that mainly interacted with Bolivia through financial, rather than physical means. The World Bank and the privatized companies it endorsed became the targets, and thus rage was harmlessly fired at an international organization, rather than targeted upon the Bolivian government. In this way, international organizations served as a scapegoat, absorbing criticism in the place of the government, which was left alone to maintain the peace.

The government of Bolivia, like many governments in region susceptible to water conflict, was not itself affected by the water scarcity. Governments have the power, resources, and authority to find and secure water in their country, and a water shortage is generally unlikely to severely affect those within a government. Rather, a water shortage is felt most acutely by those with almost no power, little money, and few resources. Water shortages hit the poorest hard, and the government is slow to respond since governmental officials are generally not impacted by such shortages. While this might seem at first consideration like a factor that is more likely to exacerbate water conflicts by allowing scarcity to rise undetected, it is ultimately a major dampener on the chances of water war. While individual citizens may protest their condition, and in extreme cases mount ail insurgency, these actions are unlikely to have a substantial effect on the country. The most powerless in a country already have much to protest about, and the addition of water scarcity is unlikely to dramatically alter the frequency or fervor of protest. The government of a nation must expect that some citizens cannot be fully provided for, and therefore protests are inevitable. The propensity of water shortages to impact this segment of the population means that the net effect of water shortage will be relatively small, reducing the necessity of the government to respond to the crisis. Even an insurgency will be mounted by those with many grievances and few resources, which makes the insurgency comparatively simple to combat. Critically, the government has little incentive to start a war over water shortages impacting those the state is already failing—their protests are inevitable, and the shortages do not impact the government. While water shortages will of course trigger mass protest if enough of the population is impacted, the tendency of water shortages to prey upon the most vulnerable makes the onset of such mass protest less likely.

The idea of water wars fits many contemporary narratives well. In an era where we are forced to face the consequences of economic growth—pollution, climate change, and unrest—water wars seems a convenient instance of our failure to properly safeguard our natural resources. While it is easy to think of local consequences of the corruption of natural resources (for example, lung cancer resulting from air pollution), it is more difficult to give examples of widespread social change spurred by pollution. Despite a litany of international conferences issuing increasingly urgent manifestos demanding dramatic change, society has changed its patterns of consumption comparatively little, with seemingly few more widespread societal (rather than local) consequences. Although global warming threatens to destroy our way of life, society has not responded to the impacts of a warmer climate. Water wars seem to make up for this lack of action, since they are a powerful social problem easily attributable to the degradation of national resources. However, they have so far failed to meaningfully transpire, thanks to the very forces—the international geopolitical order and economic growth—that would presumably cause water wars in the first place. While the degradation of natural resources is a serious problem with modern society, the lack of water wars serves as a reminder of the power of the forces of peace and prosperity that are an inherent part of the modern world.

Critical Thinking

1. Why was it expected that the growing scarcity of water in the 21st century would lead to growing conflict?
2. What factors explain the low incidence of water wars?
3. Provide an example of a water war.

Internet References

International Freshwater Treaties

http://ocid.macse.org/4fdd/treaties.php

The International Water Events Data Base

http://www.transboundarywaters.orst.edu/database/event_bar_scale.html

Article

Prepared by: Robert Weiner, *University of Massachusetts, Boston*

Taiwan's Dire Straits

JOHN J. MEARSHEIMER

Learning Outcomes

After reading this article, you will be able to:

- Understand how to apply the theory of offensive realism to the rise of China.
- Understand the Chinese approach to world order.

What are the implications for Taiwan of China's continued rise? Not today. Not next year. No, the real dilemma Taiwan will confront looms in the decades ahead, when China, whose continued economic growth seems likely although not a sure thing, is far more powerful than it is today.

Contemporary China does not possess significant military power; its military forces are inferior, and not by a small margin, to those of the United States. Beijing would be making a huge mistake to pick a fight with the American military nowadays. China, in other words, is constrained by the present global balance of power, which is clearly stacked in America's favor.

But power is rarely static. The real question that is often overlooked is what happens in a future world in which the balance of power has shifted sharply against Taiwan and the United States, in which China controls much more relative power than it does today, and in which China is in roughly the same economic and military league as the United States. In essence: a world in which China is much less constrained than it is today. That world may seem forbidding, even ominous, but it is one that may be coming.

It is my firm conviction that the continuing rise of China will have huge consequences for Taiwan, almost all of which will be bad. Not only will China be much more powerful than it is today, but it will also remain deeply committed to making Taiwan part of China. Moreover, China will try to dominate Asia the way the United States dominates the Western Hemisphere, which means it will seek to reduce, if not eliminate, the American military presence in Asia. The United States, of course, will resist mightily, and go to great lengths to contain China's growing power. The ensuing security competition will not be good for Taiwan, no matter how it turns out in the end. Time is not on Taiwan's side. Herewith, a guide to what is likely to ensue between the United States, China, and Taiwan.

In an ideal world, most Taiwanese would like their country to gain de jure independence and become a legitimate sovereign state in the international system. This outcome is especially attractive because a strong Taiwanese identity—separate from a Chinese identity—has blossomed in Taiwan over the past 65 years. Many of those people who identify themselves as Taiwanese would like their own nation-state, and they have little interest in being a province of mainland China.

According to National Chengchi University's Election Study Center, in 1992, 17.6 percent of the people living in Taiwan identified as Taiwanese only. By June 2013, that number was 57.5 percent, a clear majority. Only 3.6 percent of those surveyed identified as Chinese only. Furthermore, the 2011 Taiwan National Security Survey found that if one assumes China would not attack Taiwan if it declared its independence, 80.2 percent of Taiwanese would in fact opt for independence. Another recent poll found that about 80 percent of Taiwanese view Taiwan and China as different countries.

However, Taiwan is not going to gain formal independence in the foreseeable future, mainly because China would not tolerate that outcome. In fact, China has made it clear that it would go to war against Taiwan if the island declares its independence. The antisecession law, which China passed in 2005, says explicitly that "the state shall employ nonpeaceful means and other necessary measures" if Taiwan moves toward de jure independence. It is also worth noting that the United States does not recognize Taiwan as a sovereign country, and according to President Obama, Washington "fully supports a one-China policy."

Thus, the best situation Taiwan can hope for in the foreseeable future is maintenance of the status quo, which means de facto independence. In fact, over 90 percent of the Taiwanese surveyed this past June by the Election Study Center favored maintaining the status quo indefinitely or until some later date.

The worst possible outcome is unification with China under terms dictated by Beijing. Of course, unification could happen in a variety of ways, some of which are better than others. Probably the least bad outcome would be one in which Taiwan ended up with considerable autonomy, much like Hong Kong enjoys today. Chinese leaders refer to this solution as "one country, two systems." Still, it has little appeal to most Taiwanese. As Yuan-kang Wang reports: "An overwhelming majority of Taiwan's public opposes unification, even under favorable circumstances. If anything, longitudinal data reveal a decline in public support of unification."

In short, for Taiwan, de facto independence is much preferable to becoming part of China, regardless of what the final political arrangements look like. The critical question for Taiwan, however, is whether it can avoid unification and maintain de facto independence in the face of a rising China.

What about China? How does it think about Taiwan? Two different logics, one revolving around nationalism and the other around security, shape its views concerning Taiwan. Both logics, however, lead to the same endgame: the unification of China and Taiwan.

The nationalism story is straightforward and uncontroversial. China is deeply committed to making Taiwan part of China. For China's elites, as well as its public, Taiwan can never become a sovereign state. It is sacred territory that has been part of China since ancient times, but was taken away by the hated Japanese in 1895—when China was weak and vulnerable. It must once again become an integral part of China. As Hu Jintao said in 2007 at the Seventeenth Party Congress: "The two sides of the Straits are bound to be reunified in the course of the great rejuvenation of the Chinese nation."

The unification of China and Taiwan is one of the core elements of Chinese national identity. There is simply no compromising on this issue. Indeed, the legitimacy of the Chinese regime is bound up with making sure Taiwan does not become a sovereign state and that it eventually becomes an integral part of China.

The continuing rise of China will have huge consequences for Taiwan, almost all of which will be bad.

Chinese leaders insist that Taiwan must be brought back into the fold sooner rather than later and that hopefully it can be done peacefully. At the same time, they have made it clear that force is an option if they have no other recourse.

The security story is a different one, and it is inextricably bound up with the rise of China. Specifically, it revolves around a straightforward but profound question: How is China likely to behave in Asia over time, as it grows increasingly powerful? The answer to this question obviously has huge consequences for Taiwan.

The only way to predict how a rising China is likely to behave toward its neighbors as well as the United States is with a theory of great-power politics. The main reason for relying on theory is that we have no facts about the future, because it has not happened yet. Thomas Hobbes put the point well: "The present only has a being in nature; things past have a being in the memory only; but things to come have no being at all." Thus, we have no choice but to rely on theories to determine what is likely to transpire in world politics.

My own realist theory of international relations says that the structure of the international system forces countries concerned about their security to compete with each other for power. The ultimate goal of every major state is to maximize its share of world power and eventually dominate the system. In practical terms, this means that the most powerful states seek to establish hegemony in their region of the world, while making sure that no rival great power dominates another region.

To be more specific, the international system has three defining characteristics. First, the main actors are states that operate in anarchy, which simply means that there is no higher authority above them. Second, all great powers have some offensive military capability, which means they have the wherewithal to hurt each other. Third, no state can know the intentions of other states with certainty, especially their future intentions. It is simply impossible, for example, to know what Germany's or Japan's intentions will be toward their neighbors in 2025.

In a world where other states might have malign intentions as well as significant offensive capabilities, states tend to fear each other. That fear is compounded by the fact that in an anarchic system there is no night watchman for states to call if trouble comes knocking at their door. Therefore, states recognize that the best way to survive in such a system is to be as powerful as possible relative to potential rivals. The mightier a state is, the less likely it is that another state will attack it. No Americans, for example, worry that Canada or Mexico will attack the United States, because neither of those countries is strong enough to contemplate a fight with Uncle Sam.

But great powers do not merely strive to be the strongest great power, although that is a welcome outcome. Their ultimate aim is to be the hegemon—which means being the only great power in the system.

What exactly does it mean to be a hegemon in the modern world? It is almost impossible for any state to achieve global hegemony, because it is too hard to sustain power around the globe and project it onto the territory of distant great powers. The best outcome a state can hope for is to be a regional hegemon, to dominate one's own geographical area. The United States has been a regional hegemon in the Western Hemisphere since about 1900. Although the United States is clearly the most powerful state on the planet today, it is not a global hegemon.

States that gain regional hegemony have a further aim: they seek to prevent great powers in other regions from duplicating their feat. Regional hegemons, in other words, do not want peer competitors. Instead, they want to keep other regions divided among several great powers, so that those states will compete with each other and be unable to focus their attention and resources on them. In sum, the ideal situation for any great power is to be the only regional hegemon in the world. The United States enjoys that exalted position today.

What does this theory say about how China is likely to behave as it rises in the years ahead? Put simply, China will try to dominate Asia the way the United States dominates the Western Hemisphere. It will try to become a regional hegemon. In particular, China will seek to maximize the power gap between itself and its neighbors, especially India, Japan, and Russia. China will want to make sure it is so powerful that no state in Asia has the wherewithal to threaten it.

It is unlikely that China will pursue military superiority so it can go on a rampage and conquer other Asian countries, although that is always possible. Instead, it is more likely that it will want to dictate the boundaries of acceptable behavior to neighboring countries, much the way the United States lets other states in the Americas know that it is the boss.

An increasingly powerful China is also likely to attempt to push the United States out of Asia, much the way the United States pushed the European great powers out of the Western Hemisphere in the 19 century. We should expect China to come up with its own version of the Monroe Doctrine, as Japan did in the 1930s.

These policy goals make good strategic sense for China. Beijing should want a militarily weak Japan and Russia as its neighbors, just as the United States prefers a militarily weak Canada and Mexico on its borders. What state in its right mind would want other powerful states located in its region? All Chinese surely remember what happened in the previous two centuries when Japan was powerful and China was weak.

Furthermore, why would a powerful China accept U.S. military forces operating in its backyard? American policy makers, after all, go ballistic when other great powers send military forces into the Western Hemisphere. Those foreign forces are invariably seen as a potential threat to American security. The same logic should apply to China. Why would China feel safe with U.S. forces deployed on its doorstep? Following the logic of the Monroe Doctrine, would China's security not be better served by pushing the American military out of Asia?

Why should we expect China to act any differently than the United States did? Are Chinese leaders more principled than American leaders? More ethical? Are they less nationalistic? Less concerned about their survival? They are none of these things, of course, which is why China is likely to imitate the United States and try to become a regional hegemon.

What are the implications of this security story for Taiwan? The answer is that there is a powerful strategic rationale for China—at the very least—to try to sever Taiwan's close ties with the United States and neutralize Taiwan. However, the best possible outcome for China, which it will surely pursue with increasing vigor over time, would be to make Taiwan part of China.

Unification would work to China's strategic advantage in two important ways. First, Beijing would absorb Taiwan's economic and military resources, thus shifting the balance of power in Asia even further in Chinas direction. Second, Taiwan is effectively a giant aircraft carrier sitting off China's coast; acquiring that aircraft carrier would enhance China's ability to project military power into the western Pacific Ocean.

In short, we see that nationalism as well as realist logic give China powerful incentives to put an end to Taiwan's de facto independence and make it part of a unified China. This is clearly bad news for Taiwan, especially since the balance of power in Asia is shifting in China's favor, and it will not be long before Taiwan cannot defend itself against China. Thus, the obvious question is whether the United States can provide security for Taiwan in the face of a rising China. In other words, can Taiwan depend on the United States for its security?

Let us now consider America's goals in Asia and how they relate to Taiwan. Regional hegemons go to great lengths to stop other great powers from becoming hegemons in their region of the world. The best outcome for any great power is to be the sole regional hegemon in the system. It is apparent from the historical record that the United States operates according to this logic. It does not tolerate peer competitors.

During the 20th century, there were four great powers that had the capability to make a run at regional hegemony:

Imperial Germany from 1900 to 1918, Imperial Japan between 1931 and 1945, Nazi Germany from 1933 to 1945 and the Soviet Union during the Cold War. Not surprisingly, each tried to match what the United States had achieved in the Western Hemisphere.

How did the United States react? In each case, it played a key role in defeating and dismantling those aspiring hegemons.

An increasingly powerful China is likely to attempt to push the United States out of Asia, much the way the United States pushed the European great powers out of the Western Hemisphere.

The United States entered World War I in April 1917 when Imperial Germany looked like it might win the war and rule Europe. American troops played a critical role in tipping the balance against the Kaiserreich, which collapsed in November 1918. In the early 1940s, President Franklin Roosevelt went to great lengths to maneuver the United States into World War II to thwart Japan's ambitions in Asia and Germany's ambitions in Europe. The United States came into the war in December 1941, and helped destroy both Axis powers. Since 1945, American policy makers have gone to considerable lengths to put limits on German and Japanese military power. Finally, during the Cold War, the United States steadfastly worked to prevent the Soviet Union from dominating Eurasia and then helped relegate it to the scrap heap of history in the late 1980s and early 1990s.

Shortly after the Cold War ended, the George H. W. Bush administration's controversial "Defense Planning Guidance" of 1992 was leaked to the press. It boldly stated that the United States was now the most powerful state in the world by far and it planned to remain in that exalted position. In other words, the United States would not tolerate a peer competitor.

That same message was repeated in the famous 2002 National Security Strategy issued by the George W. Bush administration. There was much criticism of that document, especially its claims about "preemptive" war. But hardly a word of protest was raised about the assertion that the United States should check rising powers and maintain its commanding position in the global balance of power.

The bottom line is that the United States—for sound strategic reasons—worked hard for more than a century to gain hegemony in the Western Hemisphere. Since achieving regional dominance, it has gone to great lengths to prevent other great powers from controlling either Asia or Europe.

Thus, there is little doubt as to how American policy makers will react if China attempts to dominate Asia. The United States can be expected to go to great lengths to contain China and ultimately weaken it to the point where it is no longer capable of ruling the roost in Asia. In essence, the United States is likely to behave toward China much the way it acted toward the Soviet Union during the Cold War.

China's neighbors are certain to fear its rise as well, and they too will do whatever they can to prevent it from achieving regional hegemony. Indeed, there is already substantial evidence that countries like India, Japan, and Russia as well as smaller powers like Singapore, South Korea, and Vietnam are worried about China's ascendancy and are looking for ways to contain it. In the end, they will join an American-led balancing coalition to check China's rise, much the way Britain, France, Germany, Italy, Japan, and even China joined forces with the United States to contain the Soviet Union during the Cold War.

How does Taiwan fit into this story? The United States has a rich history of close relations with Taiwan since the early days of the Cold War, when the Nationalist forces under Chiang Kai-shek retreated to the island from the Chinese mainland. However, Washington is not obliged by treaty to come to the defense of Taiwan if it is attacked by China or anyone else.

Regardless, the United States will have powerful incentives to make Taiwan an important player in its anti-China balancing coalition. First, as noted, Taiwan has significant economic and military resources and it is effectively a giant aircraft carrier that can be used to help control the waters close to Chinas all-important eastern coast. The United States will surely want Taiwan's assets on its side of the strategic balance, not on China's side.

Second, America's commitment to Taiwan is inextricably bound up with U.S. credibility in the region, which matters greatly to policy makers in Washington. Because the United States is located roughly 6,000 miles from East Asia, it has to work hard to convince its Asian allies—especially Japan and South Korea—that it will back them up in the event they are threatened by China or North Korea. Importantly, it has to convince Seoul and Tokyo that they can rely on the American nuclear umbrella to protect them. This is the thorny problem of extended deterrence, which the United States and its allies wrestled with throughout the Cold War.

If the United States were to sever its military ties with Taiwan or fail to defend it in a crisis with China, that would surely send a strong signal to America's other allies in the region that they cannot rely on the United States for protection. Policy makers in Washington will go to great lengths to avoid that outcome and instead maintain America's reputation as a reliable partner. This means they will be inclined to back Taiwan no matter what.

While the United States has good reasons to want Taiwan as part of the balancing coalition it will build against China, there are also reasons to think this relationship is not sustainable over the long term. For starters, at some point in the next decade or so it will become impossible for the United States to help Taiwan defend itself against a Chinese attack. Remember that we are talking about a China with much more military capability than it has today.

In addition, geography works in China's favor in a major way, simply because Taiwan is so close to the Chinese mainland and so far away from the United States. When it comes to a competition between China and the United States over projecting military power into Taiwan, China wins hands down. Furthermore, in a fight over Taiwan, American policy makers would surely be reluctant to launch major attacks against Chinese forces on the mainland, for fear they might precipitate nuclear escalation. This reticence would also work to China's advantage.

One might argue that there is a simple way to deal with the fact that Taiwan will not have an effective conventional deterrent against China in the not-too-distant future: put America's nuclear umbrella over Taiwan. This approach will not solve the problem, however, because the United States is not going to escalate to the nuclear level if Taiwan is being overrun by China. The stakes are not high enough to risk a general thermonuclear war. Taiwan is not Japan or even South Korea. Thus, the smart strategy for America is to not even try to extend its nuclear deterrent over Taiwan.

There is a second reason the United States might eventually forsake Taiwan: it is an especially dangerous flashpoint, which could easily precipitate a Sino-American war that is not in America's interest. U.S. policy makers understand that the fate of Taiwan is a matter of great concern to Chinese of all persuasions and that they will be extremely angry if it looks like the United States is preventing unification. But that is exactly what Washington will be doing if it forms a close military alliance with Taiwan, and that point will not be lost on the Chinese people.

It is important to note in this regard that Chinese nationalism, which is a potent force, emphasizes how great powers like the United States humiliated China in the past when it was weak and appropriated Chinese territory like Hong Kong and Taiwan. Thus, it is not difficult to imagine crises breaking out over Taiwan or scenarios in which a crisis escalates into a shooting war. After all, Chinese nationalism will surely be a force for trouble in those crises, and China will at some point have the military wherewithal to conquer Taiwan, which will make war even more likely.

There was no flashpoint between the superpowers during the Cold War that was as dangerous as Taiwan will be in a Sino-American security competition. Some commentators liken Berlin in the Cold War to Taiwan, but Berlin was not sacred territory for the Soviet Union and it was actually of little strategic importance for either side. Taiwan is different. Given how dangerous it is for precipitating a war and given the fact that the United States will eventually reach the point where it cannot defend Taiwan, there is a reasonable chance that American policy makers will eventually conclude that it makes good strategic sense to abandon Taiwan and allow China to coerce it into accepting unification.

All of this is to say that the United States is likely to be somewhat schizophrenic about Taiwan in the decades ahead. On one hand, it has powerful incentives to make it part of a balancing coalition aimed at containing China. On the other hand, there are good reasons to think that with the passage of time the benefits of maintaining close ties with Taiwan will be outweighed by the potential costs, which are likely to be huge. Of course, in the near term, the United States will protect Taiwan and treat it as a strategic asset. But how long that relationship lasts is an open question.

So far, the discussion about Taiwan's future has focused almost exclusively on how the United States is likely to act toward Taiwan. However, what happens to Taiwan in the face of Chinas rise also depends greatly on what policies Taiwan's leaders and its people choose to pursue over time. There is little doubt that Taiwan's overriding goal in the years ahead will be to preserve its independence from China. That aim should not be too difficult to achieve for the next decade, mainly because Taiwan is almost certain to maintain close relations with the United States, which will have powerful incentives as well as the capability to protect Taiwan. But after that point Taiwan's strategic situation is likely to deteriorate in significant ways, mainly because China will be rapidly approaching the point where it can conquer Taiwan even if the American military helps defend the island. And, as noted, it is not clear that the United States will be there for Taiwan over the long term.

In the face of this grim future, Taiwan has three options. First, it can develop its own nuclear deterrent. Nuclear weapons are the ultimate deterrent, and there is no question that a Taiwanese nuclear arsenal would markedly reduce the likelihood of a Chinese attack against Taiwan.

Taiwan pursued this option in the 1970s, when it feared American abandonment in the wake of the Vietnam War. The United States, however, stopped Taiwan's nuclear-weapons program in its tracks. And then Taiwan tried to develop a bomb secretly in the 1980s, but again the United States found out and forced Taipei to shut the program down. It is unfortunate for

Taiwan that it failed to build a bomb, because its prospects for maintaining its independence would be much improved if it had its own nuclear arsenal.

No doubt Taiwan still has time to acquire a nuclear deterrent before the balance of power in Asia shifts decisively against it. But the problem with this suggestion is that both Beijing and Washington are sure to oppose Taiwan going nuclear. The United States would oppose Taiwanese nuclear weapons, not only because they would encourage Japan and South Korea to follow suit, but also because American policy makers abhor the idea of an ally being in a position to start a nuclear war that might ultimately involve the United States. To put it bluntly, no American wants to be in a situation where Taiwan can precipitate a conflict that might result in a massive nuclear attack on the United States.

China will adamantly oppose Taiwan obtaining a nuclear deterrent, in large part because Beijing surely understands that it would make it difficult—maybe even impossible—to conquer Taiwan. What's more, China will recognize that Taiwanese nuclear weapons would facilitate nuclear proliferation in East Asia, which would not only limit China's ability to throw its weight around in that region, but also would increase the likelihood that any conventional war that breaks out would escalate to the nuclear level. For these reasons, China is likely to make it manifestly clear that if Taiwan decides to pursue nuclear weapons, it will strike its nuclear facilities, and maybe even launch a war to conquer the island. In short, it appears that it is too late for Taiwan to pursue the nuclear option.

There was no flashpoint between the superpowers during the Cold War that was as dangerous as Taiwan will be in a Sino-American security competition.

Taiwan's second option is conventional deterrence. How could Taiwan make deterrence work without nuclear weapons in a world where China has clear-cut military superiority over the combined forces of Taiwan and the United States? The key to success is not to be able to defeat the Chinese military—that is impossible—but instead to make China pay a huge price to achieve victory. In other words, the aim is to make China fight a protracted and bloody war to conquer Taiwan. Yes, Beijing would prevail in the end, but it would be a Pyrrhic victory. This strategy would be even more effective if Taiwan could promise China that the resistance would continue even after its forces were defeated on the battlefield. The threat that Taiwan might turn into another Sinkiang or Tibet would foster deterrence for sure.

This option is akin to Admiral Alfred von Tirpitz's famous "risk strategy," which Imperial Germany adopted in the decade before World War I. Tirpitz accepted the fact that Germany could not build a navy powerful enough to defeat the mighty Royal Navy in battle. He reasoned, however, that Berlin could build a navy that was strong enough to inflict so much damage on the Royal Navy that it would cause London to fear a fight with Germany and thus be deterred. Moreover, Tirpitz reasoned that this "risk fleet" might even give Germany diplomatic leverage it could use against Britain.

There are a number of problems with this form of conventional deterrence, which raise serious doubts about whether it can work for Taiwan over the long haul. For starters, the strategy depends on the United States fighting side by side with Taiwan. But it is difficult to imagine American policy makers purposely choosing to fight a war in which the U.S. military is not only going to lose, but is also going to pay a huge price in the process. It is not even clear that Taiwan would want to fight such a war, because it would be fought mainly on Taiwanese territory—not Chinese territory—and there would be death and destruction everywhere. And Taiwan would lose in the end anyway.

Furthermore, pursuing this option would mean that Taiwan would be constantly in an arms race with China, which would help fuel an intense and dangerous security competition between them. The sword of Damocles, in other words, would always be hanging over Taiwan.

Finally, although it is difficult to predict just how dominant China will become in the distant future, it is possible that it will eventually become so powerful that Taiwan will be unable to put up major resistance against a Chinese onslaught. This would certainly be true if America's commitment to defend Taiwan weakens as China morphs into a superpower.

Taiwan's third option is to pursue what I will call the "Hong Kong strategy." In this case, Taiwan accepts the fact that it is doomed to lose its independence and become part of China. It then works hard to make sure that the transition is peaceful and that it gains as much autonomy as possible from Beijing. This option is unpalatable today and will remain so for at least the next decade. But it is likely to become more attractive in the distant future if China becomes so powerful that it can conquer Taiwan with relative ease.

So where does this leave Taiwan? The nuclear option is not feasible, as neither China nor the United States would accept a nuclear-armed Taiwan. Conventional deterrence in the form of a "risk strategy" is far from ideal, but it makes sense as long as China is not so dominant that it can subordinate Taiwan without difficulty. Of course, for that strategy to work, the United States must remain committed to the defense of Taiwan, which is not guaranteed over the long term.

Once China becomes a superpower, it probably makes the most sense for Taiwan to give up hope of maintaining its de facto independence and instead pursue the "Hong Kong strategy." This is definitely not an attractive option, but as Thucydides argued long ago, in international politics "the strong do what they can and the weak suffer what they must."

By now, it should be glaringly apparent that whether Taiwan is forced to give up its independence largely depends on how formidable China's military becomes in the decades ahead. Taiwan will surely do everything it can to buy time and maintain the political status quo. But if China continues its impressive rise, Taiwan appears destined to become part of China.

There is one set of circumstances under which Taiwan can avoid this scenario. Specifically, all Taiwanese should hope there is a drastic slowdown in Chinese economic growth in the years ahead and that Beijing also has serious political problems on the home front that work to keep it focused inward. If that happens, China will not be in a position to pursue regional hegemony and the United States will be able to protect Taiwan from China, as it does now. In essence, the best way for Taiwan to maintain de facto independence is for China to be economically and militarily weak. Unfortunately for Taiwan, it has no way of influencing events so that this outcome actually becomes reality.

When China started its impressive growth in the 1980s, most Americans and Asians thought this was wonderful news, because all of the ensuing trade and other forms of economic intercourse would make everyone richer and happier. China, according to the reigning wisdom, would become a responsible stakeholder in the international community, and its neighbors would have little to worry about. Many Taiwanese shared this optimistic outlook, and some still do.

They are wrong. By trading with China and helping it grow into an economic powerhouse, Taiwan has helped create a burgeoning Goliath with revisionist goals that include ending Taiwan's independence and making it an integral part of China. In sum, a powerful China isn't just a problem for Taiwan. It is a nightmare.

Critical Thinking

1. Why doesn't the United States recognize Taiwan as a sovereign state?
2. Will the United States go to war with China to defend Taiwan? Why or why not?
3. Do you think that China's rise as a regional hegemon in Asia is peaceful? Why or why not?

Internet References

Kissinger Institute on China and the United States
http://www.wilsoncenter.org/program/kissinger-institute-china-and-the-UnitedStates

Minister of Foreign Affairs, Republic of China
http://www.fmprc.gov.cn/eng

Minister of Foreign Affairs, Republic of China (Taiwan)
http://www.mofa.gov.tw/en

John J. Mearsheimer is the R. Wendell Harrison Distinguished Service Professor of Political Science at the University of Chicago. He serves on the Advisory Council of *The National Interest.* This article is adapted from a speech he gave in Taipei on December 7, 2013, to the Taiwanese Association of International Relations. An updated edition of his book *The Tragedy of Great Power Politics* will be published in April by W. W. Norton.

Article

Prepared by: Robert Weiner, *University of Massachusetts, Boston*

Why 1914 Still Matters

NORMAN FRIEDMAN

Learning Outcomes

After reading this article, you will be able to:

- Understand the causes of World War I.
- Understand the legacy of the war.

Today, as a century ago, the fact that war between trading nations would be ruinous does not necessarily mean that its outbreak is impossible.

Imagine that your closest trading partner is also your most threatening potential enemy. Imagine, too, that this partner is building a large navy specifically targeted at yours, hence at the overseas trade vital to you. Does that sound like the current U.S. situation with respect to China? It was certainly the British situation relative to Germany a century ago, on the eve of World War I. History never repeats, but it is often instructive to look at the mistakes of the past. The worse the mistakes, the more instructive. No one looking at the outbreak and then the course of World War I can see it as anything but a huge mistake. Hopefully we can do better.

The worst mistake, from a British point of view, was to forget that this was a maritime war. Had the British not entered the war at all, it would have been a European land war. Once Britain entered, the character of the war changed, not only because Britain was the world's dominant sea power, but also because the British Empire—including vital informal elements—was a seaborne entity, drawing much of its strength from overseas. As an island, Britain was almost impossible to invade. Centuries earlier, Sir Francis Bacon had written that he who controls the sea can take as much or as little of the war as he likes. The sea power did not have to place a mass army ashore. That was not necessarily its appropriate contribution to a coalition effort.

Our memory of World War I overwhelmingly emphasizes the blood and horror of the Western Front, to which the U.S. Army and Marines were assigned when entering the conflict in 1917. The war at sea is usually dismissed as a sideshow, at best an enabler for the more important action ashore. That view obscures the reality that the war was shaped by maritime considerations, and, at least as importantly, the potential that seaborne mobility offered the British and the Allies. The one instance of a strategic attack from the sea, Gallipoli (the Dardanelles campaign), is usually dismissed as an attempt by First Lord of the Admiralty Winston Churchill to gain publicity for the Royal Navy. In fact, it was a high-risk, high-payoff operation supported by the British cabinet for very rational reasons. That it failed does not make it a foolish bit of grandstanding. It only proves that planning and execution were extraordinarily poor. Our memory of how the war was fought obscures the fact that there were real alternatives, at least for the British.

Our present situation is more like that of the British than that of their continental allies. How well would we do in a similar situation? We were actually confronted by one during the Cold War. The U.S. Navy's Maritime Strategy was an alternative way to fight a continental war. It is still worth thinking about.

The Accidental Army

When the British entered World War I, Prime Minister Herbert Asquith expected the French and the Russians to provide the bulk of the forces on land; the British army's contribution in France was to be largely symbolic.[1] The British expected the French to hold the German army in the west while the "Russian steamroller" smashed from the east. However, Asquith casually approved War Minister Lord Herbert Kitchener's program to create massive "New Armies" (without ever being forced to explain their rationale). The British slid into creating the largest army in their history. Once that army existed, it could not be denied to the French when they found themselves in serious trouble in 1915. Once there, it could not easily be withdrawn. Most of the 800,000 British Empire troops killed in World War I died on the Western Front.

Were these horrific losses inevitable? Given the sheer depth of modern economies and the power of the defense, the war on land would surely have been a protracted bloodbath. Did it have to be a British bloodbath? Asquith was Prime Minister of the United Kingdom, not of some Franco-British combination. It was clearly in the interest of the French that the British army fought alongside theirs and helped preserve France. Was that in British interests, too? How deep should coalition parteneship cut? Could the British have fought a more maritime war? In Vietnam, in Iraq, and in Afghanistan the United States has faced the question of how far to go in support of a coalition partner.

Perhaps the saddest feature of British prewar and wartime planning was Admiral Sir John Fisher's futile attempt to point out that although (as everyone agreed) no success on the Western Front could be decisive, the Germans were extraordinarily sensitive to threats to their Baltic coast—a place accessible by sea, albeit with considerable danger. Unfortunately, Fisher made his point, both before and during the war, in an obscure, even mystical way.[2] The often-denigrated Dardanelles operation was a remnant of the abortive British maritime strategy; it was intended to help sustain Russia. Fisher's great objection was that it would swallow forces he thought could have been used more effectively in the Baltic—again, to support the Russians on what he and others thought was the decisive front.

The deeper reason for British planning failure is that almost up to the declaration of war virtually no one in London believed that there could ever be a war. It was widely accepted that, because the major economies were so closely intertwined, any war would be disastrous. The Britain of 1914 was a much more modern nation than its European partners. International finance played a larger part in the British economy than in any other. The financial sector still considers war futile: If one asks someone on Wall Street right now whether a war with China is possible, the answer is emphatically no, that would be ruinous. If the point of government is to maintain national prosperity, big wars are absurd. The British government of the years before 1914 did not, it seems, understand that those governing Germany had rather different ideas. How well do we understand how foreign governments think? Are big wars really obsolete?

Economy as Weapon = Double-Edged Sword

In effect, those in London thought that what was much later called mutual assured destruction prevailed. War fighting and therefore war planning were of little account. The British army commitment to France was much more symbolic than real, an attempt to show the French that the British would back them in the event of a crisis. This plan was accepted (though not, it seems, wholeheartedly) largely because it was far more important that prewar War Minister Richard Haldane led an influential faction in the governing Liberal Party than that the army's favored plan for deployment in France made much military sense.

The British government naturally became interested in economic attack as a means of quickly concluding any war that broke out. The Admiralty became an advocate of such warfare as a natural extension of the traditional naval economic weapon of blockade. In 1908 a prominent British economist pointed out that in a crisis the British banks, which were central to the world economic system, could attack German credit with devastating results.[3] Somewhat later the British banks pointed out that since Germany was Britain's most important trading partner, any damage would go both ways. Banking had to be omitted from the arsenal of economic weapons. It turned out that sanctions imposed on Britain's main trading partner were less than popular in the United Kingdom—and that they badly damaged the British economy which depended on trade. For example, a prohibition against trading with the enemy made it necessary to prove that every transaction was not with the enemy. It was not at all clear that the damage done to the British economy did not exceed that done to the German.

In pre-1914 Europe the single life-and-death problem for most governments was internal stability. Most thought in domestic terms. For example, the British Liberal Party resisted naval and military spending because it considered social spending vital for British stability. The tsarist government in Russia sought to create a strong peasant class as a bulwark against socialist workers (assuring grain exports, which would create the prosperous peasant class, required free access to the world grain market via the Dardanelles). However, the Austro-Hungarian government feared nationalist upheaval triggered from outside, most notably from Serbia (and was unable to promote internal reform).

German leaders thought they faced an imminent internal crisis.[4] The perceived crisis was the rise of a hostile majority in the Reichstag, the lower house of the German parliament. Although hardly comparable to the British Parliament, the Reichstag was responsible for the budget. In elections from 1890 on, the Social Democrats, whom the Kaiser and his associates considered dangerous revolutionaries, consistently won majorities of the vote, but because seats were gerrymandered they did not win a majority in the Reichstag until 1912. The German army's general staff considered itself and the army the bulwark of the regime. Although in theory the Kaiser ruled Germany, in fact he had been sidelined for several years. Army expansion, which might be associated with the sense of internal crisis, began in 1912.

The following year the nightmare became visible, as the Reichstag passed a vote of no confidence after the army exonerated an officer who had attacked a civilian in Alsace.[5] The vote did not bring down the government, because Prime Minister Theobald von Bethmann-Hollweg was responsible to the Kaiser rather than to the Reichstag. The center-left coalition shrank from rejecting the year's budget. However, there was a sense of escalating internal crisis. A member of the German General Staff told a senior Foreign Ministry official that his task for the coming year was to foment a world war, and to make it defensive for Germany so that the Reichstag would support the war.[6]

In this light, the event that precipitated the war—the assassination of the Austrian crown prince Franz Ferdinand—seems to have been much more a useful pretext than the reason the world blew up. The Kaiser was largely on the periphery of rapidly unfolding events during the crisis. He kept asking why the army was attacking France when the crisis was about Russia and Serbia. Do we understand who actually rules countries that may be hostile to us?

Internal Motivations, External Aggression

In 1912–14 the German army general staff could look back to 1870. By drawing France into a war at that time, Prussia had created the German Empire. The spoils of that war were a way of showing that it had been worthwhile, but the war was really about the internal political needs of the German state. In 1914, the general staff doubtless expected that victory would shrivel the Social Democrats (a 1907 military victory over the Hottentots in Africa had reversed their rise, though only briefly). No other military seems to have had a record of deliberately instigating war as a specific way of gaining an internal political end. After World War I, there was a general sense that the German general staff had been responsible for the war, but not to the extent that now seems apparent.

At one time a standard explanation for enmity between Britain and Germany, leading to war, was commercial rivalry. It was taken so seriously that interwar U.S. Navy war planners used British-U.S. rivalry to explain why a war might break out between the two countries. Similarly, one might see Chinese-U.S. trade rivalry as a possible cause of war. However, those concerned with commerce are too aware of how ruinous war can be. Wall Street really does prefer commercial competition to blowing apart its rivals. It has too clear an idea of what war might mean. Naval wars connected with commercial rivalry were fought before commercial and financial interests came to dominate governments. The perceived need to keep the state alive is a very different matter, and it seems to have been what propelled Germany in 1914. Do we see similar motives at work now, or in the near future? The lesson of 1914 is that others' decision to fight is far more often about internal politics than about what we may do.

The Vital Importance of Coalitions

British strategy in 1914–15 may not seem odd in itself, but it is decidedly odd in the context of other wars the British fought on the continent. Everyone in the 1914 Cabinet knew something of the Napoleonic Wars, though probably not from a strategic point of view. That was unfortunate, because they might have benefited from seeing the new war in terms of the earlier one. The British fought Napoleon as a member of a coalition. They watched their coalition partners collapse, to the point where they alone resisted Napoleon. They were forced to agree to a peace in 1801, which they rightly considered nothing more than a pause in the war—and they used that peace to consolidate what advantages they could.

Once the war against Napoleon resumed, the British wisely made it their first step to insure against invasion by blocking and then neutralizing the French and their allied fleets. Once they had been freed from the threat of invasion by the victory at Trafalgar, they could mount high-risk, high-gain operations around the periphery of Napoloeon's empire. Ultimately that meant Wellington's war on the Iberian Peninsula. Napoleon realized that he could not tolerate British resistance. Since he could not invade, he was forced into riskier and riskier operations intended to crush Britain economically. His disastrous 1812 invasion of Russia was in this category (it was intended to cut off Russian trade with Britain). The British limited their own liability on the Continent. Knowing that they could not be invaded (hence defeated), they could afford to be patient—and they won. Victory was a coalition achievement, which is why it did not matter that so many of the troops at Waterloo were not British.

World War I was shaped by the fact that Britain entered it. Until that moment, the German army staff could envisage a quick war which would end in the West with the hoped-for defeat of the French army. Once Britain was in the war, no German victory on land could be complete. Ironically, the Germans guaranteed that Britain would enter the war by building a large fleet specifically directed against it. Some current British historians have asked whether it was really worthwhile for the British of 1914 to have resisted the creation of a unified Europe under German control. They have missed the maritime point. In 1914 the British saw the Germans as a direct threat to their lives, because the Germans had been building their massive fleet. By 1914 most Britons well understood that their country lived or died by its access to the sea and to the resources of the world. The Royal Navy had worked hard for nearly 30 years to bring that message home. It resonated because it was true. In 1914 the British government would have had to fight public opinion to keep the country out of a war the Germans started.

The German decision to build a fleet seems, in retrospect, to have been remarkably casual. The fleet was completely disconnected from the war plan created by the army's general staff; it had no initial role whatsoever. The German navy came into its own only when it became clear that the army could not achieve a decision on land. Then it was not so much the big fleet (that had caught British attention prewar) but the U-boats that Admiral Alfred Tirpitz, the fleet's creator, grudgingly built. The British government might well have decided to oppose Germany in 1914 to preserve the balance of power in Europe—a historic British policy—but without the obvious threat of the German fleet its decision would not have enjoyed anything like the same level of support.

In 1939 the British again faced a continental war. Everyone in the British government had experienced World War I as a horrific bloodbath. This time the British consciously limited their liability. It helped that by 1939 they believed that the Germans could not destroy the United Kingdom by air attack (thanks to radar and modern fighters), so that as in World War I, Britain was a defensible island. Winston Churchill, who had a far more strategic viewpoint than most, certainly did not intend to surrender when the British were ejected from the continent in 1940. He understood that the overseas Empire and the overseas world could and would support Britain against Germany (which is why the Battle of the Atlantic was his greatest concern). He also understood that it would take a coalition to destroy Hitler.

During the Cold War, NATO faced a continental threat not entirely unlike that the British had faced in 1914 and in 1939. Attention was focussed on the Central Front, unfortunately so named because it was in the center between the alliance's northern and southern flanks. The U.S. Navy offered a maritime alternative, both in the 1950s and in the 1980s. Captain Peter Swartz, U.S. Navy (Retired), who chronicled the U.S. Navy's Maritime Strategy, summarized the way that a maritime power deals with a land power: It combines a coalition with its own land partner and it exploits maritime mobility to cripple the enemy army.

"Hard Thinking about the Object of War."

Not being able to end a war may seem to be a tame sort of disadvantage to the land power sweeping all before it in Europe. However, both in Napoleon's time and during World War I, the land power (France and Germany, respectively) found that it could not stop fighting. Its effort to knock the British out of the war eventually brought in enemies the land power could not handle. In Napoleon's time that was the Russians, whose territory absorbed the French army, and whose limitless mass of troops eventually helped invade France. Obviously there were many other contributions to French defeat, including Wellington's campaign in Spain, but the point is that none of that would have mattered had Napoleon been able to end the war as he liked.

In World War I the Germans found that their only leverage against the British was to attack their overseas source of strength, either at source in the United States or at sea en route to Britain. Either move was risky. Unrestricted submarine warfare against shipping led to angry reactions from the United States; in 1915–16 the German Foreign Ministry convinced the government (i.e. the general staff) to pull back. As an alternative, in 1916 the Germans organized the sabotage of munitions plants supplying the Allies, most notably Black Tom in New York Harbor. Although the U.S. government almost immediately discovered that the Germans had caused the Black Tom explosion, President Woodrow Wilson badly wanted to stay out of the war. That was not enough for the German general staff. Against Foreign Ministry opposition, it turned again in February 1917 to unrestricted submarine warfare as a way of strangling the Allies.

It was understood that resumption of such warfare would probably bring the United States into the war. With this possibility in mind, the Germans authorized their diplomats in Mexico to offer an alliance under which Mexico would regain the territory it had lost to the United States 60 years earlier: California, New Mexico, Arizona, Nevada, and Texas. Revelation of this Zimmermann Telegram helped bring the United States into the war on the Allied side. U.S. naval and industrial resources helped neutralize the German U-boat campaign in the Atlantic. The U.S. Army and Marines Corps tipped the balance of power in Europe, though it was at least as important that the British and the French became adept at all-arms warfare.

It is also possible that, in the end, the Western Front, where so much blood was spilled, was not decisive in itself. In 1918 the defense still enjoyed considerable advantages. The Germans told themselves that they could shore up their defense in the West, but in September and October 1918 their position in the south, the area in which maritime power had made Allied action possible, collapsed. Whatever they could do on the Western Front, the Germans could not spare troops to cover their southern and eastern borders. In this sense the collapse in the south (of Austria-Hungary, Turkey, and Bulgaria) may have been far more important than is generally imagined.

Maritime never meant purely naval. Success came from using land and sea forces in the right combinations. Maritime did demand hard thinking about the object of the war. In 1914, was it to preserve France or above all to defeat Germany? Because the prewar British government believed in deterrence, it never thought through this kind of question, and by the time it might have been asked, there was a huge British army in France. Withdrawal would have been difficult at best. After the disaster on the Somme in 1916, many in the British government began to ask what the British should do if they were forced to accept an unsatisfactory peace, as in 1801. Part of their answer was that phase two of the war should concentrate more on the east. That is why the British had such large forces in places like the Caucasus and the Middle East when the war ended in November 1918.[7]

A century later, we are in something like the position the British occupied in 1914. We are the world's largest trading nation, and we live largely by international trade—much of which has to go by sea. We do not have a formal empire like the British, but they and we are at the core of a commercial commonwealth which is our real source of economic strength. In a crisis our trade—our lifeblood—would be guaranteed by the U.S. and allied navies, the U.S. Navy dwarfing the others. That we depend on imports means that we have vital interests in far corners of the world. It happens that relatively few Americans understand as much, or see what happens in the Far East as central to their own prosperity. Access to our trading partners there is crucial to us, just as access to overseas trading partners (and the Empire) was a life-or-death matter for the British in 1914. Like the British in 1914, we regard war as too ruinous to be worthwhile, and we often assume that other governments take a similar view. Like the British, we are not very sensitive to the possibility that other governments' views may not match ours. A long look back at 1914 may be well worth our while.

Notes

1. Michael and Eleanor Brock, eds., H. H. Asquith: *Letters to Venetia Stanley* (Oxford, UK: Oxford University Press, 1982).
2. Holger M. Herwig, *"Luxury Fleet:" The German Imperial Navy 1888–1918* (London: Allen & Unwin, 1980).
3. Nicholas A. Lambert, *Planning Armageddon: British Economic Warfare and the First World War* (Cambridge, MA: Harvard University Press, 2012).
4. V. R. Berghahn, *Germany and the Approach of War in 1914,* second ed. (New York: St. Martin's Press, 1993).
5. Jack Beatty, *The Lost History of 1914: How the Great War Was Not Inevitable* (London: Bloomsbury, 2012).
6. David Fromkin, *Europe's Last Summer: Who Started the Great War in 1914?* (New York: Knopf, 2004).
7. Brock Millman, *Pessimism and British War Policy, 1916–1918* (London: Frank Cass, 2001).

Critical Thinking

1. Do you think that a great war is possible between the United States and China? Why or why not?
2. Why was World War I called the Great War?
3. Why did the United Kingdom declare war against Germany?

Internet References

Centenary News
http://www.centenarynews.com

National Army Museum Website
http://www.nam.ac.ukwwI

The Great War Centenary
http://www.greatwar.co.uk/events/2014-2018-www1-centenary-events-htm

Trenches on the Web
http://www.worldwar1.com

Dr. Friedman, whose "World Naval Developments" column appears monthly in *Proceedings,* is the author of *The Naval Institute Guide to World Naval Weapons Systems,* Fifth Edition, *The Fifty-Year War: Conflict and Strategy in the Cold War,* and other works. This article is based on his new book, *Fighting the Great War at Sea: Strategy, Tactics, and Technology,* forthcoming in September from the Naval Institute Press.

Article

Prepared by: Robert Weiner, *University of Massachusetts, Boston*

The Utility of Cyberpower

KEVIN L. PARKER

Learning Outcomes

After reading this article, you will be able to:

- Understand what is meant by cyberspace.
- Understand what is the relationship between realism and the defense of U.S. national interest in cyberspace.

After more than 50 years, the Korean War has not officially ended, but artillery barrages seldom fly across the demilitarized zone.[1] U.S. forces continue to fight in Afghanistan after more than 10 years, with no formal declaration of war.[2] Another conflict rages today with neither bullets nor declarations. In this conflict, U.S. adversaries conduct probes, attacks, and assaults on a daily basis.[3] The offensives are not visible or audible, but they are no less real than artillery shells or improvised explosive devices. This conflict occurs daily through cyberspace.

To fulfill the U.S. military's purpose of defending the nation and advancing national interests, today's complex security environment requires increased engagement in cyberspace.[4] Accordingly, the Department of Defense (DOD) now considers cyberspace an operational domain.[5] Similar to other domains, cyberspace has its own set of distinctive characteristics. These attributes present unique advantages and corresponding limitations. As the character of war changes, comprehending the utility of cyberpower requires assessing its advantages and limitations in potential strategic contexts.

Defining Cyberspace and Cyberpower

A range of definitions for cyberspace and cyberpower exist, but even the importance of establishing definitions is debated. Daniel Kuehl compiled 14 distinct definitions of cyberspace from various sources, only to conclude he should offer his own.[6] Do exact definitions matter? In bureaucratic organizations, definitions do matter because they facilitate clear division of roles and missions across departments and military services. Within DOD, some duplication of effort may be desirable but comes at a high cost; therefore, definitions are necessary to facilitate the rigorous analyses essential for establishing organizational boundaries and budgets.[7] In executing assigned roles, definitions matter greatly for cross-organizational communication and coordination.

No matter how important, precise definitions to satisfy all viewpoints and contexts are elusive. Consider defining the sea as all the world's oceans. This definition lacks sufficient clarity to demarcate bays or riverine waterways. Seemingly inconsequential, the ambiguity is of great consequence for organizations jurisdictionally bound at a river's edge. Unlike the sea's constant presence for millennia, the Internet is a relatively new phenomenon that continues to expand and evolve rapidly. Pursuing single definitions of cyberspace and cyberpower to put all questions to rest may be futile. David Lonsdale argued that from a strategic perspective, definitions matter little. In his view, "what really matters is to perceive the infosphere as a place that exists, understand the nature of it and regard it as something that can be manipulated and used for strategic advantage."[8] The definitions below are consistent with Lonsdale's viewpoint and suffice for the purposes of this discussion, but they are unlikely to satisfy practitioners who wish to apply them beyond a strategic perspective.

> Cyberspace: the domain that exists for inputting, storing, transmitting, and extracting information utilizing the electromagnetic spectrum. It includes all hardware, software, and transmission media used, from an initiator's input (e.g., fingers making keystrokes, speaking into microphones, or feeding documents into scanners)

to presentation of the information for user cognition (e.g., images on displays, sound emitted from speakers, or document reproduction) or other action (e.g., guiding an unmanned vehicle or closing valves).

Cyberpower: The potential to use cyberspace to achieve desired outcomes.[9]

Advantages of Wielding Cyberpower

With these definitions being sufficient for this discussion, consider the advantages of operations through cyberspace.

Cyberspace provides worldwide reach. The number of people, places, and systems interconnecting through cyberspace is growing rapidly.[10] Those connections enhance the military's ability to reach people, places, and systems around the world. Operating in cyberspace provides access to areas denied in other domains. Early airpower advocates claimed airplanes offered an alternative to boots on the ground that could fly past enemy defenses to attack power centers directly.[11] Sophisticated air defenses developed quickly, increasing the risk to aerial attacks and decreasing their advantage. Despite current cyberdefenses that exist, cyberspace now offers the advantage of access to contested areas without putting operators in harm's way. One example of directly reaching enemy decision makers through cyberspace comes from an event in 2003, before the U.S. invasion of Iraq. U.S. Central Command reportedly emailed Iraqi military officers a message on their secret network advising them to abandon their posts.[12] No other domain had so much reach with so little risk.

Cyberspace enables quick action and concentration. Not only does cyberspace allow worldwide reach, but its speed is unmatched. With aerial refueling, air forces can reach virtually any point on the earth; however, getting there can take hours. Forward basing may reduce response times to minutes, but information through fiber optic cables moves literally at the speed of light. Initiators of cyberattacks can achieve concentration by enlisting the help of other computers. By discretely distributing a virus trained to respond on command, thousands of co-opted botnet computers can instantly initiate a distributed denial-of-service attack. Actors can entice additional users to join their cause voluntarily, as did Russian "patriotic hackers" who joined attacks on Estonia in 2007.[13] With these techniques, large interconnected populations could mobilize on an unprecedented scale in mass, time, and concentration.[14]

Cyberspace allows anonymity. The Internet's designers placed a high priority on decentralization and built the structure based on the mutual trust of its few users.[15] In the decades since, the number of Internet users and uses has grown exponentially beyond its original conception.[16] The resulting system makes it very difficult to follow an evidentiary trail back to any user.[17] Anonymity allows freedom of action with limited attribution.

Cyberspace favors offense. In Clausewitz' day, defense was stronger, but cyberspace, due to the advantages listed above, currently favors the attack.[18] Historically, advantages from technological leaps erode over time.[19] However, the current circumstance pits defenders against quick, concentrated attacks, aided by structural security vulnerabilities inherent in the architecture of cyberspace.

Cyberspace expands the spectrum of nonlethal weapons. Joseph Nye described a trend, especially among democracies, of antimilitarism, which makes using force "a politically risky choice."[20] The desire to limit collateral damage often has taken center stage in NATO operations in Afghanistan, but this desire is not limited to counterinsurgencies.[21] Precision-guided munitions and small-diameter bombs are products of efforts to enhance attack capabilities with less risk of collateral damage. Cyberattacks offer nonlethal means of direct action against an adversary.[22] The advantages of cyberpower may be seductive to policymakers, but understanding its limitations should temper such enthusiasm. The most obvious limitation is that your adversary may use all the same advantages against you. Another obvious limitation is its minimal influence on nonnetworked adversaries. Conversely, the more any organization relies on cyberspace, the more vulnerable it is to cyberattack. Three additional limitations require further attention.

Cyberspace attacks rely heavily on second order effects. In Thomas Schelling's terms, there are no brute force options through cyberspace, so cyberoperations rely on coercion.[23] Continental armies can occupy land and take objectives by brute force, but success in operations through cyberspace often hinges on how adversaries react to provided, altered, or withheld information. Cyberattacks creating kinetic effects, such as destructive commands to industrial control systems, are possible. However, the unusual incidents of malicious code causing a Russian pipeline to explode and the Stuxnet worm shutting down Iranian nuclear facility processes were not ends.[24] In the latter case, only Iranian leaders' decisions could realize abandonment of nuclear technology pursuits. Similar to strategic bombing's inability to collapse morale in World War II, cyberattacks often rely on unpredictable second order effects.[25] If Rear Adm. Wylie is correct in that war is a matter of control, and "its ultimate tool . . . is the man on the scene with a gun," then operations through cyberspace can only deliver a lesser form of control.[26] Evgeny Morozov quipped, "Tweets, of course, don't topple governments; people do."[27]

Cyberattacks risk unintended consequences. Just as striking a military installation's power system may have cascading ramifications on a wider population, limiting effects through interconnected cyberspace is difficult. Marksmanship instructors teach shooters to consider their maximum range and what lies beyond their targets. Without maps for all systems, identifying maximum ranges and what lies beyond a target through cyberspace is impossible.

Defending against cyberattacks is possible. The current offensive advantage does not make all defense pointless. Even if intrusions from sophisticated, persistent attacks are inevitable, certain defensive measures (e.g., physical security controls, limiting user access, filtering and antivirus software, and firewalls) do offer some protection. Redundancy and replication are resilience strategies that can deter some would-be attackers by making attacks futile.[28] Retaliatory responses via cyberspace or other means can also enhance deterrence.[29] Defense is currently disadvantaged, but offense gets no free pass in cyberspace.

Expectations and Recommendations

The advantages and limitations of using cyberpower inform expectations for the future and several recommendations for the military.

Do not expect clear, comprehensive policy soon.[30] Articulating a comprehensive U.S. strategy for employing nuclear weapons lagged 15 years behind their first use, and the timeline for clear, comprehensive cyberspace policy may take longer.[31] Multiple interests collide in cyberspace, forcing policy makers to address concepts that traditionally have been difficult for Americans to resolve. Cyberspace, like foreign policy, exposes the tension between defaulting to realism in an ungoverned, anarchic system, and aspiring to the liberal ideal of security through mutual recognition of natural rights. Cyberspace policy requires adjudicating between numerous priorities based on esteemed values such as intellectual property rights, the role of government in business, bringing criminals to justice, freedom of speech, national security interests, and personal privacy. None of these issues is new. Cyberspace just weaves them together and presents them from unfamiliar angles. For example, free speech rights may not extend to falsely shouting fire in crowded theaters, but through cyberspace all words are broadcast to a global crowded theater.[32]

Beyond the domestic front, the Internet access creates at least one significant foreign policy dilemma. While it can help mobilize and empower dissidents under oppressive governments, it also can provide additional population control tools to authoritarian leaders.[33] The untangling of these sets of overlapping issues in new contexts is not likely to happen quickly. It may take several iterations, and it may only occur in crises. Meanwhile, the military must continue developing capabilities for operating through cyberspace within current policies.

Defend in Depth—Inner Layers

Achieving resilience requires evaluating dependencies and vulnerabilities at all levels. Starting inside the firewall and working outward, defense begins at the lowest unit level. Organizations and functions should be resilient enough to sustain attacks and continue operating. In a period of declining budgets, decision makers will pursue efficiencies through leveraging technology.[34] Therefore, prudence requires reinvesting some of the savings to evaluate and offset vulnerabilities created by new technological dependencies.[35] Future war games should not just evaluate what new technologies can provide, but also they should consider how all capabilities would be affected if denied access to cyberspace.

Beyond basic user responsibilities, forces providing defense against cyberattacks require organizations and command structures particular to their function. Martin van Creveld outlined historical evolutions of command and technological developments. Consistent with his analysis, military cyberdefense leaders should resist the technology-enabled urge to centralize and master all available information at the highest level. Instead, their organizations should act semi-independently, set low decision thresholds, establish meaningful regular information reporting, and use formal and informal communications.[36] These methods can enhance "continuous trial-and-error learning essential to collectively make sense of disabling surprises" and shorten response times.[37] Network structures may be more appropriate for this type of task than traditional hierarchical military structures.[38] Whatever the structure, military leaders must be willing to subordinate tradition and task-organize their defenses for effectiveness against cyberattacks.[39] After all, weapons "do not triumph in battle; rather, success is the product of man-machine weapon systems, their supporting services of all kinds, and the organization, doctrine, and training that launch them into battle."[40]

Defend in Depth—Outer Layers

Defending against cyberattacks takes more than firewalls. Expanding defense in depth requires creatively leveraging influence. DOD has no ownership or jurisdiction over the civilian sectors operating the Internet infrastructure and developing computer hardware and software. However, DOD systems are vulnerable to cyberattack through each of these avenues beyond their control.[41] Richard Clarke recommended federal regulation starting with the Internet backbone as the best way to overcome systemic vulnerabilities.[42] Backlash over potential legislation regulating Internet activity illustrates the problematic nature of regulation.[43] So, how can DOD effect change seemingly beyond its control? Label it "soft power" or "friendly conquest of cyberspace," but the answer lies in leveraging assets.[44]

One of DOD's biggest assets to leverage is its buying power. In 2011, DOD spent over $375 billion on contracts.[45] The military should, of course, use its buying power to insist on strict security standards when purchasing hardware and software. However, it also can use its acquisition process to reduce vulnerabilities through its use of defense contractors. Similar to detailed classification requirements, contracts should specify network security protocols for all contract firms as well as their suppliers, regardless of the services provided. Maintaining stricter security protocols than industry standards would become a condition of lucrative contracts. Through its contracts, allies, and position as the nation's largest employer, DOD can affect preferences to improve outer layer defenses.[46]

Develop an Offensive Defense

Even in defensive war, Clausewitz recognized the necessity of offense to return enemy blows and achieve victory.[47] Robust offensive capabilities can enhance deterrence by affecting an adversary's decision calculus.[48] DOD must prepare for contingencies calling for offensive support to other domains or independent action through cyberspace.

The military should develop offensive capabilities for potential scenarios but should purposefully define its preparations as defense. Communicating a defensive posture is important to avoid hastening a security-dilemma-inspired cyberarms race that may have already started.[49] Over 20 nations reportedly have some cyberwar capability.[50] Even if it is too late to slow others' offensive development, controlling the narrative remains important.[51] Just as the name Department of Defense sends a different message than its former name—War Department—developing defensive capabilities to shut down rogue cyberattackers sounds significantly better than developing offensive capabilities that "knock [the enemy] out in the first round."[52]

Do not expect rapid changes in international order or the nature of war. Without question, the world is changing, but world order does not change overnight. Nye detailed changes due to globalization and the spread of information technologies, including diffusion of U.S. power to rising nations and nonstate actors. However, he claimed it was not a "narrative of decline" and wrote, "The United States is unlikely to decay like ancient Rome or even to be surpassed by another state."[53] Adapting to current trends is necessary, but changes in the strategic climate are not as dramatic as some proclaim.

Similarly, some aspects of war change with the times while its nature remains constant. Clausewitz advised planning should account for the contemporary character of war.[54] Advances in cyberspace are changing war's character but not totally eclipsing traditional means. Sir John Slessor noted, "If there is one attitude more dangerous than to assume that a future war will be just like the last one, it is to imagine that it will be so utterly different that we can afford to ignore all the lessons of the last one."[55] Further, Lonsdale advised exploiting advances in cyberspace but not to "expect these changes to alter the nature of war."[56] Wars will continue to be governed by politics, affected by chance, and waged by people even if through cyberspace.[57]

Do not Overpromise

Advocates of wielding cyberpower must bridle their enthusiasm enough to see that its utility only exists within a strategic context. Colin Gray claimed airpower enthusiasts "all but invited government and the public to ask the wrong questions and hold air force performance to irrelevant standards of superheroic effectiveness."[58] By touting decisive, independent, strategic capabilities, airpower advocates often failed to meet such hyped expectations in actual conflicts. Strategic contexts may have occurred where airpower alone could achieve strategic effects, but more often, airpower was one of many tools employed.

Cyberpower is no different. Gray claimed, "When a new form of war is analyzed and debated, it can be difficult to persuade prophets that prospective efficacy need not be conclusive."[59] Cyberpower advocates must recognize not only its advantages, but also its limitations applied in a strategic context.

Conclusion

If cyberpower is the potential to use cyberspace to achieve desired outcomes, then the strategic context is key to understanding its utility. As the character of war changes and cyberpower joins the fight alongside other domains, military leaders must make sober judgments about what it can contribute to achieving desired outcomes. Decision makers must weigh the opportunities and advantages cyberspace presents against the vulnerabilities and limitations of operations in that domain. Sir Arthur Tedder discounted debate over one military arm or another winning wars single-handedly. He insisted, "All three arms of defense are inevitably involved, though the correct balance between them may and will vary."[60] Today's wars may involve more arms, but Tedder's concept of applying a mix of tools based on their advantages and limitations in the strategic context still stands as good advice.

Notes

1. See Chico Harlan, "Korean DMZ troops exchange gunfire," *Washington Post,* 30 October 2010, <http://www.washingtonpost.com/wp-dyn/content/article/2010/10/29/AR2010102906427.html>. Bullets occasionally fly across the demilitarized zone, but occurrences are rare.
2. See Authorization for Use of Military Force, Public Law 107–40, 107th Cong., 18 September 2001, <http://www.gpo.gov/fdsys/pkg/PLAW-107publ40/html/PLAW-107publ40.htm>. The use of military force in Afghanistan was authorized by the U.S. Congress in 2001 through Public Law 107–40, which does not include a declaration of war.
3. "DOD systems are probed by unauthorized users approximately 250,000 times an hour, over 6 million times a day." Gen. Keith Alexander, director, National Security Agency and Commander, U.S. Cyber Command (remarks, Center for Strategic and International Studies Cybersecurity Policy Debate Series: US Cybersecurity Policy and the Role of US Cybercom, Washington, DC, 3 June 2010, 5), <http://www.nsa.gov/public_info/_files/speeches_testimonies/100603_alexander_transcript.pdf>.
4. "The purpose of this document is to provide the ways and means by which our military will advance our enduring national interests . . . and to accomplish the defense objectives in the 2010 Quadrennial Defense Review." Joint Chiefs of Staff, *The National Military Strategy of the United States of America, 2011: Redefining America's Military Leadership* (Washington, DC: United States Government Printing Office [GPO], 8 February 2011), i.
5. DOD, *DOD Strategy for Operating in Cyberspace* (Washington, DC: GPO, July 2011), 5.
6. Daniel T. Kuehl, "From Cyberspace to Cyberpower: Defining the Problem," in *Cyberpower and National Security,* eds. Franklin D. Kramer, Stuart H. Starr, and Larry K. Wentz (Dulles, VA: Potomac Books, 2009): 26–28.
7. *Staff Report to the Senate Committee on Armed Services, Defense Organization: The Need for Change,* 99th Cong., 1st sess., 1985, Committee Print, 442–44.
8. David J. Lonsdale, *The Nature of War in the Information Age: Clausewitzian Future* (London: Frank Cass, 2004), 182.
9. See Joseph S. Nye, Jr., *The Future of Power* (New York: PublicAffairs, 2011), 123. This definition is influenced by the work of Nye.
10. "From 2000 to 2010, global Internet usage increased from 360 million to over 2 billion people," DOD Strategy for Operating in Cyberspace, 1.
11. Giulio Douhet, *The Command of the Air* (Tuscaloosa, AL: University of Alabama Press, 2009), 9.
12. Richard A. Clarke and Robert K. Knake, *Cyber War: The Next Threat to National Security and What to Do about It* (New York: HarperCollins Publisher, 2010), 9–10.
13. Nye, 126.
14. Audrey Kurth Cronin, "Cyber-Mobilization: The New Levée en Masse," *Parameters* (Summer 2006): 77–87.
15. Clarke and Knake, 81–84.
16. See Clarke and Knake, 84–85. Trends in the number of Internet-connected devices threaten to use up all 4.29 billion available addresses based on the original 32-bit numbering system.
17. Clay Wilson, "Cyber Crime," in *Cyberpower and National Security,* eds. Franklin D. Kramer, Stuart H. Starr, Larry Wentz (Washington, DC: NDU Press, 2009), 428.
18. Carl von Clausewitz, *On War,* ed. and trans. Michael Howard and Peter Paret (Princeton, NJ: Princeton University Press, 1976), 357; John B. Sheldon, "Deciphering Cyberpower: Strategic Purpose in Peace and War," *Strategic Studies Quarterly* (Summer 2011): 98.
19. Martin van Creveld, *Command in War* (Cambridge, MA: Harvard University Press, 1985), 231.
20. Nye, 30.
21. Dexter Filkins, "US Tightens Airstrike Policy in Afghanistan," *New York Times,* 21 June 2009, <http://www.nytimes.com/2009/06/22/world/asia/22airstrikes.html>.
22. "We will improve our cyberspace capabilities so they can often achieve significant and proportionate effects with less cost and lower collateral impact." Chairman of the Joint Chiefs of Staff *The National Military Strategy of the United States of America 2011: Redefining America's Military Leadership* (Washington, DC: GPO, 2011), 19.
23. Thomas C. Schelling, *Arms and Influence* (New Haven, CT: Yale University, 2008), 2–4.
24. For Russian pipeline, see Clarke and Knake, 93; for Stuxnet, see Nye, 127.
25. Lonsdale, 143–45.
26. Rear Adm. J.C. Wylie, *Military Strategy: A General Theory of Power Control* (Annapolis, MD: Naval Institute Press, 1989), 74.
27. Evgeny Morozov, *The Net Delusion: The Dark Side of Internet Freedom* (New York: PublicAffairs, 2011), 19.
28. Nye, 147.
29. Richard L. Kugler, "Deterrence of Cyber Attacks," *Cyberpower and National Security,* eds. Franklin D. Kramer, Stuart H. Starr, and Larry K. Wentz (Washington, DC: NDU Press, 2009), 320.
30. See United States Office of the President, *International Strategy for Cyberspace: Prosperity, Security, and Openness in a Networked World,* May 2011.
31. See Clarke and Knake, 155. International strategy for cyberspace addresses diplomacy, defense, and development in cyberspace but fails to outline relative priorities for conflicting policy interests. 31.
32. First Amendment free speech rights and their limits have been a contentious issue for decades. "Shouting fire in a crowded theater" comes from a 1919 U.S. Supreme Court case, *Schenck v. United States.* Justice Oliver Wendell Holmes' established context as relevant for limiting free speech. An "imminent lawless action" test superseded his "clear and present danger"

test in 1969, <http://www.pbs.org/wnet/supremecourt/capitalism/landmark_schenck.html>.

33. Morozov, 28.
34. "Today's information technology capabilities have made this vision [of precision logistics] possible, and tomorrow's demand for efficiency has made the need urgent." Gen. Norton Schwartz, chief of staff, U.S. Air Force, "Toward More Efficient Military Logistics," address on 29 March 2011, to the 27th Annual Logistics Conference and Exhibition, Miami, FL, <http://www.af.mil/shared/media/document/AFD-110330-053.pdf>.
35. Chris C. Demchak, *Wars of Disruption and Resilience: Cybered Conflict, Power, and National Security* (Athens, GA: University of Georgia Press, 2011), 44.
36. Van Creveld, 269–70.
37. Demchak, 73.
38. Antoine Bousquet, *The Scientific Way of Warfare: Order and Chaos on the Battlefields of Modernity* (New York: Columbia University Press, 2009), 228–29.
39. See R.A. Ratcliff, *Delusions of Intelligence: Enigma, Ultra, and the End of Secure Ciphers* (Cambridge, UK: Cambridge University Press, 2006), 229–30. Allied World War II Enigma code-breaking offers a successful example of creatively task-organizing without rigid hierarchy.
40. Colin S. Gray, *Explorations in Strategy* (Westport, CT: Praeger, 1996), 133.
41. *DOD Strategy for Operating in Cyberspace,* 8.
42. Clarke and Knake, 160.
43. Geoffrey A. Fowler, "Wikipedia, Google Go Black to Protest SOPA," *Wall Street Journal,* 18 January 2012, <http://online.wsj.com/article/SB10001424052970204555904577167873208040252.html?mod=WSJ_Tech_LEADTop>; Associated Press, "White House objects to legislation that would undermine 'dynamic' Internet," Washington Post, 14 January 2012, <http://www.washingtonpost.com/politics/courts-law/white-house-objects-to-legislation-that-would-undermine-dynamic-internet/2012/01/14/gIQAJsFcyP_story.html>.
44. "Soft power," see Nye, 81–82; "friendly conquest," see Martin C. Libicki, *Conquest in Cyberspace: National Security and Information Warfare* (Cambridge, UK: Cambridge University Press, 2007), 166.
45. U.S. Government, USASpending.gov official Web site, "Prime Award Spending Data," <http://www.usaspending.gov/explore?carryfilters=on> (18 January 2012). "2011" refers to the fiscal year.
46. DOD Web site, "About the Department of Defense," <http://www.defense.gov/about> (18 January 2012). DOD employs 1.4 million active, 1.1 million National Guard/Reserve, 718,000 civilian personnel.
47. Clausewitz, 357.
48. Kugler, "Deterrence of Cyber Attacks," 335.
49. "Many observers postulate that multiple actors are developing expert [cyber] attack capabilities." Ibid., 337.
50. Clarke and Knake, 144.
51. "Narratives are particularly important in framing issues in persuasive ways." Nye, 93–94.
52. Quote from Gen. Robert Elder as commander of Air Force Cyber Command. See Clarke and Knake, 158; Defense Tech, "Chinese Cyberwar Alert!" 15 June 2007, <http://defensetech.org/2007/06/15/chinese-cyberwar-alert>.
53. Nye, 234.
54. Clausewitz, 220.
55. John Cotesworth Slessor, *Air Power and Armies* (Tuscaloosa, AL: University of Alabama Press, 2009), iv.
56. Lonsdale, 232.
57. Clausewitz, 89.
58. Gray, 58.
59. Colin S. Gray, *Modern Strategy* (Oxford, UK: Oxford University Press, 1999), 270.
60. Arthur W. Tedder, *Air Power in War* (Tuscaloosa: University of Alabama Press, 2010), 88.

Critical Thinking

1. What is the greatest threat to U.S. cybersecurity?
2. Why is it so difficult to defend U.S. cyberspace?
3. What recommendations would you make to defend U.S. cyberspace?

Internet References

Department of Defense Strategy for Operating in Cyberspace
http://www.defense.gov/news/d20110714.cyber.pdf

Economist debates cyberwar
http://www.economist.com/debate/overview/256

National Security Agency
https://www.nsa.gov

Kevin L. Parker, U.S. Air Force, is the commander of the 100th Civil Engineer Squadron at RAF Mildenhall, United Kingdom. He holds a BS in civil engineering from Texas A&M University, an MA in human resource development from Webster University, and an MS in military operational art and science and an MPhil in military strategy from Air University. He has deployed to Saudi Arabia, Kyrgyzstan, and twice to Iraq.

Lt. Col. Kevin Parker, "The Utility of Cyberpower," *Military Review,* May/June 2014, pp 26–33. HQ. Department of the Army, US Army Combined Arms Center.

Article

Prepared by: Robert Weiner, *University of Massachusetts, Boston*

Turkey at a Tipping Point

JENNY WHITE

Learning Outcomes

After reading this article, you will be able to:

- Understand the relationship between Kemalism and Turkish National Identity.
- Discuss the impact of Erdoğan on the Turkish political system.

Something substantially different is shaping up in today's Turkey. Given the many variables in play, no one can be sure what the country will look like in 10 years. The recent autocratic turn of the pious former prime minister and now president, Recep Tayyip Erdoğan, cannot be explained simply as a form of Islamic radicalization. After more than a decade of economic growth and social reform under the Justice and Development Party (AKP), Muslim and Turkish identities have been transformed to such an extent that it is nearly impossible to assign people to one end or the other of a secular-Islamist divide, particularly that half of the population that is under 30. Many young people have heterogeneous identities, composed of seemingly contradictory positions and affiliations. Turkey is now split along more complex lines, pitting Sunni against Sunni, Sunni against Alevi (a heterodox Shia sect that makes up more than 10 percent of the population), and both pious and secular nationalists against Kurds. It could be argued that a lust for power and profit on the part of one man and his inner circle, rather than a wider cohort, has driven recent events as much as religion. This is no novelty in the world of dictators, which may well be the direction Turkey is taking.

Part of the answer to what is happening in the present lies in the past, in Kemalist practices (the legacy of Mustafa Kemal Atatürk, who founded the modern Turkish state in 1923) that still powerfully shape social and political life today. Erdoğan, threatened by recent street protests and the actions of a rival Islamic movement, has returned to the fearmongering and aggressive political paternalism that were ingrained in the Turkish psyche for much of the twentieth century, making them powerful tools for social manipulation. Kemalism has been largely dethroned, but the levers of power it developed remain in place. In the absence of Kemalist symbolism, AKP rule has taken on an Ottoman and Sunni Muslim veneer.

What is fundamentally different, though, is that Erdoğan has begun, for the first time, to dismantle the democratic structures that, creaky and biased though they were, provided a balance of power among institutions. Under Erdoğan, these institutions, from universities and the media to police, prosecutors, and judges, have been forced to answer not to a party, but essentially to one man who has taken control of most mechanisms of rule. This is a new and worrisome development, out of step with the AKP's (and Erdoğan's) accomplishments over the previous decade. Those who claim to have seen this coming could have done so only by closing their eyes to what the party accomplished—and what these newest developments put at risk.

David or Goliath?

From 2002 until 2011, the AKP attracted a wide variety of voters, drawn to its economic program, global outlook, revival of Turkey's European Union accession process, and introduction of much-needed reforms, which included placing the military under civilian control. The party profited from a reservoir of public sympathy and support after the military in 2007 and the Constitutional Court in 2008 threatened to bring the government down for alleged anti-secular activities. The AKP represented David against the military Goliath that had ousted several governments since 1960.

Once in power, the AKP reached out to minorities and former national enemies like Greece and Armenia. It broke nationalist taboos by acknowledging, to some degree, the 1915 Armenian massacres and the slaughter of Alevis at Dersim in 1937 and 1938, while pursuing a solution to the division of Cyprus, Kurdish cultural rights, and peace with the separatist Kurdistan Workers' Party (PKK).

Per capita income doubled on the AKP's watch, although unemployment remained near 10 percent, with youth unemployment much higher and women's labor force participation just 29 percent. An improved economy, social welfare, and new roads and subways brought votes, while opposition parties were ineffectual. This combination continues to be successful: About half of the population consistently votes for the AKP (43 percent in March local elections and 52 percent in the August presidential election). In other words, the AKP appears to have done well by the country, and there is no other party voters trust to keep the train on the rails.

A noticeable change in direction occurred in 2011. In a general election that June, the AKP won just under 50 percent of the vote, giving it a majority of 326 seats in the 550-seat parliament, and empowering Erdoğan to centralize power. He replaced independent thinkers in the party with loyalists who often lacked the requisite experience or expertise. The military was brought to heel through a series of trials (known as the Ergenekon and Sledgehammer cases) and the subsequent imprisonment of hundreds of high-ranking officers accused of plotting coups. In July 2011, the chief of the general staff and the commanders of the land, sea, and air forces resigned en masse; they were replaced by more tractable men. Once the threat of a military coup and dissenting voices within the party were removed, the AKP's message became narrower, focused on a romanticized notion of Ottoman Sunni brotherhood, and more intolerant. Erdoğan began to see enemies and threats everywhere, mistaking dissent and protest against government policies for coup attempts.

For most of its rule, the AKP had worked in tandem with the Hizmet movement led by the Muslim cleric Fethullah Gülen, who has lived in self-imposed exile in Pennsylvania since 1997. Hizmet excelled at setting up well-regarded schools and businesses in Turkey and abroad, with the aim of developing what Gülen called a "golden generation" of youth equipped with business and science skills and Muslim ethics, who could staff state agencies. For every embassy the AKP government opened abroad—dozens in sub-Saharan Africa alone—Hizmet would set up local schools and businesses. But relations between the AKP and Hizmet began to fray several years ago.

Hizmet is widely thought to have a heavy presence in the Turkish police and security services. In December 2013, Erdoğan accused it of being behind prosecutors and police who tried to arrest close members of his circle on corruption charges. He claimed that the investigation was a coup attempt, and that Hizmet had created a "parallel state." He transferred or fired thousands of police officers and prosecutors in a successful attempt to derail the charges. The AKP also closed down Hizmet's lucrative prep schools in Turkey and brought Bank Asya, which is associated with Gülen, to its knees by orchestrating a massive withdrawal of deposits. Each side, proclaiming its Sunni piety, has vowed to destroy the other.

The constitution is designed to protect the rights of the state, not the individual.

Out of Touch

In response to this perceived coup attempt, the AKP curtailed civil liberties, banning YouTube and Twitter after they were used to circulate taped evidence from the corruption investigation. Recently passed laws allow intrusive government surveillance and arrests of citizens for thought crimes. Given the jailing and harassment of journalists and protesters, and the impunity of the police in using violence, little is now possible in the way of freedom of speech. Erdoğan has revived the Kemalist threat paradigm, using the same language, railing against outside and inside enemies, and presenting himself in his campaign ads and speeches as the heroic savior of the nation, the patriarchal father protecting the honor of his national family and keeping the dangerous chaos of liberalism at bay.

By pulling the levers of suspicion and social polarization, Erdoğan appeals to the conservative nationalist core of his supporters, but he is out of touch with a large part of the population. There is a growing disconnect between the twenty-first-century aspirations of both pious and secular youth, who grew up in the AKP environment of great promise, and the twentieth-century values and practices of Turkey's leadership, which cannot bend to meet that promise and is preoccupied with serving its own interests. The AKP raked in enormous profits through rampant development all over the country, despoiling environments, neighborhoods, and archaeological sites. The 2013 demonstrations began as a peaceful sit-in to save Gezi Park in Istanbul's Taksim Square, then grew into a nationwide protest against the disproportionate police violence used to break it up.

The Gezi events occurred around the same time that enormous crowds filled Cairo streets to show their approval of the Egyptian army's coup against President Mohamed Morsi. Erdoğan, who felt a kinship with the Muslim Brotherhood and Morsi, clearly viewed the Gezi protests in light of the events in Egypt, convinced that the protesters were plotting to overthrow him. He responded with an all-out crackdown, including arrests of protesters under draconian terrorism laws. It is not only secular youth, however, who have taken up the call of environmentalism and other social justice issues. There has been a convergence in lifestyle and aspirations between secular and pious youth, who have developed a taste for making their own choices and demanding accountability.

Erdoğan's increasing volatility and consolidation of power have opened fissures in the AKP edifice. Party members uncomfortable with his policies dare not speak up. Many hoped that Abdullah Gül, when he stepped down from the presidency in August, would capitalize on his popularity and legitimacy by leading a moderate branch of the party, but he has disappeared from the headlines.

Even the conservative provincial folk who make up a large part of the AKP's core constituency have recoiled from the gloves-off exercise of raw power by Erdoğan and his circle, which even religious pretexts can no longer disguise. Earlier this year, many citizens were shocked by the callousness with which Erdoğan and his advisers treated family members waiting for news of their missing relatives after a mine disaster in the western town of Soma, in which 301 miners were killed. Despite media censorship, a photo of an Erdoğan aide kicking a miner went viral, as did a video of a large crowd booing the prime minister. Erdoğan was forced to take refuge in a market, where he was caught on camera punching another miner.

Another wild card is the recruitment of Turks by Islamic State (ISIS) jihadists to join the group's fighters in Iraq and Syria. Many of its recruits hail from nearby countries like Iraq and Saudi Arabia; they have moved freely across Turkey's borders and taken up residence in its cities and border towns. Turkey's largely Sunni and Alevi population has no affinity with ISIS's puritan Salafist creed and in the past has been suspicious of foreigners, including Arabs. But the weakening of physical borders as a result of the AKP's dream of a Muslim union of states in former Ottoman lands, and the breakdown of firm national and Muslim identities and proliferation of alternative practices beyond "Turkish Islam," have opened cracks in Turkish society in which ISIS can establish roots.

Sèvres Syndrome

Over the past decade, Washington slowly and somewhat reluctantly came to the realization that Turkey was no longer the pliant, army-led Kemalist ally of Cold War years, but had become a self-possessed nation with a booming economy, proactive foreign policy, global political and economic reach, and a headstrong and openly pious Muslim prime minister. Pundits initially warned that the Islam-rooted AKP was moving the country away from the West and toward the Islamic East, but that view dissipated when it became clear that Turkey was pursuing interests in Europe, sub-Saharan Africa, South America, and Asia, not just the Middle East. The new Turkish leaders imagined themselves walking in the footsteps not of Atatürk, the war hero and first president of the nation, but of the Ottomans, lords of a world empire. When the Middle East imploded in the 2011 Arab uprisings and their turbulent aftermath, Turkey seemed to be the one stable Muslim-majority country left standing in the region.

This new brand of Turkey emerged in sharp contrast to the crisis-ridden country of earlier decades. Although the Kemalist state oversaw free and fair elections that became the expected standard, the country was micromanaged socially and politically by elites positioned in state institutions and by the military, which carried out several coups when it felt that national unity was threatened by nonconforming identities and ideologies. This aggressive defensiveness, which some scholars call Turkey's Sèvres Syndrome, is a century-long hangover from the dismemberment of the Ottoman Empire by the Europeans, formalized by the 1923 Treaty of Sèvres.

Since then, in schoolbooks and a variety of rituals from grade school to adulthood, Turks have learned to be militant, to know who their enemies are, and to be suspicious of outsiders. Polls show that a majority of Turks not only lead the world in disliking the United States, but they dislike pretty much everyone else too, Muslim countries included. That hostility extends to next-door neighbors with different religious beliefs or lifestyles. A continual drumbeat of acts of intolerance against Armenians, Greek Christians, Protestants, Kurds, Alevis, Roma, Jews, and others has left deep tears in the social fabric.

Citizenship, in the sense of a contract between the nation-state and its people, was poorly developed. Schoolchildren were taught that the ideal quality was unquestioning obedience to the state, the highest expression of which would be to sacrifice their lives for it. There was little mention of what the state would provide for its citizens, aside from protection against the ever-present threat posed by what were called inside and outside enemies, the bogeymen of the nation-state. The current constitution, written under military oversight following a 1980 coup, is designed to protect the rights of the state, not the individual. Kemalism's message was one of unceasing embattlement, buttressed by conspiracy theories, and nurturing a deep-seated belief that a strong patriarchal state (*Devlet Baba*, or Father State, in popular parlance) and army were necessary in order to protect the national family and its citizen children from outsiders still hell-bent on destroying them.

Muslim Nationalism

Non-Muslim citizens and other ethnic minorities like the Kurds suffered greatly under Kemalist nationalist policies that defined them as pawns manipulated by outside powers to undermine Turkish national unity. Although Kemalists promoted a secular lifestyle, their policies were based on a religio-racial understanding of Turkishness that was contingent on being Muslim. Yet Kemalist Islam did not require piety and, indeed, eyed it with suspicion; for many years, the headscarf was barred from government offices and universities (the ban was lifted in 2013). Until the 1990s, the headscarf and other overt demonstrations of piety were associated with the rural poor and urban migrants from the countryside, both romanticized and disdained.

The Kemalist state ran a tight Islamic ship. The Presidency of Religious Affairs controlled mosques, religious teaching, and public expressions of faith. State laicism was not secularism so much as state-controlled Sunni Islam. Other faiths and forms of Islamic worship, such as the heterodox Alevi sect and officially banned but proliferating Sufi orders, coexisted in the shadows and gained adherents, including some politicians. In the 1980s, under the leadership of Necmettin Erbakan, political parties with a clear Islamist bent began to make headway in elections, but were continually closed down by the courts, only to reopen under other names. Erdoğan, Gül, and other dissidents broke away from Erbakan's Welfare Party after his government was forced out in 1997, and in 2001 they founded the AKP, which they claimed was not Islamic, but rather a secular (not laicist) party run by pious Muslims. That is, Muslimhood was a personal attribute of individual politicians, not a party ideology. The party would make policy based on pragmatic considerations, not Islam. It aimed to represent all sectors of Turkish society. And for a time, it did.

Erdoğan began to see enemies and threats everywhere, mistaking dissent and protest for coup attempts.

In the mid-1980s, Prime Minister Turgut Özal had opened Turkey's economy to the world market, unleashing provincial entrepreneurs who had been left out of state-supported industrial development. These businessmen tended to be pious, and their newly acquired wealth and dominance in social and political networks led to the rise of an Islamic bourgeoisie. Under the AKP, they have developed alternative definitions of the nation and the citizen based on a post-Ottoman rather than a republican model.

Such changes have allowed the new pious elites to experiment with expressions of Muslimhood and national identity that would not have been possible before. Muslim nationalism is based on a cultural ideal of Turkishness, rather than blood-based Turkish ethnicity. It imagines the nation with more flexible Ottoman postimperial boundaries, instead of the historically embattled republican borders. The founding moment for this ideology is not the 1923 establishment of the nation-state, but the 1453 conquest of Constantinople by the Turks, which is reenacted, visually depicted in public places, and commemorated in festivities, sometimes displacing Kemalist national rituals.

This shift has created quite a different understanding of Turkish national interests, freeing the AKP to engage with Turkey's non-Muslim minorities, open borders to Arab states by waiving visa requirements, and make global alliances and pursue economic and political interests without concern for the ethnic identity of its interlocutors or the role they played in republican history—for instance, in relations with former enemies Greece and Armenia. When it was first elected, the AKP systematically began to break down military tutelage and reach out to non-Muslims and Kurds, returning confiscated properties and allowing use of previously banned non-Turkish languages. Erdoğan began to negotiate a peace deal with the Kurdish PKK, which the government classifies as a terrorist organization, after three decades of fighting and more than 40,000 dead.

The ban on three letters of the alphabet used in Kurdish—q, w, and x—was eliminated. Education in the Kurdish language was allowed in private institutions, though not in public ones. Place names of villages and regions were restored to their Kurdish or Alevi originals. Tunceli, for instance, would once again become Dersim, reminding everyone of the state massacre of Alevis that occurred there in the 1930s (Erdoğan blamed it on the secular, Kemalist Republican People's Party, which was in power at the time).

There has been a convergence in lifestyle and aspirations between secular and pious youth.

New Identities

Kemalism as a nationalist ideology has been pushed to the margins, although nationalism itself is alive and thriving in new forms. The concept of what it means to be Turkish, which was shaped by ideological indoctrination in schools, has become more malleable in recent years, up for reinterpretation in a marketplace of identities browsed by a burgeoning middle class that is young, globalized, and desires to be modern. For the first time in republican history, an Islamic identity is associated with upward mobility. Islam is a faith, but also a lifestyle choice with its own fashions, leisure options, musical styles, and media that mirror secular society. If they choose to work, pious young women can now find jobs and arenas of activism and professional development open to them, especially since the lifting of the headscarf ban.

The 2013 protests began in response to the government's attempt to turn Gezi Park, one of central Istanbul's last parks in a city with less than 2 percent public green space, into a mall. The police violently put down the protests, but instead of making them fade away, this response provoked a spontaneous, nationwide series of mass demonstrations. Mostly young and secular, and including many women, the Gezi protesters are another product of the changes in Turkish society since the 1980s. They are global, playful, and consumerist. Turkishness is a personal attribute for them, just as the AKP suggested that Muslimhood was a personal attribute. They represent themselves, not an ideological position, a party, or a scheming foreign power. It was the first time in Turkish history that such masses of people—many with contradictory or competing interests—came together without any ideological or party organization.

The emergence of these new publics, even if only briefly, heralded an important step in Turkey's transformation away from twentieth-century values and incomplete political structures, toward a more tolerant democratic order and a civic nationalism based on citizenship rather than blood or group membership. But young people and women have little place in a political system dominated by older males. They find outlets in a civil society and in the street, but are unlikely for at least the next decade to have an impact on the system that Erdoğan is consolidating under himself—unless that system changes dramatically to permit independent voices, which at this juncture seems doubtful.

The rigidity of the political system is heightened by a widely shared majoritarian understanding of democracy in which the electoral winners get to determine what is allowed and what is banned in social life according to the norms of their community, with no room for nonconforming practices or ideas. This is true whether the issue is banning alcohol consumption or banning the veil. As Erdoğan told the Gezi protesters: If you don't agree with my decisions, win an election.

Kurdish Crisis

In October, nationwide protests by Kurdish citizens broke out against the government's refusal to help protect the Kurdish town of Kobani, just across the border in Syria, against an ISIS onslaught. The protests turned violent, leaving 40 people dead. The reluctance to act reflected Turkish perceptions that Syrian President Bashar al-Assad's survival and the strengthening of Kurdish nationalist aspirations in Syria are greater dangers to Turkey's national integrity than ISIS. The Turkish government (as well as many of its nationalist constituents who will be casting votes in the June 2015 general election) perceives the PKK as an existential threat, though Ankara is in peace talks with jailed PKK leader Abdullah Öcalan and on good terms with the Kurdistan Democratic Party (PDK) in Iraq. Indeed, Iraqi Kurdistan has become a lucrative trade partner.

If Kobani falls, the peace negotiations may be a dead letter; but one could argue that they are already on life support. The PKK appears to be experiencing a struggle for supremacy between the still-popular Öcalan and top military commander Cemil Bayık. On Ankara's side, nationalist factions in government and the military may be pushing against any accommodation with the Kurds, while others advocate continuing the talks. In October, the negotiations were proceeding in Ankara at the same moment as Turkish planes were bombing PKK militants in eastern Turkey in retaliation for the killing of three soldiers.

Turkey sought to enlist a Syrian Kurdish group, the Democratic Union Party (PYD), to help topple Assad, but was rebuffed. If there is to be an autonomous Kurdish region in Syria (which could benefit Turkey by buffering it from the Syrian war), Ankara would prefer that it not be run by the unpredictable PYD, an ally of the PKK. Ankara's recent decision to allow *peshmerga* fighters from Iraqi Kurdistan to cross into Kobani via Turkey, while rejecting international pressure to arm the PYD, is an awkward compromise. Turkey trusts the peshmerga, but Iraqi Kurds and the PYD/PKK are rivals for power, not friends.

Nevertheless, Turkey had to do something to avert another wave of refugees. In the first week after ISIS assaulted Kobani, 140,000 Syrians fled into Turkey in two days alone—a 10 percent increase in the refugee population of 1.4 million. Turkey feels it does not get enough international aid or respect for carrying this burden. Officials fear that any further influx, combined with rising unrest among the Kurds and increasing anti-refugee sentiment, could lead to major social instability. ISIS is a threat, but Ankara sees no good outcome from confronting it. The international coalition fighting ISIS seems to have no strategic goals to resolve the situation in Syria. Turkish public opinion outside of the Kurdish areas is strongly against involvement in Syria, and suspicion of the PKK is widespread. ISIS is fighting both Assad and the PYD, which seem to be the more immediate evils.

Turkey's broader foreign policy is in tatters, as illustrated in October by its humiliatingly decisive loss in a bid for a nonpermanent seat on the United Nations Security Council. The AKP's support for the Muslim Brotherhood and Hamas, both considered threats by Saudi Arabia, the Gulf states (with the exception of Qatar), Egypt, and other regimes in the region, has led not to a Sunni *Pax Ottomana,* but rather to an attenuation of diplomatic ties with these countries.

Open Wounds

Turkey is at a tipping point, held in the balance between those seeking to loosen the reins of heavy-handed paternalistic governance and those unsettled by the chaos of liberalism and desiring order and prosperity (the AKP demonized the Gezi protesters as hoodlums destroying property). Pulling the sectarian lever, however, nourishes extremism.

Within the new context of Muslim nationalism, these tensions have dangerous implications. ISIS penetration of Turkish borders is made possible partly because geographic boundaries in practice have become nearly irrelevant. Although Turkish opinion polls show widespread revulsion against ISIS, it could be argued that part of the population might be vulnerable to recruitment because boundaries of identity are also in flux. In the new post-Ottoman, globalized, commercialized environment of today's Turkey, a choosing Muslim does not have to see himself as a Turkish Muslim, and being a Turk no longer means being bounded by the borders of the nation-state. ISIS recruits are primed to embrace jihadist life by the deep structure of Turkish society, which requires obedience to a patriarchal hierarchy and submergence of selfhood, casting the citizen as self-sacrificing hero.

All of this is destabilizing Turkey internally, ripping open wounds that had partly healed after a decade of reforms. Those wounds are now vulnerable to infection by outside ideologies and actors. Erdoğan, in the meantime, is dismantling Turkey's checks and balances. Surrounded by yes-men, he has moved into his newly constructed thousand-room presidential palace in Ankara. Recently he railed against "those Lawrences" (of Arabia) in the Middle East who, he claimed, are trying to do again today what they did with the Treaty of Sèvres after World War I. Preoccupied with imagined enemies, Turkey's leader is blind to the real threat inside the gates.

Critical Thinking

1. What is the relationship between the Turkish government and the Kurds?
2. How has Turkey's foreign policy changed recently?
3. How has the national identity of Turkey changed, if at all?

Internet References

Justice and Development Party
www.akparti.org.tr/english

People's Democratic Party
https://hdpenglish.wordpress.com

Republic of Turkey, Ministry of Foreign Affairs
www.mfa.gov.tr/default.en.mfa

Jenny White is a professor of anthropology at Boston University. Her latest book is *Muslim Nationalism and the New Turks* (Princeton University Press, 2012).

Article

Prepared by: Robert Weiner, *University of Massachusetts, Boston*

The New Russian Chill in the Baltic

MARK KRAMER

Learning Outcomes

After reading this article, you will be able to:

- Understand the effect of the conflict in Ukraine on the Baltic States.
- Understand the Russian military provocations in the Baltics.

In late February and March 2014, shortly after the violent overthrow of Ukrainian President Viktor Yanukovych, Russian President Vladimir Putin sent troops to occupy the Crimean Peninsula, which had long been part of Ukraine. Putin's subsequent annexation of Crimea sparked a bitter confrontation with Western governments and stoked deep anxiety in Central and Eastern Europe about the potential for Russian military encroachments elsewhere. Nowhere has this anxiety been more acute than in Poland and the three Baltic countries—Lithuania, Latvia, and Estonia—where fears have steadily mounted as Russia has helped to fuel a civil war in eastern Ukraine while undertaking a series of military provocations in the Baltic region.

Estonia, Latvia, and Lithuania were forcibly annexed by the Soviet Union in 1940 and remained an involuntary part of it until 1991, when they were finally able to regain their independence. Their relations with post-Soviet Russia have often been tense, even though Russian troops were withdrawn on schedule from Baltic territory by 1994. As a deterrent against possible threats from Russia, all three Baltic countries pressed hard to gain membership in NATO, a status that would entitle them to protection from the United States and other alliance members. Initially, the NATO governments were skeptical about bringing in the Baltic states, but in November 2002 the allied leaders invited Estonia, Latvia, and Lithuania to join. The three were formally admitted into the alliance in 2004.

Poland, for its part, had joined NATO several years earlier. After Communist rule came to an end in Poland in 1989, a broad consensus emerged among Polish elites and the public that membership in NATO would be crucial for the country's long-term security vis-à-vis Russia and other potential threats. In the early 1990s, leaders in Washington and other NATO capitals were wary of adding new members to an alliance that already included 16 countries. Over time, however, NATO shifted in favor of enlargement, and in 1997 the member states agreed to invite Poland and two other former Warsaw Pact countries (Hungary and the Czech Republic) to join. The three formally gained membership in 1999.

The subsequent entry of Estonia, Latvia, and Lithuania into NATO brought most of the Baltic region under the alliance's auspices. The only exceptions have been Finland and Sweden, both of which have chosen thus far to remain outside military alliances, as they have since 1945. However, one of the by-products of the Russia-Ukraine confrontation and the recent spate of Russian military provocations in northern Europe has been a surge of public discussion in both Finland and Sweden about the need for closer links with NATO and even possible membership in the alliance—a step that once would have been unthinkable.

Prior Misgivings

Well before the conflict between Russia and Ukraine erupted in 2014, concerns had arisen in the Baltic region about Russia's intentions. The August 2008 Russia-Georgia war, which saw the Russian Army quickly overwhelm and defeat the much smaller Georgian military and then carve off two sizable parts of Georgia's territory (the self-declared independent republics of South Ossetia and Abkhazia), stirred doubts in the Baltic countries and Poland about the willingness of the United States and other key NATO members to defend them against Russian military pressure or intervention. Even though Georgia was not a NATO member and thus had not received any guarantee of allied protection, the televised images of Russian forces sweeping across Georgian territory and overrunning Georgian military positions came as a jolt to both elites and the wider public in the Baltic countries. Their misgivings were reinforced by the conspicuous maneuvers undertaken by Russian ground forces along the Russian-Estonian border in late 2008 and by the provocative nature of Russia's "Zapad 2009" military exercises with Belarus in September 2009, which involved simulations of rapid offensive operations against NATO.

To allay misgivings in the Baltic countries, NATO leaders in December 2009 authorized the preparation of contingency plans for the reinforcement and defense of the whole Baltic region against an unspecified enemy. Contingency planning known as Eagle Guardian already existed for the defense of Poland, but until 2009 neither the United States nor Germany had wanted to produce additional blueprints to defend the Baltic states, for fear that such an effort would damage relations with Russia if it became publicly known. Polish officials initially expressed concern that the decision to include the Baltic countries would dilute Eagle Guardian, but they were eventually willing to embrace the expanded contingency plan, provided that Poland was treated separately and that US bilateral military support would increase.

Russian actions came perilously close to provoking a military confrontation or a collision with a passenger aircraft.

The new plan designated a minimum of nine NATO divisions—from the United States, Britain, Germany, and Poland—for combat operations to repulse an attack against Poland or the Baltic countries. US and German policy makers tried to keep the revised contingency planning secret, but some details began leaking to the press in early 2010. Soon thereafter, the unauthorized release of thousands of classified US State Department documents on the WikiLeaks website, including many items pertaining to Eagle Guardian and the concerns that led to its expansion, revealed the alliance's planning for all to see.

The public disclosure of NATO's behind-the-scenes deliberations and revised Eagle Guardian plans in late 2010 and early 2011 spawned hyperbolic commentary in the Russian press and drew a harsh reaction from the Russian Foreign Ministry. High-ranking officials claimed to be "bewildered" and "dismayed" that NATO, after issuing countless "proclamations of friendship," would be treating Russia as "the same old enemy in the Cold War." The Russian ambassador to NATO, Dmitri Rogozin—a notorious hard-liner—denounced the "sinister manipulations and intrigues" of the allied governments and accused them of engaging in "warmongering," "odious discrimination," "hateful anti-Russian propaganda," and "flagrant hypocrisy."

The ensuing tensions, coming at an early stage of Barack Obama's presidency, tarnished his administration's much-ballyhooed "reset" of relations with Moscow and eroded NATO's credibility in its dealings with Russia, including its repeated statements that "NATO does not view Russia as a threat." Perhaps if Dmitri Medvedev had stayed on as Russian president (as the Obama administration expected), the damage from the disclosures would have abated relatively quickly and would not have hindered closer ties via the NATO-Russia Council. But with the return of Putin as president in 2012 and the Russian government's growing predilection for flamboyant anti-Western rhetoric and policies, the adverse impact of the revelations persisted.

Among other things, the Russian Army stepped up its military exercises, including simulations of attacks against the Baltic countries and Poland. Russia's Zapad 2011 and Zapad 2013 exercises with Belarus, which were given wide publicity in the Russian and Belarusian media, featured simulated preventive nuclear strikes against Poland and large-scale thrusts toward the Baltic countries. In March 2012, Russian combat aircraft also began to conduct simulated attacks against military sites in Sweden. The Russian authorities' shift to a more belligerent posture throughout the Baltic region began when Medvedev was still president, and any prospect for a rapprochement between NATO and Russia disappeared after Putin returned to the presidency, lending an even sharper edge to the two sides' military rivalry vis-à-vis the Baltic countries and Poland.

NATO sought to offset Russian military activities in the region by carrying out major maneuvers of its own in 2012 and 2013, especially Exercise Steadfast Jazz in the Baltic countries and Poland in early November 2013, which included more than 1,000 mechanized infantry, 2,000 other troops, 3,000 command-and-control personnel, 40 combat aircraft, 2 submarines, and 15 surface vessels. All the NATO countries as well as Finland, Sweden, and Ukraine (then still headed by Yanukovych) took part in Steadfast Jazz, which was tied to plans for a NATO Response Force capable of deploying thousands of allied troops to "defend all member states" in the Baltic region against external attack at very short notice. Before the exercise began, Russian Deputy Defense Minister Anatoly Antonov complained that it would mark a return to the "chill of the Cold War." Although the results of Steadfast Jazz and other joint exercises helped to calm nerves among Baltic leaders, apprehension in the region about Russia's intentions was mounting long before Put in authorized the annexation of Crimea or began fueling a civil war in eastern Ukraine.

Grave Threat

No sooner had the Russian government embarked on the takeover of Crimea in late February and early March 2014 than senior officials in Estonia, Latvia, and Lithuania began warning about the "grave threat" to their own countries. At an emergency meeting of European Union leaders in Brussels in early March, Lithuanian President Dalia Grybauskaite warned that "Russia today is dangerous.. . . They are trying to rewrite the borders established after the Second World War in Europe." The vice speaker of Lithuania's parliament, Petras Auštrevičius, concurred: "Russia is presenting a clear threat, and, knowing the Russian leadership, there is a great risk they might not stop with Ukraine. There is a clear risk of an extension of [Russia's military] activities." Estonian President Toomas Hendrik Ilves emphasized that "no one in [the Baltic] countries can safely assume that Russia's predatory designs will end with the seizure of Crimea."

No country would see much point in belonging to an alliance that refused to protect its members against external aggression.

Baltic leaders' concerns about the prospect of Russian aggression intensified after Putin delivered a bellicose speech before the Russian parliament on March 18, 2014, announcing the annexation of Crimea and proclaiming a duty to protect ethnic Russian populations in other countries. Senior Baltic officials and military commanders urged the United States and other leading NATO countries to reaffirm and strengthen Article 5 of the North Atlantic Treaty, which stipulates that "an armed attack against one or more of them in Europe or North America shall be considered an attack against them all" and obliges every NATO country to take "such action as it deems necessary, including the use of armed force," to repulse an attack against another NATO member state. The commander-in-chief of the Estonian Defense Forces, Major General Riho Terras, declared that although Estonia faced no immediate military threat, the events in Crimea demonstrated that the Russian Army has "a very credible capability" of "doing various things" elsewhere in Europe, especially in countries like Estonia with large ethnic Russian minorities. "It is very important," Terras warned, "that we [the members of NATO] now seriously think about defense plans based on Article 5."

Terras's comments were echoed a week later by Estonian Prime Minister Taavi Rõivas, who said that, in light of the annexation of Crimea and Russia's conspicuous military activities in the Baltic region, it would be "extremely important for the alliance," especially the United States, to deploy "boots on the ground" in the Baltic countries in order to "increase the NATO presence and defend all allies" against any possible encroachments. Only through a robust and lasting troop presence, he implied, could NATO counter "external threats" and ensure the "security and well-being" of alliance members. Lithuanian Defense Minister Juozas Olekas expressed much the same view, arguing that the "very active" Russian troop movements in Kaliningrad Oblast (the Russian exclave on the Baltic Sea) along the border with Lithuania necessitated the deployment of "NATO ground forces in the [Baltic] region and visits from NATO navies" to deter "aggression from the East." Terras's predecessor as Estonia's commander-in-chief, General Ants Laaneots, addressed the issue even more bluntly in an interview with the Estonian press a few weeks later: "Putin has brought sense back to European minds regarding military dangers. I am happy that NATO and above all the EU members [of the alliance] have woken up after twenty years of self-delusion in the field of security."

These comments by Baltic officials were in line with broader public opinion in the Baltic countries and Poland. A survey in Poland in March 2014, as Russia's takeover of Crimea was unfolding and Russian military forces in and around the Baltic Sea were engaging in unscheduled large-scale maneuvers, indicated that 59 percent of Polish adults viewed Russia as a "threat to Poland's security." Surveys in Estonia and Lithuania in late March turned up even higher shares of respondents—74 percent of Estonians and 68 percent of Lithuanians—who saw Russia as the "greatest threat" to their countries and to the "whole of Europe."

After a civil war erupted in eastern Ukraine in the late spring of 2014 with Russia's active support of separatist rebels, anxiety in the Baltic countries and Poland steadily intensified. Polish Foreign Minister Radoslaw Sikorski warned in August 2014 that Putin "has moved beyond all civilized norms" and "thinks he's facing a bunch of degenerate weaklings [in NATO]. He thinks we wouldn't go to war to defend the Baltics. You know, maybe he's right." Sikorski was hardly alone in this view. Nearly every senior official in Poland and the Baltic countries expressed great unease. Political leaders and military commanders in the region increasingly warned that Putin was intent on undermining NATO's resolve to protect their countries. The Latvian and Estonian governments noted with consternation that Russian diplomats had stepped up efforts to give Russian passports and higher pensions to ethnic Russians in Latvia and Estonia. They urged the United States and other NATO countries to take full account of the "overriding danger" posed by Russia and the "evident desire by Moscow authorities to reestablish domination over their former empire in Europe."

Surging Provocations

Russia's annexation of Crimea and sponsorship of an armed insurgency in eastern Ukraine were accompanied by a surge of Russian military provocations in and around the Baltic region. Some of these actions were targeted at the Baltic countries and Poland, whereas others were aimed at Sweden, Finland, and

Norway. Although most of the incidents were little more than shows of strength and bravado, some proved highly dangerous. In a few cases, Russian actions came perilously close to provoking a military confrontation between Russia and one or more NATO countries or a collision with a passenger aircraft.

For more than a decade after the Soviet Union collapsed, Russian military forces engaged in relatively few exercises and kept a very low profile. Long training flights for Russian combat aircraft nearly ceased, and sea patrols by naval vessels and submarines were drastically curtailed. The situation began to change gradually in the early 2000s as Russia's economy started to recover from the steep output decline that followed the disintegration of the Soviet economy. By 2006, Russian military forces had returned to higher levels of readiness and resumed activities beyond Russia's borders, including lengthy training flights through international airspace. The extent of these activities was not quite as sizable as in the Soviet era, but on a gradually increasing number of occasions from 2006 through 2013 NATO fighter aircraft intercepted Russian military planes as they approached Polish and Baltic airspace. According to NATO's Combined Air Operations Center in Üdem, Germany, allied aircraft scrambled to intercept Russian combat planes roughly 45 times a year in the Baltic region from 2011 to 2013.

This pattern changed dramatically in 2014, as tensions mounted over Crimea and eastern Ukraine. Russian military activities of all sorts in northern Europe, especially harassing NATO countries and Sweden, precipitously increased. Russian warships intruded into Baltic countries' territorial waters, including one occasion when Russian vessels engaged in live-firing exercises that severely disrupted civilian shipping throughout the region. Russian fighter and bomber aircraft repeatedly buzzed warships and other naval vessels from NATO countries and Sweden in the Baltic and North Seas, carried out simulated attacks against NATO countries and Sweden (as well as a simulated volley of air-launched cruise missile targeting North America), intruded into NATO countries' airspace, and undertook armed missions against US and Swedish reconnaissance aircraft, forcing them to take evasive maneuvers.

NATO fighter aircraft in the region were scrambled to intercept Russian planes more than 130 times in 2014, roughly triple the number of interceptions in 2013. On many occasions, NATO fighters intercepted formations of Russian bombers and tanker aircraft as they approached or entered Baltic and Norwegian airspace. After the largest such incident, in late October 2014, NATO's Allied Operations Command reported that "the bomber and tanker aircraft from Russia did not file flight plans or maintain radio contact with civilian air traffic control authorities and they were not using onboard transponders. This poses a potential risk to civil aviation as civilian air traffic control cannot detect these aircraft or ensure there is no interference with civilian air traffic."

The quantity and provocative nature of Russian aerial incursions over northern Europe in 2014 marked a sharp departure from the pattern of earlier years. A report published in late 2014 by the European Leadership Network (a London-based think tank) highlighted the magnitude of the difference in the Baltic region from January to September 2014, when "the NATO Air Policing Mission conducted 68 'hot' identification and interdiction missions along the Lithuanian border alone, and Latvia recorded more than 150 incidents of Russian planes approaching its airspace." Also, "Estonia recorded 6 violations of its airspace in 2014, as compared to 7 violations overall for the entire period between 2006 and 2013." This pattern continued in late 2014 and 2015, far exceeding the number of incidents since the height of the Cold War.

Reckless Endangerment

Of particular concern to Polish and Baltic leaders were the seemingly deliberate efforts by Russian military forces to provoke armed clashes or to endanger civilian passenger aircraft. Provocations directed against NATO countries, Sweden, and Finland occurred throughout 2014 and early 2015 but were particularly frequent in the aftermath of the controversy surrounding the downing of Malaysia Airlines flight MH17 by Russian-backed insurgents in eastern Ukraine. In early September 2014, two days after Obama traveled to Estonia and pledged strong support to the three Baltic countries, Russian state security forces kidnapped at gunpoint an Estonian Internal Security Service officer, Eston Kohver, from a border post on Estonian territory and spirited him to Moscow. As of February 2015, Kohver was still being held without trial in Moscow's notorious Lefortovo Prison on charges of espionage.

Apprehension in the region about Russia's intentions was mounting long before Putin authorized the annexation of Crimea.

The same month Kohver was abducted, Russian strategic nuclear bombers carried out simulated cruise missile attacks against North America; Russian fighters buzzed a Canadian frigate in the Black Sea while Russian naval forces engaged in maneuvers nearby; Russian medium-range bombers intruded into Swedish airspace to test the reactions of air defense forces; and Russian warships seized a Lithuanian fishing vessel in international waters of the Barents Sea and brought it to Murmansk in defiance of the Lithuanian government's protests. The next month, Russian military forces not only kept up their aerial incursions in the Baltic region (including a mission against Swedish surveillance aircraft that was deemed "unusually provocative") but also dispatched submarines on a prolonged series of intrusions into Swedish territorial waters, causing sharp bilateral tensions and nearly provoking an armed confrontation at sea. The Swedish navy received authorization to use force if necessary to bring the submarines to the surface, but a 10-day search for the intruders proved unsuccessful.

Equally disturbing was the apparent willingness—indeed eagerness—of the Russian authorities to deploy military forces in ways that endangered civilian air traffic in northern Europe. The most egregious incident of this sort occurred in March 2014, when a Russian reconnaissance plane that was not transmitting its position nearly collided with an SAS passenger airliner carrying 132 people. A fatal collision was avoided only because of the alertness and skillful reaction of the SAS pilots. Other such incidents occurred in the spring and early summer. Even after the MH17 incident in July 2014 drew opprobrium from around the world, Russian military aircraft continued to pose dangers to civilian airliners. The frequency and audacity of the incidents left no doubt that they were deliberate.

By the end of 2014, it had become clear that, as the European Leadership Network report stated, the "Russian armed forces and security agencies seem to have been authorized and encouraged to act in a much more aggressive way toward NATO countries, Sweden, and Finland." Against a backdrop of large-scale Russian military exercises and force redeployments in the Baltic region and elsewhere in 2014 and early 2015, the long series of Russian military provocations has raised troubling questions about Moscow's intentions.

Boots on the Ground?

The surge of tensions over Ukraine and Russian military provocations in the Baltic region spurred officials from Poland and the three Baltic countries to push for a strong show of resolve by NATO and a concrete reaffirmation of Article 5. The US government moved relatively quickly to allay some of these concerns, announcing in March 2014 that it would send six F-15C fighters and two KC-135 tanker aircraft to the headquarters of NATO's Baltic Air Policing Mission at the Šiauliai air base in Lithuania, joining the four F-15Cs that had been on patrol since the mission was established in 2004 when the Baltic countries (which lack their own combat aircraft) entered the alliance. The United States also deployed twelve F-15s and F-16s to Poland to assist air defense operations there and augmented the US naval presence in the Baltic Sea. Subsequently, Denmark, France, and Britain sent additional fighter planes to Šiauliai to expand the air policing mission further and relieve some of the aircraft already on patrol. NATO also expanded its surveillance of the Baltic region with extra flights of allied Airborne Warning and Control Systems planes, which provided broad coverage around the clock.

These incremental increases of NATO's military presence in the Baltic region, and a decision by alliance foreign ministers in April 2014 to "suspend all practical civilian and military cooperation between NATO and Russia," were welcomed by the Baltic and Polish governments, but they urged the United States and other large NATO countries to go further with defense preparations. Officials in Warsaw, Tallinn, Riga, and Vilnius sought the permanent stationing of allied ground and air forces on Polish and Baltic territory. Estonian Prime Minister Rõivas's plea for "boots on the ground" was echoed by other leaders in the region, who hoped that their requests would be endorsed by the NATO governments at a summit meeting in Wales on September 4–5, 2014. For years, such a step had been precluded by the NATO-Russia Founding Act, signed by Russian and NATO leaders at a Paris summit in May 1997. To mitigate Moscow's aversion to the enlargement of NATO, the Founding Act established conditions for the deployment of allied troops on the territory of newly admitted member states:

> NATO reiterates that in the current and foreseeable security environment, the Alliance will carry out its collective defense and other missions by ensuring the necessary interoperability, integration, and capability for reinforcement rather than by additional permanent stationing of substantial combat forces. Accordingly, it will have to rely on adequate infrastructure commensurate with the above tasks. In this context, reinforcement may take place, when necessary, in the event of defense against a threat of aggression.

In accordance with this provision, the United States and other allied countries had always eschewed any prolonged deployment of "substantial" military forces on the territory of new NATO members in the Baltic region.

In the lead-up to the Wales summit, the Polish and Baltic governments argued that international circumstances had fundamentally changed since 1997 and that NATO should no longer be bound by anything in the Founding Act. Referring to the clause stating that NATO and Russia no longer regarded each other as enemies, General Terras of Estonia contended in May 2014 that the whole document had become obsolete: "Russia sees NATO as a threat, and therefore NATO should not view Russia as a friendly, cooperative country. That is very clear. The threat assessment of NATO needs to fit the current realistic circumstances." Terras returned to this theme a few months later, just before the Wales summit:

> No one now believes that friendly relations between NATO and Moscow can be reestablished. Russia today regards NATO as an enemy, and this must facilitate changes [in NATO's force posture]. Some changes have already taken place, and the NATO summit must give out a clear message to the allies and to Russia that NATO

is the world's most powerful military organization and is willing to do everything to protect its member states, including increasing its presence in areas bordering Russia.

Terras and other senior military and political officials in the Baltic region also argued that in light of Russia's actions in Ukraine and elsewhere, NATO should move ahead as expeditiously as possible with concrete military preparations for "defense against a threat of aggression," as stipulated in the Founding Act.

Limited Measures

Many officials in the United States and other NATO countries were sympathetic to the arguments of Polish and Baltic leaders, but the US government ultimately decided not to proceed with long-term deployments of "substantial" military forces in the Baltic region. US officials at the Wales summit did make an effort to address Baltic and Polish concerns, not least by joining with all the other NATO allies in vowing to uphold Article 5: "The Alliance poses no threat to any country. But should the security of any Ally be threatened we will act together and decisively, as set out in Article 5 of the Washington Treaty." The summit declaration made clear that this warning was meant for Russia.

In the military sphere, however, the summit mostly just endorsed and extended the relatively limited measures that had been adopted earlier in the year to expand the Baltic Air Policing Mission and to bolster Western naval forces in the Baltic Sea. Although the Wales summit participants welcomed the fact that more than 200 military exercises had been held in Europe in 2014, the reality was that few of these exercises were of any appreciable size. Moreover, although they endorsed the deployment of "ground troops in the eastern parts of the Alliance for training and exercises," they made clear that these troops were stationed there solely "on a rotational basis," not permanently. The allied leaders did adopt a Readiness Action Plan to enlarge the long-planned NATO Response Force (from 13,000 to 20,000 troops) and to put it on a higher state of readiness, with a "Spearhead Force" of up to several thousand troops and reinforcements that could be deployed to the Baltic region within a few days. Whether those projections will actually materialize in 2015 and 2016 remains to be seen, however. Even if the proposals are fully implemented, they fall well short of what the Baltic countries and Poland had been seeking.

In the months following the Wales summit, the NATO governments tried to fulfill several of the pledges they had adopted, most notably with the establishment of multinational command-and-control centers in the Baltic countries, Poland, Bulgaria, and Romania, consisting of "personnel from Allies on a rotational basis" who are to "focus on planning and exercising collective defense." At a February 2015 meeting in Brussels, NATO defense ministers pledged to increase the size of the Response Force to 30,000, including a Spearhead Force of 5,000. NATO military planners and individual governments took other concrete steps, including the upgrading of infrastructure and the prepositioning of weaponry and support equipment, to enhance the alliance's capacity to uphold Article 5 in the Baltic region.

Nevertheless, in the absence of large, permanent deployments of US and other NATO ground and air forces in the Baltic countries and Poland, doubts about the collective defense of the region are bound to persist. Terras highlighted this problem when he noted that although he himself did not doubt NATO's willingness to carry out its defense commitments, "the real question is whether Putin believes that Article 5 works." He warned, "We should not give any option of miscalculation for President Putin."

Gloomy Scenarios

Russia's actions in Crimea and eastern Ukraine, and the risky nature of Russian military operations and exercises in and around the Baltic region in 2014 and 2015, have sparked acute unease in Poland and the Baltic countries. Although the available evidence does not indicate that the Russian authorities will attack a NATO member state, Putin does seem intent on undermining NATO by raising doubts about the credibility of Article 5. Despite the strong pledges of support offered at the Wales summit, some uncertainty remains about what would happen if Russia undertook a limited military probe against one or more of the Baltic states. Certain European members of NATO might hinder a timely response, but if that were to happen the United States and some other NATO member states would likely act outside the alliance's command structure to defend the Baltic states, as envisaged in NATO's contingency defense planning. They would undoubtedly try to avoid escalation to all-out war against Russia, not least because that would require the NATO countries to fight in a region in which they would be at a serious geographic disadvantage.

However, if the United States and its allies failed to uphold Article 5 in the Baltic region and refrained from intervening against Russian military forces, this would gravely damage the credibility of all of NATO's defense commitments. No country would see much point in belonging to an alliance that refused to protect its members against external aggression. If Putin were fool-hardy enough to risk all-out war by embarking on military action in the Baltic region, NATO would have no fully reliable or attractive military and diplomatic options. But the worst option of all would be to do nothing and allow Russian military expansion to proceed unchecked.

These gloomy scenarios seem improbable for now, but the very fact that they are being discussed seriously in NATO circles as well as in Warsaw and the Baltic capitals is a sign of how gravely Russia's actions in Ukraine and elsewhere have affected the post—Cold War European security order. Peace and security in the Baltic region, which only a decade ago appeared more robust than ever, now seem all too precarious.

Critical Thinking

1. Why would Sweden and Finland consider joining NATO?
2. Should Poland and the Baltics be defended separately or together from external attacks?
3. What can the U.S. do to protect the Baltics from Russia?

Internet References

Latvian Foreign Ministry
mfa.gov.lv

Lithuanian Ministry of Foreign Affairs
urm.lt

Republic of Estonia, Ministry of Foreign Affairs
vm.el

Russian Ministry of Foreign Affairs
government.ru

Mark Kramer is director of the Cold War Studies program and a senior fellow of the Davis Center for Russian and Eurosian Studies at Harvard University.

Article Prepared by: Robert Weiner, *University of Massachusetts, Boston*

Putin's Foreign Policy

The Quest to Restore Russia's Rightful Place

FYODOR LUKYANOV

Learning Outcomes

After reading this article, you will be able to:

- Discuss the basic elements of Putin's foreign policy.
- Identify the reasons for Putin's distrust of the West.

In February, Moscow and Washington issued a joint statement announcing the terms of a "cessation of hostilities" in Syria—a truce agreed to by major world powers, regional players, and most of the participants in the Syrian civil war. Given the fierce mutual recriminations that have become typical of U.S.–Russian relations in recent years, the tone of the statement suggested a surprising degree of common cause. "The United States of America and the Russian Federation . . . [are] seeking to achieve a peaceful settlement of the Syrian crisis with full respect for the fundamental role of the United Nations," the statement began. It went on to declare that the two countries are "fully determined to provide their strongest support to end the Syrian conflict."

What is even more surprising is that the truce has mostly held, according to the UN, even though many experts predicted its rapid failure. Indeed, when Russia declared in March that it would begin to pull out most of the forces it had deployed to Syria since last fall, the Kremlin intended to signal its belief that the truce will hold even without a significant Russian military presence.

The ceasefire represents the second time that the Russians and the Americans have unexpectedly and successful cooperated in Syria, where the civil war has pitted Moscow (which acts as the primary protector and patron of Syrian President Bashar al-Assad) against Washington (which has called for an end to Assad's rule). In 2013, Russia and the United States agreed on a plan to eliminate Syria's chemical weapons, with the Assad regime's assent. Few believed that arrangement would work either, but it did.

These moments of cooperation highlight the fact that, although the world order has changed beyond recognition during the past 25 years and is no longer defined by a rivalry between two competing superpowers, it remains the case that when an acute international crisis breaks out, Russia and the United States are often the only actors able to resolve it. Rising powers, international institutions, and regional organizations frequently cannot do anything—or don't want to. What is more, despite Moscow's and Washington's expressions of hostility and contempt for each other, when it comes to shared interests and common threats, the two powers are still able to work reasonably well together.

And yet, it's important to note that these types of constructive interactions on discrete issues have not changed the overall relationship, which remains troubled. Even as it worked with Russia on the truce, the United States continued to enforce the sanctions it had placed on Russia in response to the 2014 annexation of Crimea, and a high-level U.S. Treasury official recently accused Russian President Vladimir Putin of personal corruption.

The era of bipolar confrontation ended a long time ago. But the unipolar moment of U.S. dominance that began in 1991 is gone, too. A new, multipolar world has brought more uncertainty into international affairs. Both Russia and the United States are struggling to define their proper roles in the world. But one thing that each side feels certain about is that the other side has overstepped. The tension between them stems not merely from events in Syria and Ukraine but also from a continuing disagreement about what the collapse of the Soviet Union meant for the world order. For Americans and other Westerners, the legacy of the Soviet downfall is simple: the United States won the Cold War and has taken its rightful place as the world's sole superpower, whereas post-Soviet Russia has failed to integrate itself as a regional power in the Washington-led postwar liberal international order. Russians, of course, see things differently. In their view, Russia's subordinate position is the illegitimate result of a never-ending U.S. campaign to keep Russia down and prevent it from regaining its proper status.

In his annual address to the Russian legislature in 2005, Putin famously described the disappearance of the Soviet Union as a "major geopolitical disaster." That phrase accurately captures the sense of loss that many Russians associate with the post-Soviet era. But a less often noted line in that speech conveys the equally crucial belief that the West misinterpreted the end of the Cold War. "Many thought or seemed to think at the time that our young democracy was not a continuation of Russian statehood, but its ultimate collapse," Putin said. "They were mistaken." In other words: the West thought that Russia would forever going forward play a fundamentally diminished role in the world. Putin and many other Russians begged to differ.

In the wake of the 2014 Russian reclamation of Crimea and the launch of Russia's direct military intervention in Syria last year, Western analysts have frequently derided Russia as a "revisionist" power that seeks to alter the agreed-on post–Cold War consensus. But in Moscow's view, Russia has merely been responding to temporary revisions that the West itself has tried to make permanent. No genuine world order existed at the end of the twentieth century, and attempts to impose U.S. hegemony have slowly eroded the principles of the previous world order, which was based on the balance of power, respect for sovereignty, noninterference in other states' internal affairs, and the need to obtain the UN Security Council's approval before using military force.

By taking action in Ukraine and Syria, Russia has made clear its intention to restore its status as a major international player. What remains unclear is how long it will be able to maintain its recent gains.

No World Order

In January 1992, a month after the official dissolution of the Soviet Union, U.S. President George H. W. Bush announced in his State of the Union address: "By the grace of God, America won the Cold War." Bush put as fine a point as possible on it: "The Cold War didn't 'end'—it was won."

Russian officials have never made so clear a statement about what, exactly, happened from their point of view. Their assessments have ranged from "we won" (the Russian people overcame a repressive communist system) to "we lost" (the Russians allowed a great country to collapse). But Russian leaders have all agreed on one thing: the "new world order" that emerged after 1991 was nothing like the one envisioned by Mikhail Gorbachev and other reform-minded Soviet leaders as a way to prevent the worst possible outcomes of the Cold War. Throughout the late 1980s, Gorbachev and his cohort believed that the best way out of the Cold War would be to agree on new rules for global governance. The end of the arms race, the reunification of Germany, and the adoption of the Charter of Paris for a New Europe aimed to reduce confrontation and forge a partnership between the rival blocs in the East and the West.

But the disintegration of the Soviet Union rendered that paradigm obsolete. A "new world order" no longer meant an arrangement between equals; it meant the triumph of Western principles and influence. And so in the 1990s, the Western powers started an ambitious experiment to bring a considerable part of the world over to what they considered "the right side of history." The project began in Europe, where the transformations were mainly peaceful and led to the emergence and rapid expansion of the EU. But the United States led 1990–1991 Gulf War introduced a new dynamic: without the constraints of superpower rivalry, the Western powers seemed to feel emboldened to use direct military intervention to put pressure on states that resisted the new order, such as Saddam Hussein's Iraq.

Soon thereafter, NATO expanded eastward, mainly by absorbing countries that had previously formed a buffer zone around Russia. For centuries, Russian security strategy has been built on defense: expanding the space around the core to avoid being caught off guard. As a country of plains, Russia has experienced devastating invasions more than once; the Kremlin has long seen reinforcing "strategic depth" as the only way to guarantee its survival. But in the midst of economic collapse and political disorder in the immediate post-Soviet era, Russia could do little in response to EU consolidation and NATO expansion.

The West misinterpreted Russia's inaction. As Ivan Krastev and Mark Leonard observed last year in these pages, Western powers "mistook Moscow's failure to block the post–Cold War order as support for it." Beginning in 1994, long before Putin appeared on the national political stage, Russian President Boris Yeltsin repeatedly expressed deep dissatisfaction with what he and many Russians saw as Western arrogance. Washington, however, viewed such criticism from Russia as little more than a reflexive expression of an outmoded imperial mentality, mostly intended for domestic consumption.

From the Russian point of view, a critical turning point came when NATO intervened in the Kosovo war in 1999. Many Russians—even strong advocates of liberal reform—were appalled by NATO'S bombing raids against Serbia, a European country with close ties to Moscow, which were intended to force the Serbs to capitulate in their fight against Kosovar separatists. The success of that effort—which also led directly to the downfall of the Serbian leader Slobodan Milosevic the following year—seemed to set a new precedent and provide a new template. Since 2001, NATO or its leading member states have initiated military operations in Afghanistan, Iraq, and Libya. All three campaigns led to various forms of regime change and, in the case of Iraq and Libya, the deterioration of the state.

In this sense, it is not only NATO's expansion that has alarmed Russia but also NATO's transformation. Western arguments that NATO is a purely defensive alliance ring hollow: it is now a fighting group, which it was not during the Cold War.

Victors and Spoils

As the United States flexed its muscles and NATO became a more formidable organization, Russia found itself in a strange position. It was the successor to a superpower, with almost all of the Soviet Union's formal attributes, but at the same time, it had to overcome a systemic decline while depending on the mercy (and financial support) of its former foes. For the first dozen or so years of the post-Soviet era, Western leaders assumed that Russia would respond to its predicament by becoming part of what can be referred to as "wider Europe": a theoretical space that featured the EU and NATO at its core but that also incorporated countries that were not members of those organizations by encouraging them to voluntarily adopt the norms and regulations associated with membership. In other words, Russia was offered a limited niche inside Europe's expanding architecture. Unlike Gorbachev's concept of a common European home where the Soviet Union would be a co-designer of a new world order, Moscow instead had to give up its global aspirations and agree to obey rules it had played no part in devising. European Commission President Romano Prodi expressed this formula best in 2002: Russia would share with the EU "everything but institutions." In plain terms, this meant that Russia would adopt EU rules and regulations but would not be able to influence their development.

For quite a while, Moscow essentially accepted this proposition, making only minimal efforts to expand its global role. But neither Russian elites nor ordinary Russians ever accepted the image of their country as a mere regional power. And the early years of the Putin era saw the recovery of the Russian economy—driven to a great extent by rising energy prices but also by Putin's success in reestablishing a functioning state—with a consequent increase in Russia's international influence. Suddenly, Russia was no longer a supplicant; it was a critical emerging market and an engine of global growth.

Meanwhile, it became difficult to accept the Western project of building a liberal order as a benign phenomenon when a series of so-called color revolutions in the former Soviet space, cheered on (at the very least) by Washington, undermined governments that had roots in the Soviet era and reasonably good relations with Moscow. In Russia's opinion, the United States and its allies had convinced themselves that they had the right, as moral and political victors, to change not only the world order but also the internal orders of individual countries however they saw fit. The concepts of "democracy promotion" and "transformational diplomacy" pursued by the George W. Bush administration conditioned interstate relations on altering any system of government that did not match Washington's understanding of democracy.

The Iron Fist

In the immediate post-9/11 era, the United States was riding high. But in more recent years, the order designed by Washington and its allies in the 1990s has come under severe strain. The many U.S. failures in the Middle East, the 2008 global financial crisis and the subsequent recession, mounting economic and political crises in the EU, and the growing power of China made Russia even more reluctant to fit itself into the Western-led international system. What is more, although the West was experiencing growing difficulties steering its own course, it never lost its desire to expand—pressuring Ukraine, for example, to align itself more closely with the EU even as the union appeared to be on the brink of profound decay. The Russian leadership came to the conclusion that Western expansionism could be reversed only with an "iron fist," as the Russian political scientist Sergey Karaganov put it in 2011.

The February 2014 ouster of Ukrainian President Viktor Yanukovych by pro-Western forces was, in a sense, the final straw for Russia. Moscow's operation in Crimea was a response to the EU's and NATO's persistent eastward expansion during the post–Cold War period. Moscow rejected the further extension of Western influence into the former Soviet space in the most decisive way possible—with the use of military force. Russians had always viewed Crimea as the most humiliating loss of all the territories left outside of Russia after the disintegration of the Soviet Union. Crimea has long been a symbol of a post-Soviet unwillingness to fight for Russia's proper status. The return of the peninsula righted that perceived historical wrong, and Moscow's ongoing involvement in the crisis in Ukraine has made the already remote prospect of Ukrainian membership in NATO even more unlikely and has made it impossible to imagine Ukraine joining the EU anytime soon.

The Kremlin has clearly concluded that in order to defend its interests close to Russia's borders, it must play globally. So having drawn a line in Ukraine, Russia decided that the next place to put down the iron fist would be Syria. The Syrian intervention was aimed not only at strengthening Assad's position but also at forcing the United States to deal with Moscow on a more equal footing. Putin's decision to begin pulling Russian forces out of Syria in March did not represent a reversal; rather, it was a sign of the strategy's success. Moscow had demonstrated its military prowess and changed the dynamics of the conflict but had avoided being tied down in a Syrian quagmire.

Identity Crisis

There is no doubt that during the past few years, Moscow has achieved some successes in its quest to regain international stature. But it's difficult to say whether these gains will prove lasting. The Kremlin may have outmaneuvered its Western rivals in some ways during the crises in Ukraine and Syria, but it still faces the more difficult long-term challenge of finding a credible role in the new, multipolar environment. In recent years, Russia has shown considerable skill in exploiting the West's missteps, but Moscow's failure to develop a coherent economic strategy threatens the long-term sustainability of its newly restored status.

As Moscow has struggled to remedy what it considers to be the unfair outcome of the Cold War, the world has changed dramatically. Relations between Russia and the United States no longer top the international agenda, as they did 30 years ago. Russia's attitude toward the European project is not as important as it was in the past. The EU will likely go through painful transformations in the years to come, but mostly not on account of any actions Moscow does or does not take.

Russia has also seen its influence wane on its southern frontier. Historically, Moscow has viewed Central Asia as a chessboard and has seen itself as one of the players in the Great Game for influence. But in recent years, the game has changed. China has poured massive amounts of money into its Silk Road Economic Belt infrastructure project and is emerging as the biggest player in the region. This presents both a challenge and an opportunity for Moscow, but more than anything, it serves as a reminder that Russia has yet to find its place in what the Kremlin refers to as "wider Eurasia."

Simply put, when it comes to its role in the world, Russia is in the throes of an identity crisis. It has neither fully integrated into the liberal order nor built its own viable alternative. That explains why the Kremlin has in some ways adopted the Soviet model—eschewing the communist ideology, of course, but embracing a direct challenge to the West, not only in Russia's core security areas but far afield, as well. To accompany this shift, the Russian leadership has encouraged the idea that the Soviet disintegration was merely the first step in a long Western campaign to achieve total dominance, which went on to encompass the military interventions in Yugoslavia, Iraq, and Libya and the color revolutions in post-Soviet countries—and which will perhaps culminate in a future attempt to pursue regime change in Russia itself. This deep-rooted view is based on the conviction that the West not only seeks to continue geopolitical expansion in its classical form but also wants to make everyone do things its way, by persuasion and example when possible, but by force when necessary.

Even if one accepts that view of Western intentions, however, there is not much Moscow can do to counter the trend by military means only. Influence in the globalized world is increasingly determined by economic strength, of which Russia has little, especially now that energy prices are falling. Economic weakness can be cloaked by military power or skillful diplomacy, but only for a short time.

Angry or Focusing?

Putin and most of those who are running the country today believe that the Soviet collapse was hastened by perestroika, the political reform initiated by Gorbachev in the late 1980s. They dread a recurrence of the instability that accompanied that reform and perceive as a threat anything and anyone that might make it harder to govern. But the Kremlin would do well to recall one of the most important lessons of perestroika. Gorbachev had ambitious plans to create a profoundly different relationship with the West and the rest of the world. This agenda, which the Kremlin dubbed "new political thinking," was initially quite popular domestically and was well received abroad as well. But as Gorbachev struggled and ultimately failed to restart the Soviet economy, "new political thinking" came to be seen as an effort to compensate for—or distract attention from—rapid socioeconomic decline by concentrating on foreign policy. That strategy didn't work then, and it's not likely to work now.

It's doubtful that the Kremlin will make any significant moves on the Russian economy before 2018, when the next presidential election will take place, in order to avoid any problems that could complicate Putin's expected reelection. Russia's economy is struggling but hardly in free fall; the country should be able to muddle through for another two years. But the economic agenda will inevitably rise to the fore after the election, because at that point, the existing model will be close to exhausted.

Turbulence will almost certainly continue to roil the international system after the 2018 election, of course, so the Kremlin might still find opportunities to intensify Russia's activity on the world stage. But without a much stronger economic base, the gap between Russian ambitions and Russian capacities will grow. That could inspire a sharper focus on domestic needs—but it could also provoke even more risky gambling abroad.

"Russia is not angry; it is focusing." So goes a frequently repeated Russian aphorism, coined in 1856 by the foreign minister of the Russian empire, Alexander Gorchakov, after Russia had lowered its international profile in the wake of its defeat in the Crimean War. The situation today is in some ways the opposite: Russia has regained Crimea, has enhanced its international status, and feels confident when it comes to foreign affairs. But the need to focus is no less urgent—this time on economic development. Merely getting angry will accomplish little.

Critical Thinking

1. Should the next U.S. President try to "reset" relations with Russia? Why or why not?
2. Why did Russia intervene in the Syrian civil war?
3. Is there a new "Cold War" between Russia and the U.S.?

Internet References

Collective Security Treaty Organization
www.odkb.gov.ru/start/index_aengl.htm

Embassy of the Russian federation to the U.S.
www.RussianEmbassy.org/

Permanent Mission of the Russian federation to the United Nations
Russiaun.nu/en

The Eurasian Economic Union
https://en.wikipedia.org/wiki/Eurasian_Economic_Union

Fyodor Lukyanov is an editor in chief of *Russia in Global Affairs*, Chair of the Presidium of the Council on Foreign and Defense Policy, and a research professor at the National Research University Higher School of Economics, Moscow.

Article

Prepared by: Robert Weiner, *University of Massachusetts, Boston*

How to Prevent an Iranian Bomb

The Case for Deterrence

MICHAEL MANDELBAUM

Learning Outcomes

After reading this article, you will be able to:

- Explain how the Cold War strategy of deterrence can be applied to Iran.
- Identify the basic elements of the nuclear deal with Iran.

The Joint Comprehensive Plan of Action (JCPOA), reached by Iran, six other countries, and the European Union in Vienna in July, has sparked a heated political debate in the United States. Under the terms of the agreement, Iran has agreed to accept some temporary limits on its nuclear program in return for the lifting of the economic sanctions the international community imposed in response to that program. The Obama administration, a chief negotiator of the accord, argues that the deal will freeze and in some ways set back Iran's march toward nuclear weapons while opening up the possibility of improving relations between the United States and the Islamic Republic, which have been bitterly hostile ever since the 1979 Iranian Revolution. The administration further contends that the agreement includes robust provisions for the international inspection of Iran's nuclear facilities that will discourage and, if necessary, detect any Iranian cheating, triggering stiff penalties in response.

Critics of the deal, by contrast, argue that it permits Iran to remain very close to obtaining a bomb, that its provisions for verifying Iranian compliance are weak, and that the lifting of the sanctions will give Iranian leaders a massive windfall that they will use to support threatening behavior by Tehran, such as sponsoring global terrorism, propping up the Syrian dictator Bashar al-Assad, and backing Hezbollah in its conflict with Israel (a country that the Iranian regime has repeatedly promised to destroy).

The American political conflict came to a head in September, when Congress had the chance to register its disapproval of the accord—although the president had enough support among Democrats to prevent a vote on such a resolution. Despite the conflict, however, both the deal's supporters and its critics agreed that the United States should prevent Iran from getting a bomb. This raises the question of how to do so—whether without the deal, after the deal expires, or if the Iranians decide to cheat. Stopping Iranian nuclear proliferation in all three situations will require Washington to update and adapt its Cold War policy of deterrence, making Tehran understand clearly in advance that the United States is determined to prevent, by force if necessary, Iranian nuclearization.

A Credible Threat

The English political philosopher Thomas Hobbes noted in *Leviathan* that "covenants, without the sword, are but words." Any agreement requires a mechanism for enforcing it, and the Iranian agreement does include such a mechanism: in theory, if Iran violates the agreement's terms, the economic sanctions that the accord removes will "snap back" into place. By itself, however, this provision is unlikely to prevent Iranian cheating. The procedures for reimposing the sanctions are complicated and unreliable; even if imposed, the renewed sanctions would not cancel contracts already signed; and even as the sanctions have been in place, Iran's progress toward a bomb has continued. To keep nuclear weapons out of Tehran's hands will thus require something stronger—namely, a credible threat by the United States to respond to significant cheating by using force to destroy Iran's nuclear infrastructure.

The term for an effort to prevent something by threatening forceful punishment in response is "deterrence." It is hardly a novel policy for Washington: deterring a Soviet attack on the

United States and its allies was central to the American conduct of the Cold War.

Deterring Iran's acquisition of nuclear weapons now and in the future will have some similarities to that earlier task, but one difference is obvious: Cold War deterrence was aimed at preventing the use of the adversary's arsenal, including nuclear weapons, while in the case of Iran, deterrence would be designed to prevent the acquisition of those weapons. With the arguable exception of Saddam Hussein's Iraq, the United States has not previously threatened war for this purpose and has in fact allowed a number of other countries to go nuclear, including the Soviet Union, China, Israel, India, Pakistan, and North Korea. Does the Iranian case differ from previous ones in ways that justify threatening force to keep Iran out of the nuclear club?

It does. An Iranian bomb would be more dangerous, and stopping it is more feasible. The Soviet Union and China were continent-sized countries that crossed the nuclear threshold before the U.S. military had the capacity for precision air strikes that could destroy nuclear infrastructure with minimal collateral damage. Israel and India, like the United Kingdom and France before them, were friendly democracies whose possession of nuclear armaments did not threaten American interests. Pakistan is occasionally friendly, is a putative democracy, and crossed the nuclear threshold in direct response to India's having done so. The United States is hardly comfortable with the Pakistani nuclear arsenal, but the greatest danger it poses is the possibility that after a domestic upheaval, it could fall into the hands of religious extremists—precisely the kind of people who control Iran now.

North Korea presents the closest parallel. In the early 1990s, the Clinton administration was ready to go to war to stop Pyongyang's nuclear weapons program, before signing an agreement that the administration said would guarantee that the communist regime would dismantle its nuclear program. North Korea continued its nuclear efforts, however, and eventually succeeded in testing a nuclear weapon during the presidency of George W. Bush. Since then, North Korea has continued to work on miniaturizing its bombs and improving its missiles, presumably with the ultimate aim of being able to threaten attacks on North America. It is worth noting that in 2006, two experienced national security officials wrote in *The Washington Post* that if Pyongyang were ever to achieve such a capability, Washington should launch a military strike to destroy it. One of the authors was William Perry, who served as secretary of defense in the Clinton administration; the other was Ashton Carter, who holds that position today.

Bad as the North Korean bomb is, an Iranian one would be even worse. For in the case of North Korea, a long-standing policy of deterrence was already in place before it acquired nuclear weapons, with the United States maintaining a strong peacetime military presence on the Korean Peninsula after the end of the Korean War in 1953. For this reason, in the years since Pyongyang got the bomb, its neighbors have not felt an urgent need to acquire nuclear armaments of their own—something that would be likely in the case of Iranian proliferation.

Nor would the Iranian case benefit from the conditions that helped stabilize the nuclear standoff between the United States and the Soviet Union. A Middle East with multiple nuclear-armed states, all having small and relatively insecure arsenals, would be dangerously unstable. In a crisis, each country would have a powerful incentive to launch a nuclear attack in order to avoid losing its nuclear arsenal to a first strike by one of its neighbors. Accordingly, the chances of a nuclear war in the region would skyrocket. Such a war would likely kill millions of people and could deal a devastating blow to the global economy by interrupting the flow of crucial supplies of oil from the region.

But if an Iranian bomb would be even worse than a North Korean bomb, preventing its emergence would be easier. A U.S. military strike against North Korea would probably trigger a devastating war on the Korean Peninsula, one in which the South would suffer greatly. (South Korea's capital, Seoul, is located within reach of North Korean artillery.) This is one of the reasons the South Korean government has strongly opposed any such strike, and the United States has felt compelled, so far, to honor South Korea's wishes. In the Middle East, by contrast, the countries that would most likely bear the brunt of Iranian retaliation for a U.S. counterproliferation strike—Saudi Arabia and Israel, in particular—have made it clear that, although they are hardly eager for war with Iran, they would not stand in the way of such a strike.

A Limited Aim

Deterring Iran's acquisition of nuclear weapons by promising to prevent it with military action, if necessary, is justified, feasible, and indeed crucial to protect vital U.S. interests. To be effective, a policy of deterrence will require clarity and credibility, with the Iranian regime knowing just what acts will trigger retaliation and having good reason to believe that Washington will follow through on its threats.

During the Cold War, the United States was successful in deterring a Soviet attack on its European allies but not in preventing a broader range of communist initiatives. In 1954, for example, the Eisenhower administration announced a policy of massive retaliation designed to deter communist provocations, including costly conventional wars like the recent one in Korea, by promising an overpowering response. But the doctrine lacked the credibility needed to be effective, and a decade later, the United States found itself embroiled in another, similar war in Vietnam.

In the case of Iran, the aim of deterrence would be specific and limited: preventing Iran's acquisition of nuclear weapons. Still, a policy of deterrence would have to cope with two difficulties. One is the likelihood of Iranian "salami tactics"—small violations of the JCPOA that gradually bring the Islamic Republic closer to a bomb without any single infraction seeming dangerous enough to trigger a severe response. The other is the potential difficulty of detecting such violations. The Soviet Union could hardly have concealed a cross-border attack on Western Europe, but Iran is all too likely to try to develop the technology needed for nuclear weapons clandestinely (the United States believes it has an extensive history of doing so), and the loopholes in the agreement's inspection provisions suggest that keeping track of all of Iran's bomb-related activities will be difficult.

As for credibility—that is, persuading the target that force really will be used in the event of a violation—this posed a major challenge to the United States during the Cold War. It was certainly credible that Washington would retaliate for a direct Soviet attack on North America, but the United States also sought to deter an attack on allies thousands of miles away, even though in that case, retaliation would have risked provoking a Soviet strike on the American homeland. Even some American allies, such as French President Charles de Gaulle, expressed skepticism that the United States would go to war to defend Europe. The American government therefore went to considerable lengths to ensure that North America and Western Europe were "coupled" in both Soviet and Western European eyes, repeatedly expressing its commitment to defend Europe and stationing both troops and nuclear weapons there to trigger the U.S. involvement in any European conflict.

In some ways, credibly threatening to carry out a strike against Iran now would be easier. Iran may have duplicated, dispersed, and hidden the various parts of its nuclear program, and Russia may sell Tehran advanced air defense systems, but the U.S. military has or can develop the tactics and munitions necessary to cause enough damage to lengthen the time Iran would need to build a bomb by years, even without the use of any ground troops. The Iranians might retaliate against Saudi Arabia or Israel (whether directly or through their Lebanese proxy, Hezbollah), or attack American military forces, or sponsor acts of anti-American terrorism. But such responses could do only limited damage and would risk further punishment.

The problems with deterring Iran's acquisition of nuclear weapons are not practical but rather political and psychological. Having watched American leaders tolerate steady progress toward an Iranian bomb over the years, and then observed the Obama administration's avid pursuit of a negotiated agreement on their nuclear program, Iran's ruling clerics may well doubt that Washington would actually follow through on a threat to punish Iranian cheating. U.S. President Barack Obama initially embraced the long-standing American position that Iran should not be permitted to have the capacity to enrich uranium on a large scale, then abandoned it. He backed away from his promise that the Syrian regime would suffer serious consequences if it used chemical weapons. He made it the core argument in favor of the JCPOA that the alternative to it is war, implying that American military action against Iran is a dreadful prospect that must be avoided at all costs. Moreover, neither he nor his predecessor responded to Iran's meddling in Iraq over the past decade, even though Tehran's support for Shiite militias there helped kill hundreds of U.S. troops. The mullahs in Tehran may well consider the United States, particularly during this presidency, to be a serial bluffer.

Doubt Not

All of this suggests that in order to keep Iran from going nuclear, the JCPOA needs to be supplemented by an explicit, credible threat of military action. To be credible, such a threat must be publicly articulated and resolutely communicated. The Obama administration should declare such a policy itself, as should future administrations, and Congress should enshrine such a policy in formal resolutions passed with robust bipartisan support. The administration should reinforce the credibility of its promise by increasing the deployment of U.S. naval and air forces in the Persian Gulf region and stepping up the scope and frequency of military exercises there in conjunction with its allies. As in Europe during the Cold War, the goal of U.S. policy should be to eliminate all doubts, on all sides, that the United States will uphold its commitments.

The debate about the Iran nuclear deal has become politically polarized, but a policy of deterrence should not be controversial, since all participants in the debate have endorsed the goal of preventing an Iranian bomb. In addition, a robust policy of deterrence would help address some of the shortcomings of the JCPOA without sacrificing or undermining its useful elements. And since the deterrence policy could and should be open ended, it would help ease worries about the provisions of the accord that expire after 10 or 15 years. As during the Cold War, the policy should end only when it becomes obsolete—that is, when Iran no longer poses a threat to the international community. Should the Islamic Republic evolve or fall, eliminating the need for vigilant concern about its capabilities and intentions, the United States could revisit the policy. Until then, deterrence is the policy to adopt.

Critical Thinking

1. Is the international verification scheme of the Iranian nuclear deal working?
2. Why did the U.S. negotiate a nuclear deal with Iran?
3. What is the effect of the nuclear deal on the balance of power in the Middle East?

Internet References

International Atomic Energy Agency
https://www.iaea.org/

The Iranian Nuclear Deal
http://apps.washington.post.com/g/documents/world/full-tex

Treaty on the Non-Proliferation of Nuclear Weapons
www.un.org/disarmament/wmd/nuclear/npt

Michael Mandelbaum is Christian A. Herter professor of American Foreign Policy at the Johns Hopkins School of Advanced International Studies and the author of the forthcoming book *Mission Failure: America and the World in the Post–Cold War Era.*

Article

Prepared by: Robert Weiner, *University of Massachusetts, Boston*

Getting What We Need with North Korea

LEON V. SIGAL

Learning Outcomes

After reading this article, you will be able to:

- Understand what is involved in negotiating the denuclearization of North Korea.
- Discuss the relationship between China and North Korea.

While Washington's chattering classes were all atwitter about North Korean nuclear testing and rocket launching and China's backing for UN sanctions against Pyongyang in recent months, U.S. diplomats were tiptoeing to the negotiating table.

Any chance of a nuclear deal with North Korea depends on giving top priority to stopping the North's arming even if that means having Pyongyang keep the handful of weapons it has for the foreseeable future. Success will also require probing Kim Jong Un's seriousness about ending enmity, starting with a peace process on the Korean peninsula.

The revelation that Washington was willing to talk to Pyongyang without preconditions was a surprise to those who had not been tracking the evolution of U.S. policy closely. The Department of State confirmed that the United States held talks in New York last fall and rejected a proposal to begin negotiating a peace treaty. "To be clear, it was the North Koreans who proposed discussing a peace treaty," department spokesman John Kirby said on February 21. "We carefully considered their proposal, and made clear that denuclearization had to be part of any such discussion. The North rejected our response."[1]

Intriguingly, the revelation came on the eve of Chinese Foreign Minister Wang Yi's visit to Washington. Four days earlier, while signaling China's support for UN sanctions, Wang had made a more negotiable proposal of his own: "As chair country for the six-party talks, China proposes talks toward both achieving denuclearization and replacing the armistice agreement with a peace treaty." The proposal, Wang said, was intended to "find a way back to dialogue quickly."[2]

Wang's proposal was consistent with the September 19, 2005, six-party joint statement, which called for "the directly related parties" to "negotiate a permanent peace regime on the Korean Peninsula at an appropriate separate forum."[3] Those parties included the three countries with forces on the peninsula—North Korea, South Korea, and the United States—and China. They, along with Japan and Russia, agreed in six-party talks in September 2005 on the aim of "denuclearization of the Korean Peninsula" to be negotiated in parallel with a peace process in Korea and bilateral U.S.–North Korean and Japanese–North Korean talks on political and economic normalization.

Wang's initiative was also a way to bridge the gap between Washington and Pyongyang. North Korea has long sought a peace treaty. Its position hardened, however, after Washington, backed by Seoul and Tokyo, demanded preconditions—"pre-steps" in diplomatic parlance—to demonstrate its commitment to denuclearization before talks could begin. In response, Pyongyang began insisting that a peace treaty had to precede any denuclearization.

The Chinese proposal is a testament that sanctions are unlikely to curb North Korean nuclear and missile programs and that negotiation, however difficult, is the only realistic way forward. So is Washington's newfound openness to talks with Pyongyang.

Many in Washington and Seoul, however, still contend that negotiation is pointless if North Korea remains unwilling to give up the handful of crude nuclear weapons it has. That premise ignores the potential danger that an unbounded weapons program in North Korea poses to U.S. and allied security.

It also ignores the possibility that Pyongyang may be willing to suspend its nuclear and missile programs if its security concerns are satisfied. That was the gist of its January 9, 2015, offer of "temporarily suspending the nuclear test over which the U.S. is concerned" if the United States "temporarily suspends joint military exercises in South Korea and its vicinity this year."[4]

Like most opening bids, it was unacceptable. Instead of probing it further, however, Washington rejected it out of hand—within hours—and publicly denounced it as an "implicit threat."[5] That was a mistake Washington would not repeat in the fall.

Unofficial contacts later that January indicated that Pyongyang was prepared to suspend not just nuclear testing, but also missile and satellite launches and fissile material production. In return, the North was willing to accept a toning-down of the scale and scope of U.S.–South Korean exercises instead of the cancellation it had sought. This underscored the need for reciprocal steps to improve both sides' security.

Those contacts might have opened the way to talks at that time, but the initiative was squelched in Washington. Instead, U.S. officials continued to insist Pyongyang had to take unilateral steps to demonstrate its commitment to denuclearizing and ruled out reciprocity by Washington. Their stance was based on the flawed premise that the North alone had failed to live up to past agreements.[6] As Daniel Russel, assistant secretary of state for East Asian and Pacific affairs, put it on February 4, "North Korea does not have the right to bargain, to trade or ask for a pay-off in return for abiding by international law."[7] This crime-and-punishment approach, however warranted by North Korean flouting of international law, has never stopped North Korea from arming in the past, and it is unrealistic to think it would work now.

Tiptoeing Toward Talks

Last September 18, U.S. negotiator Sung Kim dropped Washington's preconditions for talks while still insisting that the agenda would be pre-steps North Korea would have to take to reassure Washington before formal negotiations could begin. "When we conveyed to Pyongyang that we are open to dialogue to discuss how we can resume credible and meaningful negotiations, of course we meant it. It was not an empty promise. We are willing to talk to them," Kim said. "And frankly for me, whether that discussion takes place in Pyongyang, or some other place, is not important. I think what's important is for us to be able to sit down with them and hear directly from them that they are committed to denuclearization and that if and when the six-party talks resume, they will work with us in meaningful and credible negotiations towards verifiable denuclearization."[8] In short, Washington would sit down with Pyongyang without preconditions in order to discuss U.S. preconditions for negotiations. That opened the door to contacts with the North Koreans in the New York channel in November.

In a November 3 interview, North Korean Foreign Ministry official Jong Tong Hak hinted at what the North might be proposing behind the scenes in New York. He said a permanent peace settlement on the Korean peninsula first required a North Korean–U.S. "peace agreement," perhaps a declaration committing the sides to negotiate peace. That was an advance. It was accompanied by a step backward from previous North Korean positions: "If the American government is serious about respecting the sovereignty of the DPRK [the Democratic People's Republic of Korea] and ending its ongoing hostile policy against the DPRK then it can be solved very easily between the two sides."[9] The apparent exclusion of South Korea made that proposal a nonstarter even if the North had been ready to suspend its nuclear and missile programs.

Sung Kim reiterated the U.S. position on November 10. "I think for us it's pretty straightforward: If [the North Koreans are] willing to talk about the nuclear issue and how we can move towards meaningful productive credible negotiations, [the United States would be] happy to meet with them anytime, anywhere," he said. He went on to respond to Jong obliquely: "It's not that we have no interest in seeking a permanent peace regime, peace mechanism or peace treaty. But I think they have the order wrong. Before we can get to a peace mechanism to replace the armistice, I think we need to make significant progress on the central issue of denuclearization."[10] The armistice agreement, signed in July 1953, established a cease-fire in the Korean War "until a final peaceful settlement is achieved."[11] That has yet to happen.

The Obama administration deserves praise for agreeing to meet in New York to explore what the North Koreans had in mind and not to reject a peace process out of hand. Disappointingly, North Korea proved unready to discuss denuclearization, which is stymieing talks for now.

The latest sanctions will squeeze Pyongyang but not enough to compel it to knuckle under and accept Washington's preconditions for negotiating.

The Limits of Sanctions

North Korea's January 6 nuclear test and February 7 satellite launch spurred more-stringent sanctions at the UN Security Council and in Washington. Even worse, the sanctions revived dreams of a North Korean collapse in Seoul, dreams that jeopardize a peace process.

Sanctions might have helped bring Iran around to negotiating, but North Korea is no Iran. It is far more autarkic and less dependent on trade with the rest of the world. It has no big-ticket items such as oil that require access to the global banking system to transact business.

An offer to ease sanctions may be of some utility in negotiations with Pyongyang, as it was with Tehran. The latest sanctions will squeeze Pyongyang but not enough to compel it to knuckle under and accept Washington's preconditions for negotiating. If anything, Pyongyang's nuclear advances have enhanced its leverage and given it greater confidence to proceed with negotiations on its own terms.

Wang and U.S. Secretary of State John Kerry acknowledged as much in their February 23 joint press conference announcing their agreement to move ahead on sanctions and negotiations. "China would like to emphasize that the Security Council resolution cannot provide a fundamental solution to the Korean nuclear issue. To really do that, we need to return to the track of dialogue and negotiation. And the secretary and I discussed this many times, and we agree on this," Wang said. Kerry echoed him, saying that the goal "is not to be in a series of cycling, repetitive punishments. That doesn't lead anywhere. The goal is to try to get Kim Jong Un and the DPRK to recognize that . . . it can rejoin the community of nations, it can actually ultimately have a peace agreement with the United States of America that resolves the unresolved issues of the Korean peninsula, if it will come to the table and negotiate the denuclearization."[12] Once again, news reports focused on China's willingness to endorse sanctions without paying attention to the U.S. commitment to negotiations.

Kirby, the State Department spokesman, improved that formulation on March 3: "We haven't ruled out the possibility that there could sort of be some sort of parallel process here. But—and this is not a small 'but'—there has to be denuclearization on the peninsula and work through the six-party process to get there."[13]

Many in Washington may question whether Beijing will enforce UN-mandated sanctions. By the same token, many in Beijing may wonder whether Washington will keep its commitment to negotiate.

Focus on the Urgent

North Korea's January 6 nuclear test, its fourth, was nothing to disparage. Even if it was neither a hydrogen bomb nor a boosted energy device, the test likely advanced Pyongyang's effort to develop a compact nuclear warhead that it can deliver by missile.

That is not all. The North has restarted its reactor at Yongbyon, which is working fitfully to generate more plutonium. It also is moving to complete a new reactor and has expanded its uranium-enrichment capacity. It has paraded two new longer-range missiles, the Musudan and KN-08, which it has yet to test-launch, and it is developing its first solid-fueled missile, the short-range Toksa.

That makes stopping the North's nuclear and missile programs a matter of urgency. Doing so should take priority over eliminating the handful of nuclear weapons Pyongyang already has, however desirable that may be. Such a negotiating approach is also more likely to bear fruit, given Kim Jong Un's goals.

What Is in it for Kim?

For nearly three decades, Pyongyang has sought to reconcile—end enmity—with Washington, Seoul, and Tokyo or, in the words of the 1994 Agreed Framework, "move toward full political and economic normalization."[14] To that end, it was prepared to suspend its weapons program or ramp it up if the other parties thwarted the reconciliation effort. Although North Korea's nuclear and missile brinkmanship is well understood, it is often forgotten that, from 1991 to 2003, North Korea reprocessed no fissile material and conducted very few test launches of medium-or long-range missiles. It suspended its weapons programs again from 2007 to early 2009.

U.S. negotiators need to probe whether an end to enmity remains Kim Jong Un's aim. He is not motivated by economic desperation, as many in Seoul and Washington believe. On the contrary, his economy has been growing over the past decade. Yet, he has publicly staked his rule on improving his people's standard of living, unlike his father and predecessor, Kim Jong Il. To deliver on his pledge, he needs to divert investment from military production to civilian goods.

That was the basis of his so-called *byungjin*, or "strategic line on carrying out economic construction and building nuclear armed forces simultaneously under the prevailing situation,"[15] meaning as long as U.S. "hostile policy" persists.

To curb military spending, Kim needs a calm international environment. Failing that, he will strengthen his deterrent, reducing the need for greater spending on conventional forces—a Korean version of U.S. President Dwight Eisenhower's bigger bang for the buck.

Pushback from the military on the budget may explain what prompted him to have his defense minister executed last spring.[16] It may also account for Kim Jong Un's exaggerated claims about testing an "H-bomb" in January. By crediting the party and the government, not the National Defense Commission, for the test, he was putting the military in its place.[17] The role that nuclear weapons play in putting a cap on defense spending was explicit in his March 9 claim of a "miniaturized" warhead deliverable by missile, which he called "a firm guarantee for making a breakthrough in the drive for economic construction and improving the people's standard of living on the basis of the powerful nuclear war deterrent."[18]

If Kim Jong Un still wants a fundamentally transformed relationship with his enemies or a calmer international climate in order to improve economic conditions in his country, a peace process is his way forward.

Testing whether North Korea means what it says about a peace process is also in the security interests of the United States and its allies.

Probing for Peace

Testing whether North Korea means what it says about a peace process is also in the security interests of the United States and its allies, especially now that North Korea has nuclear weapons.

North Korea's March 2010 sinking of a South Korean corvette, the *Cheonan*, in retaliation for the fatal November 2009 South Korean shooting up of a North Korean naval vessel in the contested waters of the West (Yellow) Sea showed that steps taken by each side to bolster deterrence can cause armed clashes. So did North Korea's November 2010 artillery barrage on Yeonpyeong Island in reprisal for South Korea's live-fire exercise. A peace process could reduce the risk of such clashes.

Negotiating a peace treaty is a formidable task. To be politically meaningful, it would require normalization of diplomatic, social, and economic relations and rectification of land and sea borders, whether those borders are temporary, pending unification, or permanent. To be militarily meaningful, it would require changes in force postures and war plans that pose excessive risks of unintended war on each side of the demilitarized zone (DMZ) separating the two Koreas. That would mean, above all, redeployment of the North's forward-deployed artillery and short-range missiles to the rear, putting Seoul out of range. Yet, to the extent Pyongyang would see that redeployment as weakening its deterrent against attack, it might be more determined to keep its nuclear arms.

A peace treaty is unlikely without a more amicable political environment. One way to nurture that environment is a peace process, using a series of interim peace agreements as stepping stones to a treaty. Such agreements, with South Korea and the United States as signatories, would constitute token acknowledgment of Pyongyang's sovereignty. In return, North Korea would have to take a reciprocal step by disabling and then dismantling its nuclear and missile production facilities.

A first step could be a "peace declaration." Signed by North Korea, South Korea, and the United States and perhaps China, Japan, and Russia, such a document would declare an end to enmity by reiterating the language of the October 12, 2000, U.S.–North Korean joint communiqué stating that "neither government would have hostile intent toward the other" and confirming "the commitment of both governments to make every effort in the future to build a new relationship free from past enmity." It could also commit the three parties to commence a peace process culminating in the signing of a peace treaty. The declaration could be issued at a meeting of the six foreign ministers.

A second step long sought by Pyongyang is the establishment of a "peace mechanism" to replace the Military Armistice Commission set up to monitor the ceasefire at the end of the Korean War. This peace mechanism could serve as a venue for resolving disputes such as the 1994 North Korea downing of a U.S. reconnaissance helicopter that strayed across the DMZ or the 1996 incursion by a North Korean spy submarine that ran aground in South Korean waters while dropping off agents. The peace mechanism would include the United States and the two Koreas.

The peace mechanism also could serve as the venue for negotiating a series of agreements on specific confidence-building measures, whether between the North and South, between the North and the United States, or among all three parties. A joint fishing area in the West Sea, as agreed in principle in the October 2007 North–South summit meeting, is one. Naval confidence-building measures, such as "rules of the road" and a navy-to-navy hotline, are also worth pursuing.

Lacking satellite reconnaissance, North Korea has conducted surveillance by infiltrating agents into the South. An "open skies" agreement allowing reconnaissance flights across the DMZ by both sides might reduce that risk. In October 2000, Kim Jong Il offered to end exports, production, and deployment of medium- and longer-range missiles. In return, he wanted the United States to launch North Korean satellites, along with other compensation. A more far-reaching arrangement might be to set up a joint North-South watch center that could download real-time data from U.S. or Japanese reconnaissance satellites. It is unclear how much such confidence-building measures will reduce the risk of inadvertent war, but they would provide political reassurance of an end to enmity.

A Starting Point

Before the sides can get to a peace process, they need to take steps to rebuild some trust. For Washington, that means verifiable suspension of all of North Korea's nuclear tests, missile and satellite launches, and fissile material production.

For Pyongyang, that means an easing of what it calls U.S. "hostile policy," starting with a toning-down of joint military exercises, partial relaxation of sanctions, and some commitment to initiating a peace process. Such reciprocal steps could lead to resumption of parallel negotiations among the six parties as envisioned in their September 2005 joint statement.

If the two sides can avoid deadly clashes triggered by the current joint military exercises, they may get back to exploring the only realistic off-ramp from the current impasse: reciprocal steps to open the way to negotiations that would address denuclearization and a peace process in Korea. That, however, would require a change of heart in Pyongyang, Seoul, and Washington. As the Rolling Stones put it, "You can't always get what you want/But if you try sometimes, well you just might find/You get what you need."

Endnotes

1. "U.S. Rejected Peace Talks before Last Nuclear Test," Reuters, February 21, 2016.
2. Lee Je-hun, "Could Wang's Two-Track Proposal Lead to a Breakthrough?" *Hankyoreh*, February 19, 2016.
3. U.S. Department of State, "Six-Party Talks, Beijing, China," n.d., http://www.state.gov/p/ eap/regional/c15455.htm (text of the joint statement of the fourth round of six-party talks on September 19, 2005).
4. Korean Central News Agency (KCNA), KCNA Report, January 10, 2015, www.kcna. co.jp/item2015/201501/news10/20150110-12ee.htm.
5. Marie Harf, transcript of U.S. Department of State daily briefing, January 12, 2015, http://www.state.gov/r/pa/prs/dpb/2015/01/235866.htm.
6. On the history of reneging by various parties, see Leon V. Sigal, "How to Bring North Korea Back Into the NPT," in *Nuclear Proliferation and International Order*, ed. Olaf Njolstad (London: Routledge, 2010), pp. 65–82.
7. "US: No Sign Yet NKorea Serious on Nuke Talks," Associated Press, February 4, 2015.
8. Chang Jae-soon and Roh Hyo-dong, "U.S. Nuclear Envoy Willing to Hold Talks with N. Korea in Pyongyang," Yonhap, September 19, 2015.
9. "N. Korea Accuses U.S. of 'Nuclear Blackmail,'" Associated Press, November 4, 2015.
10. Chang Jae-soon, "Amb. Sung Kim: U.S. 'Happy to Meet' With N. Korean 'Anytime, Anywhere,'" Yonhap, November 11, 2015.
11. Armistice Agreement for the Restoration of the South Korean State, North Korea–U.S., July 27, 1953, 4 U.S.T. 234.
12. U.S. Department of State, remarks of Secretary of State John Kerry and Chinese Foreign Minister Wang Yi, Washington, DC, February 23, 2016, http://www.state.gov/secretary/remarks/2016/02/253164.htm.
13. John Kirby, transcript of U.S. Department of State daily briefing, March 3, 2016, http://www.state.gov/r/pa/prs/dpb/2016/03/253948.htm.
14. Bureau of Arms Control, U.S. Department of State, "Agreed Framework between the United States and the Democratic People's Republic of Korea," October 21, 1994, http://2001-2009.state.gov/t/ac/rls/or/2004/31009.htm.
15. KCNA, "Report on Plenary Meeting of WPK Central Committee," March 31, 2013, www.kcna.co.jp/item/201303/news31/20130331-24ee.htm.
16. "N. Korean Ex-Army Chief 'Locked Horns with Technocrats,'" *Chosun Ilbo*, May 15, 2015, http://english.chosun.com/site/data/html_ dir/2015/05/15/2015051500971.html.
17. KCNA, "DPRK Proves Successful in H-Bomb Test," January 6, 2016, www.kcna. co.jp/2016/201601/news06/20160106-12ee.htm; KCNA, "WPK Central Committee Issues Order to Conduct First H-Bomb Test," January 6, 2016, www.kcna.co.jp/item/2016/201601/news06/20160106-11ee.htm.
18. KCNA, "Kim Jong-un Guides Work for Mounting Nuclear Warheads on Ballistic Rockets," March 9, 2016, http://www.kcna.kp/kcna.user.special.getArticlePage.kcmsf;jsessionid=823D154834DB0E5032E668C39EDE74B3.

Critical Thinking

1. Why does North Korea want the bomb?
2. How does North Korea's nuclear weapons capability affect the regional balance of power in East Asia?
3. What can the United States do to stop North Korea's nuclear weapons program?

Internet References

38 North
38north.org/

Arms Control Association
https://www.armscontrol.org.

International Atomic Energy Agency
https://www.iaea.org

U.S. Korea Institute at SAIS
uskoreainstitute.org/

Leon V. Sigal is director of the Northeast Asia Cooperative Security Project at the Social Science Research Council in New York and author of *Disarming Strangers: Nuclear Diplomacy with North Korea* (1998). A portion of this article draws from a piece that appeared in the Nautilus Institute for Security and Sustainability's NAPSNet Policy Forum.

Article

Prepared by: Robert Weiner, *University of Massachusetts, Boston*

A New Era for Nuclear Security

MARTIN B. MALIN AND NICKOLAS ROTH

Learning Outcomes

After reading this article, you will be able to:

- Explain what was accomplished at the last Nuclear Security Summit in 2016.
- List what steps had been taken by the international community to provide for the physical security of usable nuclear materials.

The 2016 Nuclear Security summit was a pivotal moment for the decades-long effort to secure nuclear material around the globe. More than 50 national leaders gathered in Washington for the last of four biennial meetings that have led to significant progress in strengthening measures to reduce the risk of nuclear theft.

These summits have played a critical role in nurturing that progress by elevating the political salience of nuclear security and providing a forum for world leaders to announce new commitments, share information, and hold one another accountable for following through on promised actions.

The international community is now entering the post-summit era, in which nuclear security will probably receive less-regular high-level political attention than it has in recent years. Yet, there is still critical work to be done to reduce the danger that nuclear weapons or the materials needed to make them could end up in the hands of a terrorist organization such as the Islamic State. Governments still do not agree on what nuclear security priorities are most pressing or how best to sustain the momentum generated by the summits. As the era of summitry recedes, will states continue improving measures to prevent nuclear theft and sabotage, or will the summits turn out to have been a high-water mark for nuclear security efforts?

Progress at the 2016 Summit

Over the course of the summit process, the participating states committed themselves to dozens of cooperative initiatives seeking to strengthen aspects of nuclear security, reduced vulnerabilities in their security systems, and pledged to continue joint efforts through multilateral groups and international institutions. The 2016 summit, held March 31–April 1 in Washington, marked progress on all of these fronts.

Like the 2010 summit in Washington, the 2012 summit in Seoul, and the 2014 summit in The Hague, this year's meeting produced a consensus-based communiqué. At the three most recent summits, smaller groups of participants also produced a series of joint statements and group commitments, or "gift baskets."[1] At this year's summit, all but three states participated in at least one of 18 gift baskets or nine joint statements, which covered a range of areas, including insider threats, transport security, minimization of the use of highly enriched uranium (HEU), and cybersecurity.[2] Among the most important outcomes of the recent summit was the establishment of a contact group, which will meet annually to discuss nuclear security.

Some of the major accomplishments of the summit are listed below.

Strengthening the commitment to nuclear security. China and India joined 36 states that had signed on to an important 2014 summit initiative on strengthening nuclear security implementation.[3] Members of this group committed to "meet the intent" of International Atomic Energy Agency (IAEA) nuclear security principles and recommendations, conduct self-assessments, host periodic peer reviews of their nuclear security, and ensure that "management and personnel with accountability for nuclear security are demonstrably competent," along with several other actions. This was an important commitment for China and India, demonstrating a measure of transparency and reassurance on nuclear security. Prior to the 2016 summit, neither country had been open to participating in such initiatives, although both nuclear-armed states face terrorist threats.[4]

The summit process also helped to build support for a foundational and legally binding international nuclear security instrument. After more than a decade, the 2005 amendment to the Convention on the Physical Protection of Nuclear Material (CPPNM) reached the required number of ratifications to enter into force in May. The amendment outlines nuclear security principles and requires states to establish rules and regulations for physical protection.

It also requires a review conference five years after entry into force and, if members choose to have them, additional review conferences at intervals of at least five years.[5] The amended CPPNM, now officially known as the Convention on the Physical Protection of Nuclear Material and Nuclear Facilities, could be a helpful tool for states to hold one another accountable for maintaining physical protection and strengthening norms.

Reducing nuclear security vulnerabilities. In addition to announcing new commitments, the summits were occasions for states to report on steps they had taken to remove or eliminate HEU or plutonium, convert reactors, improve physical protection, strengthen regulation, and contribute support to the IAEA or other international nuclear security work.

At the recent summit, Japan and the United States announced the completion of a commitment they made in 2014 to remove more than 500 kg of nuclear weapons-usable material from Japan.[6] Argentina announced it had eliminated the last of its HEU, making it the 18th state to clean out all of its nuclear weapons-usable material since the beginning of the summit process. Indonesia declared it had eliminated all of its fresh HEU and planned to get rid of all its HEU in 2016.

China announced the opening of its nuclear security center of excellence. Since 2010, China has worked with the United States to build the center as a hub for training, bilateral and multilateral best practice exchanges, and technology demonstration.[7] The center will help China test and strengthen its own nuclear security measures and will provide a venue for cooperation with others in the region and beyond.

The White House reported that 20 states hosted or invited peer review missions through the IAEA or from other states. Many other states announced that they had strengthened nuclear security laws or regulations, upgraded physical security, or updated the list of threats against which their nuclear facilities must be protected.

Continuing the dialogue. An important new gift basket created a nuclear security contact group that will convene annually on the margins of the IAEA General Conference. The contact group will carry forward the consultative element of the summit process, providing a forum for senior government officials to meet and discuss current efforts, evaluate progress on previously made commitments, and identify future priorities. If states buy into the idea of the contact group and take action to strengthen it, the group, whose membership is open to states that did not participate in the summits, could be an important vehicle for sustaining international nuclear security cooperation.

The summit also produced statements on bilateral nuclear security discussions between key countries. For example, China and the United States agreed to increase cooperation on nuclear terrorism prevention and conduct an annual dialogue on nuclear security.

In addition, summit participants agreed to action plans for the IAEA, the United Nations, Interpol, the Global Partnership against the Spread of Weapons and Materials of Destruction, and the Global Initiative to Combat Nuclear Terrorism (GICNT). The plans outline the roles these organizations will play in supporting ongoing nuclear security discussions now that the summits have ended.

Gaps and Missed Opportunities

In their communiqué, the participants in the 2016 summit pledged to "continuously strengthen nuclear security at national, regional, and global levels."[8] Striving for continuous improvement is the right way to frame the challenge of providing effective and sustainable nuclear security. Unfortunately, summit participants missed important opportunities to give added momentum to the effort. The following issues continue to require attention.

Still no global standard for nuclear security. Although the amended CPPNM establishes general security principles, it lacks specific standards or guidelines and applies only to materials in civilian use. UN Security Council Resolution 1540 requires states to provide "appropriate effective" protection for all materials, among other relevant measures, but does not specify what constitutes appropriate effective protection.[9] IAEA recommendations, to which dozens of states have now publicly subscribed, provide somewhat more specificity, but their implementation is voluntary. Although the summit process certainly helped produce a shared understanding of the importance of nuclear security, it fell short of producing a consensus on a meaningful minimum global standard.

If a global standard was beyond reach during the summits, a public commitment to stringent nuclear security measures among the states possessing the biggest stocks of HEU and plutonium would have been a consequential step. Although China's and India's endorsements of the initiative on strengthening nuclear security implementation was an important development, Russia's absence from the summit and Russia's and Pakistan's refusal to sign that statement is a significant gap in the patchwork of nuclear security commitments.

Furthermore, the summit outcomes were not comprehensive. Although the summit communiqués explicitly covered "all" nuclear material, most of the concrete progress from the meetings focused on civilian materials, largely ignoring the roughly ⅘ of the world's remaining HEU and plutonium that is controlled by military organizations.[10]

A mixed picture on implementation. Nuclear facilities in many countries still are not protected against the full range of threats. States with large stocks of nuclear weapons-usable material still contend with corruption and extremism.[11] On the ground, security upgrades remain urgently needed in many spots around the world. One indication of the extent of the inconsistent application of physical protection measures is that, after all of the high-level attention since the 2010 summit, at least six countries—Argentina, Brazil, the Netherlands, Slovakia, Spain, and Sweden—still do not have armed guards at their nuclear facilities.[12]

The collapse of U.S.–Russian bilateral cooperation is particularly alarming. Without Russian and U.S. commitments to rebuilding their bilateral nuclear security relationship, it will be impossible for the two states that possess roughly 80 percent of the world's weapons-usable nuclear material to reassure one another that their nuclear security is sound.

Slippage of consolidation and minimization goals. The Obama administration put laudable effort into cleaning out HEU and plutonium from many countries and minimizing the use of HEU elsewhere. Yet, political obstacles will likely make substantial additional progress more difficult than in the past, in particular for the hundreds of kilograms of HEU in Belarus and South Africa. Conversion of additional HEU-fueled research reactors to use low-enriched uranium fuel, particularly but not only in Russia, is hampered by technical challenges and political inattention. Moreover, summit participants failed to reach agreement, even in principle, on stopping or reversing the buildup of separated plutonium.[13]

Continuing culture of complacency in some countries. The summits put the notion of nuclear security culture on the agenda for many countries where it previously had been neglected. Nevertheless, workers, managers, policy officials, and even national leaders in many places still dismiss the threat of terrorist theft or sabotage as remote or implausible.[14] Many organizations handling nuclear weapons, HEU, or separated plutonium do not have specific programs focused on strengthening security culture. The IAEA has still not published its nuclear security culture self-assessment guide.[15] The summit process helped spark interest in strengthening security culture, but much more work is needed.

Need for morerobust channels for dialogue. The political momentum created by the summits will not likely be re-created through other organizations, although the contact group, IAEA ministerial meetings, a review conference for the amended CPPNM, and other forums certainly will provide important opportunities for discussion, reporting on progress, and further cooperation.

The recent summit's action plans did not significantly expand or strengthen the global nuclear security architecture. The IAEA has assumed greater responsibility for convening high-level discussions on nuclear security and has intensified its nuclear security efforts since the first summit. Yet, the agency still deals only with civilian material and has no authority to require states to take any action on nuclear security.[16] The nuclear security capacities of the UN and Interpol are even less robust, and the multilateral groupings, the GICNT and Global Partnership, remain unchanged by the action plans the summit participants produced.

Finally, Russia's absence from the recent summit may bode ill for the successful implementation of the summit action plans. Moscow's leadership and cooperation in all of the organizations referenced in the action plans will be necessary for many key nuclear security steps.

In the interest of promoting cooperation, the summits frequently focused on plucking low-hanging fruit.

Progress in the Post-Summit Era

In the interest of promoting cooperation, the summits frequently focused on plucking low-hanging fruit, while failing to advance more-difficult discussions of threats and persistent challenges. Governments must focus not only on what is most feasible but also on what is most urgently needed in light of the evolving threats they face.[17]

Nuclear security efforts should have a clear goal: ensuring that all nuclear weapons and the materials that could be used to make them, wherever they are in the world, are effectively and sustainably secured against the full range of threats that terrorists and thieves might plausibly pose.[18] Building an international consensus around such a goal will be a major challenge for the next U.S. president and for like-minded leaders.

The 2016 summit communiqué alludes to the goal of continuous improvement. Achieving that goal will require work on several fronts. Here are some of the most important areas of focus.[19]

Building up the commitment to stringent nuclear security standards. A legally binding set of international standards for nuclear security is unfortunately out of reach for the present. Yet, a group of states like-minded emanating from within the contact group or a special working group of the GICNT could develop a set of principles and guidelines that they pledge to apply to all stocks of nuclear weapons, HEU, and plutonium and invite other states to join them. Such a commitment should include the provision of well-trained, well-equipped on-site guard forces; comprehensive measures to protect against insider threats; control and accounting systems that can detect and localize any theft of weapons-usable nuclear material; protections against cyberthreats that are integrated with other nuclear security measures; effective nuclear security rules and regulations and independent regulators capable of enforcing them; regular and realistic testing of nuclear security systems, including force-on-force exercises; a robust program for enhancing security culture; and regular assessments of the evolving threat of theft or sabotage. Following the example of the initial group of adherents, the accumulation of international support for more-comprehensive standards could grow over time.

In the meantime, leading states that are bound by the amended CPPNM should push to universalize the treaty, and the states that have joined the initiative on strengthening nuclear security implementation initiative should encourage others to commit to implement IAEA recommendations and accept peer review.

Implementing effective and sustainable security measures on the ground. Commitments to stringent standards are meaningful only if they translate into real improvements. Bilateral cooperation can help spur the actions that are needed. The United States should expand nuclear security cooperation with China, India, and Pakistan, sharing additional information on security arrangements without revealing sensitive information that would increase vulnerability to terrorist attack. The United States also will need to make a priority of discussions with a wide range of countries on enhancing their own nuclear security, providing resources when needed.

Despite tensions over Ukraine and other issues, Russia and the United States should agree to a package of cooperation that includes nuclear energy initiatives, which are of particular interest to Russia, and nuclear security initiatives, which are of particular interest to the United States. Although it is unlikely in the current political environment, one mechanism for achieving this goal would be to restart the U.S.–Russian Nuclear Energy and Nuclear Security Working Group, which facilitated dialogue from 2009 until it was suspended in 2014 because of tensions between the two countries. Cooperation should no longer be based on a donor–recipient relationship but on an equal partnership with ideas and resources coming from both sides.[20]

Increasing efforts to reduce the number of sites where nuclear weapons and weapons-usable materials are stored. Today, there are fewer locations where HEU and plutonium can be stolen because of removals motivated by the summit process. The consolidation process must continue. Stringent security requirements can help to incentivize the process of consolidation, as can well-funded programs for conversion of HEU-fueled reactors and removal of material. Russia and the United States, as the countries whose nuclear stockpiles are dispersed in the largest number of buildings and bunkers, should each develop a national-level plan for accomplishing their military and civilian nuclear objectives with the smallest practicable number of locations. The United States and other interested countries should ensure that plutonium and HEU bulk processing facilities do not spread to other countries or expand in number or scale of operations and that no more plutonium is separated than is used, bringing global plutonium stocks down over time.

Establishing a nuclear security culture that does not tolerate complacency about threats and vulnerabilities. Every country with relevant materials and facilities should have a program in place to assess and strengthen security culture, and all nuclear managers and security-relevant staff should receive regular information, appropriate to their role, on evolving threats to nuclear security. At the same time, interested countries should launch initiatives to combat complacency, including a shared database of security incidents and lessons learned; detailed reports and briefings on the nuclear terrorism threat; discussions among intelligence agencies, on which most governments rely for information about the threats to their country; and an expanded program of nuclear theft and terrorism exercises.

Building up channels for dialogue. Countries must continue to share information and devise plans to meet current nuclear security challenges. The IAEA ministerial-level meetings on nuclear security will provide an important forum. If parties to the amended CPPNM elect to meet every five years to review progress, this process could create important opportunities to place high-level pressure on states to step up nuclear security commitments and implementation.

A more comprehensive scope of cooperation, including on military materials, could take place in multilateral forums. The GICNT, cochaired by Russia and the United States and still valued by both, consists of more than 80 states committed to the group's statement of principles, which includes improving measures that reduce the risk of nuclear theft such as accounting, control, and protection of nuclear and radiological materials. The group has not focused on these preventive approaches so far, but it should in the future.[21] This summer represents the GICNT's 10th anniversary, which would be an excellent time to announce the creation of a GICNT working group focused specifically on strengthening security for nuclear materials and facilities. The GICNT could also be a useful forum for Russia and the United States to expand nuclear security cooperation.

The contact group created at the nuclear security summit this year holds promise for facilitating dialogue, sharing information, and germinating joint activities. Its openness to all IAEA members has the advantage of potentially attracting states beyond the ring of past summit participants. Its size and heterogeneity, however, may limit the depth and effectiveness of the discussions. The contact group should select an executive committee of member state representatives—perhaps former summit hosts plus Russia, if it chooses to join—to establish and coordinate its agenda for discussion.

Finally, summit-level nuclear security meetings could be continued on the side of Group of 20 meetings, perhaps once every four years. This would sustain the kind of executive-level political attention to nuclear security that summits provided.

The nuclear security summits periodically pressed participants to commit themselves to new and stronger measures for preventing nuclear terrorism. They facilitated a process of stocktaking and reporting on the concrete actions participants had taken. Moreover, they were a vehicle for forging stronger international collaboration on bolstering nuclear security around the globe. States must continue to build on the progress they made through the summit process. If they do, the 2016 summit will mark the beginning, rather than the end, of a new era of continuous improvement in nuclear security.

Endnotes

1. For a comprehensive assessment of progress in fulfilling commitments from the summits prior to 2016, see Michelle Cann, Kelsey Davenport, and Jenna Parker, "The Nuclear Security Summit: Progress Report on Joint Statements," Arms Control Association and Partnership for Global Security, March 2015, https://www.armscontrol.org/reports/2015/The-Nuclear-Security-Summit-Progress-Report-on-Joint-Statements.
2. The three countries that did not join gift baskets were Gabon, Pakistan, and Saudi Arabia. For a list of gift baskets and joint statements from the 2016 summit, see "2016 Washington Summit," Nuclear Security Matters, n.d., http://nuclearsecuritymatters.belfercenter.org/2016-washington-summit.
3. "Strengthening Nuclear Security Implementation," March 25, 2014, http://www.state.gov/documents/organization/235508.pdf. Thirty-five countries signed the 2014 statement. Jordan joined in late 2015.
4. See Rajeswari Pillai Rajagopalan, "India and the Nuclear Security Summit," Nuclear Security Matters, April 26, 2016, http://nuclearsecuritymatters.belfercenter.org/blog/india-and-nuclear-security-summ; Hui Zhang, "China Makes Significant Nuclear Security Pledges at 2016 Summit," Nuclear Security Matters, April 8, 2016, http://nuclearsecuritymatters.belfercenter.org/blog/china-makes-significant-nuclear-security-pledges-2016-summit.
5. For background on the amended Convention on the Physical Protection of Nuclear Material, see "Convention on the Physical Protection of Nuclear Material," International Atomic Energy Agency (IAEA), n.d., https://www.iaea.org/publications/documents/conventions/convention-physical-protection-nuclear-material. For an argument that the review conferences envisioned in the amendment could help drive nuclear security progress, see Jonathan Herbach and Samantha Pitts-Kiefer, "More Work to Do: A Pathway for Future Progress on Strengthening Nuclear Security," *Arms Control Today*, October 2015.
6. "Joint Statement by the Leaders of Japan and the United States on Contributions to Global Minimization of Nuclear Material," April 1, 2016, http://nuclearsecuritymatters.belfercenter.org/files/nuclearmatters/files/joint_statement_by_the_leaders_of_japan_and_the_united_states_on_contrib.pdf.
7. Office of the Press Secretary, The White House, "U.S.–China Joint Statement on Nuclear Security Cooperation," March 31, 2016, https://www.whitehouse.gov/the-press-office/2016/03/31/us-china-joint-statement-nuclear-security-cooperation.
8. "Nuclear Security Summit 2016 Communiqué," April 1, 2016, http://nuclearsecuritymatters.belfercenter.org/files/nuclearmatters/files/nuclear_security_summit_2016_communique.pdf?m=1460469255.
9. See Matthew Bunn, "Appropriate Effective Nuclear Security and Accounting— What Is It?" (presentation, "Appropriate Effective" Material Accounting and Physical Protection—Joint Global Initiative/UNSCR 1540 Workshop," Nashville, TN, July 18, 2008), http://belfercenter.ksg.harvard.edu/files/bunn-1540-appropriate-effective50.pdf.
10. For a discussion of security for military materials, see Des Browne, Richard Lugar, and Sam Nunn, "Bridging the Military Nuclear Materials Gap," Nuclear Threat Initiative (NTI), 2015, http://www.nti.org/media/pdfs/NTI_report_2015_e_version.pdf. The 2016 summit communiqué reaffirmed that states had a fundamental responsibility "to maintain at all times effective security of all nuclear and other radioactive material, including nuclear materials used in nuclear weapons." See "Nuclear Security Summit 2016 Communiqué."
11. For a more complete discussion of the threats some countries with nuclear material face, see Matthew Bunn et al., "Preventing Nuclear Terrorism: Continuous Improvement or Dangerous Decline?" Belfer Center for Science and International Affairs, Harvard Kennedy School, March 2016, pp. 39–52, http://belfercenter.ksg.harvard.edu/files/PreventingNuclearTerrorism-Web.pdf.
12. For country information on physical protection, see the 2016 NTI Nuclear Security Index for sabotage, http://ntiindex.org/wp-content/uploads/2016/03/2016-NTI-Index-Data-2016.03.25.zip. Belgium has only recently added armed guards to its nuclear facilities. The Swedish regulator has ordered that facilities post armed guards by February 2017. See Steven Mufson, "Brussels Attacks Stoke Fears About Security of Belgian Nuclear Facilities," *The Washington Post*, March 25, 2016; "Swedish Regulator Orders Tighter Security at Nuclear Plants," Reuters, February 5, 2016, http://www.reuters.com/article/sweden-nuclear-security-idUSL8N15K3SS.
13. The 2014 summit communiqué states, "We encourage States to minimise their stocks of [highly enriched uranium] and to keep their stockpile of separated plutonium to the minimum level, both as consistent with national requirements." "The Hague Nuclear Security Summit Communiqué," March 25, 2014,

http://www.state.gov/documents/organization/237002.pdf. In 2016, there was no mention of plutonium in the communiqué.

14. Matthew Bunn and Eben Harrell surveyed nuclear experts in states with nuclear weapons-usable material and found that some respondents did not find certain threats credible, despite extensive evidence to the contrary. See Matthew Bunn and Eben Harrell, "Threat Perceptions and Drivers of Change in Nuclear Security Around the World: Results of a Survey," Belfer Center for Science and International Affairs, Harvard Kennedy School, March 2014, http://belfercenter.ksg.harvard.edu/files/surveypaperfulltext.pdf.
15. IAEA, "Self-Assessment of Nuclear Security Culture in Facilities and Activities That Use Nuclear and/or Radioactive Material: Draft Technical Guidance," July 2, 2014, http://www-ns.iaea.org/downloads/security/security-series-drafts/tech-guidance/nst026.pdf.
16. See Trevor Findlay, "Beyond Nuclear Summitry: The Role of the IAEA in Nuclear Security Diplomacy After 2016," Belfer Center for Science and International Affairs, Harvard Kennedy School, March 2014, http://belfercenter.hks.harvard.edu/files/beyondnuclearsummitryfullpaper.pdf.
17. For a discussion of how the threat of nuclear terrorism has evolved over time, see Bunn et al., "Preventing Nuclear Terrorism," pp. 14–26, 133–143.
18. Ibid., p. 96.
19. For the recommendations on which this section draws, see Bunn et al., "Preventing Nuclear Terrorism," pp. 96–133.
20. For a more complete description of the end of nuclear security cooperation, see Nickolas Roth, "Russian Nuclear Security Cooperation: Rebuilding Equality, Mutual Benefit, and Respect," Deep Cuts Commission, June 2015, http://deepcuts.org/files/pdf/Deep_Cuts_Issue_Brief4_US-Russian_Nuclear_Security_Cooperation1.pdf.
21. Global Initiative to Combat Nuclear Terrorism (GICNT), "Fact Sheet," n.d., http://www.gicnt.org/content/downloads/sop/GICNT_Fact_ Sheet_ June2015.pdf. Although the GICNT terms of reference state that its activities do not involve "military nuclear programs of the nuclear weapon states party to the Treaty on the Nonproliferation of Nuclear Weapons," the group's statement of principles, which is the only document GICNT members are required to endorse, contains no such exclusion. See Bureau of International Security and Nonproliferation, U.S. Department of State, "Terms of Reference for Implementation and Assessment," November 20, 2006, http://2001-2009.state.gov/t/isn/rls/other/76421.htm; GICNT, "Statement of Principles," 2015, http://gicnt.org/content/downloads/sop/Statement_of_Principles.pdf.

Critical Thinking

1. What is the role of the International Atomic Energy Agency in providing for physical nuclear security?
2. Why is it important that China and India cooperate to provide for physical nuclear security?
3. What are the gaps that still need to be dealt with in the field of physical nuclear security?

Internet References

Arms Control Association
https:armscontrol.org

Bulletin of the Atomic Scientists
www.thebulletin.org

Non-Proliferation Policy Education Center
www.npolicy.org/

Nuclear Security Summit 2016
www.nss2016.org/

Martin B. Malin is an executive director of the Project on Managing the Atom at Harvard Kennedy School's Belfer Center for Science and International Affairs. From 2000 to 2007, he was director of the Program on Science and Global Security at the American Academy of Arts and Sciences.

Nickolas Roth is a research associate at the Project on Managing the Atom. Parts of this article draw from the authors' article with Matthew Bunn and William H. Tobey in 2016 titled "Preventing Nuclear Terrorism: Continuous Improvement or Dangerous Decline?"

Unit 6

UNIT

Prepared by: Robert Weiner, *University of Massachusetts, Boston*

Ethics and Values

This unit contains several articles which deal with the question of the rule of law, both on the national and international level. The year 2015 marked 800 years since the signing of the first Magna Carta or Great Charter. The Magna Carta not only established the rule of law in England, but also established a model for the rule of law around the world. A definition of a rule of law state includes a liberal democracy which is characterized by free and fair elections, a competitive party system, a real rather than a nominal constitution, a free press, and an impartial judicial system which is appointed and allowed to function by the government without intimidation and political interference. An objective judicial system is critical in rooting out corruption in states like China. There have been a series of show trials in China, involving select high-level officials, in an effort to maintain the political legitimacy of the ruling elite. Kleptocracy not only exists in China but is a widespread phenomenon in a number of states, ranging from Ukraine to Nigeria. This unit contains an interesting discussion of corruption in Latin America, especially Brazil. Kleptocracy invariably occurs in a repressive regime marked by human rights violations and a loss of legitimacy of the government.

Furthermore, a rule of law state is necessary to eliminate racism in such liberal democracies as the United States, where unarmed black males fall victim to a discriminatory system, in a society which is becoming more diverse.

The year 2015 also marked 100 years, since the Armenian genocide took place in Ottoman Turkey. Genocide is a term which was invented by the Polish lawyer and linguist Raphael Lemkin, who almost single-handedly persuaded the international community in 1948 to adopt the Convention on the Prevention and Punishment of the Crime of Genocide. The term genocide is based on the Greek word "gens" which means people, and the Latin word "cide" which means killing. Until 1948, genocide was known as the "crime with no name," as famously stated by Winston Churchill. Genocide was finally recognized as a crime under international law after the Holocaust of World War II. The Genocide Convention lists various acts of genocide, which are designed to destroy in whole or in part a group of people based on race, ethnicity, religion, or nationality. The Convention was criticized for only focusing on these four groups, and not including political, economic, or social groups. Moreover, the Convention did not contain a definition of "group" and left it up to subsequent international criminal courts to come up with a definition of group. International criminal courts have wrestled for years with figuring out whether membership in a group can be defined by objective or subjective factors, or both. Furthermore, the Convention did not explain what was meant by "in part." The courts have tried to determine whether "in part" refers to a substantial or significant part of a group, sometimes arriving at contradictory opinions. Finally, an international judicial body, as mentioned in the Convention, did not exist at the time the Convention was adopted, and took 50 years to appear as the International Criminal Court in 1998.

Although the international community hoped that genocide would never again take place, it has occurred again and again since the Second World War, For example, in Rwanda in 1994 genocide resulted in the deaths of at least 800,000 to 1 million Tutsis and moderate Hutus. In Srebrenica, Bosnia in 1995, 7,000–8,000 Bosniak men and boys were massacred. Based on the genocides which occurred in Rwanda and Bosnia, a considerable amount of case law has been produced to provide a solid legal foundation for the prevention and punishment of this most heinous of crimes against humanity, as long as states have the political will to deal with such mass atrocities. The Genocide Convention has been virtually incorporated into the Statute of the International Criminal Court (ICC) which, after decades of negotiations, finally came into existence in 1998. A number of countries have also enacted national laws, using the Genocide Convention as a model. Although human rights advocates have been disappointed by the narrow legal basis of the Genocide Convention, the expansion of the definition of crimes against humanity has filled in the gap that has been left by the Convention. In 2016, for the very first time, a non-state actor, ISIL, was accused of committing genocide.

Article

Prepared by: Robert Weiner, *University of Massachusetts, Boston*

Xi's Corruption Crackdown: How Bribery and Graft Threaten the Chinese Dream

JAMES LEUNG

Learning Outcomes

After reading this article, you will be able to:

- Explain the scope of corruption in China.
- Understand the relationship between the Party and corruption.

In a series of speeches he delivered shortly after taking office in 2012, Chinese President Xi Jinping cast corruption as not merely a significant problem for his country but an existential threat. Endemic corruption, he warned, could lead to "the collapse of the [Chinese Communist] Party and the downfall of the state." For the past two years, Xi has carried out a sweeping, highly publicized anticorruption campaign. In terms of sheer volume, the results have been impressive: according to official statistics, the party has punished some 270,000 of its cadres for corrupt activities, reaching into almost every part of the government and every level of China's vast bureaucracy. The most serious offenders have been prosecuted and imprisoned; some have even been sentenced to death.

The majority of the people caught up in Xi's crackdown have been low- or midlevel party members and functionaries. But corruption investigations have also led to the removal of a number of senior party officials, including some members of the Politburo, the group of 25 officials who run the party, and, in an unprecedented move, to the expulsion from the party and arrest of a former member of the Politburo's elite Standing Committee.

Xi's campaign has proved enormously popular, adding a populist edge to Xi's image and contributing to a nascent cult of personality the Chinese leader has begun to build around himself. And it has the quiet support of the aristocratic stratum of "princelings," the children and grandchildren of revolutionary leaders from the Mao era. They identify their interests with those of the country and consider Xi to be one of their own. But there has been pushback from other elites within the system, some of whom believe the campaign is little more than a politically motivated purge designed to help Xi solidify his own grip on power. Media organizations in Hong Kong have reported that Xi's two immediate predecessors, Jiang Zemin and Hu Jintao, have asked him to dial back the campaign. And some observers have questioned the campaign's efficacy: in 2014, despite Xi's efforts, China scored worse on Transparency International's Corruption Perceptions Index than it had in 2013. Even Xi himself has expressed frustration, lamenting a "stalemate" in his fight to clean up the system while pledging, in grandiose terms, not to give up: "In my struggle against corruption, I don't care about life or death, or ruining my reputation," he reportedly declared at a closed-door Politburo meeting last year.

There is no doubt that Xi's campaign is in part politically motivated. Xi's inner circle has remained immune, the investigations are far from transparent, and Xi has tightly controlled the process, especially at senior levels. Chinese authorities have placed restrictions on foreign media outlets that have dared to launch their own investigations into corruption, and the government has detained critics who have called for more aggressive enforcement efforts.

But that doesn't mean the campaign will fail. The anticorruption fight is only one part of Xi's larger push to consolidate his authority by establishing himself as "the paramount leader within a tightly centralized political system," as the China expert Elizabeth Economy has written. So far, Xi seems capable of pulling off that feat. Although this power grab poses other risks, it puts Xi in a good position to reduce corruption significantly—if not necessarily in a wholly consistent, apolitical manner.

This might seem paradoxical: after all, too much central power has been a major factor in creating the corruption epidemic. That is why, in the long term, the fate of Xi's anticorruption fight will depend on how well Xi manages to integrate it into a broader economic, legal, and political reform program. His vision of reform, however, is not one that will free the courts, media, or civil society, or allow an opposition party that could check the ruling party's power. Indeed, Xi believes that Western-style democracy is at least as prone to corruption as one-party rule. Rather, Xi's vision of institutional reform involves maintaining a powerful investigative force that is loyal to an honest, centralized leadership. He seems to believe that, over the course of several years, consistent surveillance and regular investigations will change the psychology of bureaucrats, from viewing corruption as routine, as many now do, to viewing it as risky—and, finally, to not even daring to consider it.

Stamping out graft, bribery, and influence peddling could very well help China's leaders maintain the political stability they fear might slip away as economic growth slows and geopolitical tensions flare in Asia. But if Xi's fight against corruption becomes disconnected from systemic reforms, or devolves into a mere purge of political rivals, it could backfire, inflaming the grievances that stand in the way of the "harmonious society" the party seeks to create.

I'll Scratch Your Back . . .

One school of thought holds that corruption is a deeply rooted cultural phenomenon in China. Some political scientists and sociologists argue that when it comes to governance and business, the traditional Chinese reliance on guanxi—usually translated as "connections" or "relationships"—is the most important factor in explaining the persistence and scope of the problem. The comfort level that many Chinese citizens have with the guanxi system might help explain why it took so long for public outrage to build up to the point where the leadership was forced to respond. But all cultures and societies produce a form of guanxi, and China's version is not distinct enough to explain the depth and severity of the corruption that inflicts the Chinese system today. The main culprits are more obvious and banal: one-party rule and state control of the economy. The lack of firm checks and balances in a one-party state fuels the spread of graft and bribery; today, no Chinese institution is free of them. And state control of resources, land, and businesses creates plenty of opportunities for corruption. In the past three decades, the Chinese economy has become increasingly mixed. According to Chinese government statistics, the private sector now accounts for around two-thirds of China's GDP and employs more than 70 percent of the labor force. And the Chinese economy is no longer isolated; it has been integrated into the global market. Nevertheless, the private sector is still highly dependent on the government, which not only possesses tremendous resources but also uses its regulatory and executive power to influence and even control private businesses.

When it comes to government purchasing and contracting and the sale of Chinese state assets (including land), bidding and auctioning processes are extremely opaque. Officials, bureaucrats, and party cadres exploit that lack of transparency to personally enrich themselves and to create opportunities for their more senior colleagues to profit in exchange for promotions. Midlevel officials who oversee economic resources offer their superiors access to cheap land, loans with favorable terms from state-owned banks, government subsides, tax breaks, and government contracts; in return, they ask to rise up the ranks. Such arrangements allow corruption to distort not just markets but also the workings of the party and the state.

Similar problems also exist in government organizations that do not directly control economic resources, such as China's military. To win promotion, junior military officers routinely bribe higher-ranking ones with gifts of cash or luxury goods. Last year, the authorities arrested Xu Caihou, a retired general who had served as a member of the Politburo and had been the vice chair of the Central Military Commission. In his house, they discovered enormous quantities of gold, cash, jewels, and valuable paintings—gifts, the party alleged, from junior officers who sought to advance up the chain of command. After the party expelled him, Xu confessed, according to Chinese state media; a few months later, he died, reportedly of cancer.

Direct state ownership, however, is hardly a prerequisite for self-dealing. The immense regulatory power that Chinese authorities hold over the private sector also helps them line their own pockets. In highly regulated industries, such as finance, telecommunications, and pharmaceuticals, relatives of senior government officials often act as "consultants" to private businesspeople seeking to obtain the licenses and approvals they need to operate. Zheng Xiaoyu, the former head of the State Food and Drug Administration, accepted around $850,000 in bribes from pharmaceutical companies seeking approval for new products. In 2007, after more than 100 people in Panama died after taking contaminated cough syrup that Zheng had approved, he was tried on corruption charges; he was found guilty and executed a few months later.

Corruption has also infected law enforcement and the legal system. Organized criminal groups pay police officers to protect their drug and prostitution rings. Criminal suspects and their relatives often bribe police officers to win release from jail or to avoid prosecution. If that fails, they can try their luck with prosecutors and judges. And of course, since China's judiciary is not independent, there are always party and government officials who might be able and willing to intervene in a case—for the right price. Authorities allege that Zhou Yongkang, a former member of the party's Standing Committee who oversaw legal and internal security affairs, personally intervened in many court cases after accepting bribes. Zhou was arrested, charged, and expelled from the party last year and is currently awaiting trial—the first time in decades that the state has pursued a criminal case against a former member of the Standing Committee.

As China's domestic markets have grown, multinational companies and banks have learned that getting access means knowing whose palms to grease. Many firms have taken to hiring the children of senior government officials, sometimes even paying their tuition at Western universities. Others have opted for a more direct route, paying hefty "consulting" fees to middlemen in order to participate in stock offerings or to win preferential treatment in bidding for government contracts. This environment has discouraged some multinational companies from investing and conducting business in China, especially those constrained by U.S. anticorruption laws.

Meanwhile, officials have taken advantage of loose financial controls and a lack of transparency to safeguard their illicit profits. Many officials hold a number of Chinese passports, often under different names but with valid visas, and use them to travel abroad and stash their money in foreign bank accounts.

But corruption is hardly limited to official circles and big business; every aspect of society feels its effects. Consider education. To give their child a shot at getting into one of the relatively small number of high-quality Chinese primary and secondary schools and universities, parents often have to bribe admissions officers or headmasters. Similarly, the scarcity of good hospitals and well-trained medical personnel has led to the practice of supplying doctors or medical administrators with a hongbao—a "red packet" of cash—to secure decent treatment.

Keep It Clean

Faced with this far-reaching problem, Xi has promised more than a mere Band-Aid, envisioning a long-term process of systemic reform. The first phase has been the heavily stage-managed crackdown of the past two years. So far, the campaign has contained an element of populism: it has targeted only officials, bureaucrats, and major business figures whom the party suspects of corrupt dealings; no ordinary Chinese people have felt the sting.

The campaign seeks not only to punish corruption but to prevent it as well. In late 2012, the party published a set of guidelines known as the "eight rules and six prohibitions," banning bureaucrats from taking gifts and bribes; attending expensive restaurants, hotels, or private clubs; playing golf; using government funds for personal travel; using government vehicles for private purposes; and so on.

The government has also required all officials and their immediate family members to disclose their assets and income, to make it harder to hide ill-gotten gains. At the same time, the party has sought to reduce incentives for graft by narrowing the income gaps within the system. In the last year, it raised the salaries and retirement benefits of military officers, law enforcement personnel, and other direct government employees, while sharply cutting the higher salaries enjoyed by top managers of state owned enterprises.

Still, to date, Xi's campaign has been chiefly an enforcement effort. Investigations are led by the party's Central Commission for Discipline Inspection (CCDI), which sends inspection teams to examine every ministry and agency and every large state-owned enterprise. The teams enjoy the unlimited power to investigate, detain, and interrogate almost anyone, but mainly government officials, the vast majority of whom are party members. Once the teams believe they have gathered sufficient evidence of wrongdoing, the CCDI expels suspects from the party and then hands them over to the legal system for prosecution.

Xi has declared that no corrupt official will be spared, no matter how high his position. In practice, however, the CCDI has chosen its targets very carefully, especially at senior levels. The decision to go after Zhou was heralded as setting a new precedent—since the late 1980s, the party has followed an unspoken rule against purging a member or former member of the Standing Committee. And yet Zhou's removal and prosecution remain unique; they appear to have been less a signal of things to come than a shot across the bow, intended to scare off any potential opposition to Xi within the leadership. Zhou was vulnerable because he was retired and no longer had direct control or power. Also, Zhou had backed a group of senior party officials who had challenged Xi's power and authority early in his tenure; among them was Bo Xilai, the influential party chief of Chongqing, who in 2013 was brought down by a scandal involving corruption and a murder plot in which his wife participated. Finally, Zhou and his immediate family members were particularly flagrant in their corrupt pursuits, which made him an easy target. Some media reports have indicated that authorities are investigating the family members of other retired Standing Committee members. But so far, no ranking member of the "red aristocracy" has yet been targeted, and all the highest-level targets, including Zhou and

Xu, have been part of a single loose political network. Apparently, there are still lines Xi is not willing to cross.

It is also worth noting that although Xi has allowed investigations of the country's key military institutions, he has yet to make any major personnel changes within the Commission for Discipline Inspection of the Central Military Commission, the armed forces' equivalent of the CCDI. Xi still needs more time to consolidate his control over the military and its institutions.

A number of other elements of Xi's campaign are also problematic, because they present opportunities for abuse and run contrary to the spirit of the legal reforms that Xi is pursuing. Xi claims that he wants to improve due process and reduce abusive police and judicial practices. But the CCDI itself does not always follow standard legal procedures. For example, Chinese law allows police to detain a suspect for only seven days without formally charging him, unless the police obtain express permission from legal authorities to extend the detention. The CCDI, on the other hand, has kept suspects in custody for far longer periods without seeking any approval and without issuing any formal charges, giving the appearance of a separate standard.

Meanwhile, with its newfound authority, the CCDI is gradually becoming the most powerful institution within the party system. Unless the party balances and limits the agency's power and influence, the CCDI could grow unaccountable and become a source of the very kinds of conduct it is supposed to combat.

Perhaps the biggest potential obstacle to the success of the campaign is strong resistance to it within the bureaucratic system. Xi has launched a direct attack on the interests of many entrenched bureaucrats and officials; even those who have escaped prosecution have watched their prosperity and privilege shrink. Many officials might also resent the idea that there is something fundamentally wrong with the way they are accustomed to conducting themselves. They may feel that they deserve the benefits they get through graft; without their work, after all, nothing would get done—the system wouldn't function.

Early in Xi's tenure, some officials seemed to believe that although the days of flagrant self-dealing were over, it would still be possible to exploit their positions for profit; they would just need to be a bit more subtle about it. In 2013, *The New York Times,* citing Chinese state media, reported that a new slogan had become popular among government officials: "Eat quietly, take gently, and play secretly." But that sense of confidence has evaporated as it has become clear that Xi is serious about cracking down. During the past two years, party members and state bureaucrats have become extremely cautious about running afoul of the new ethos, although many are quietly seething about the situation. This has interfered with the traditional wheel-greasing function of corruption and contributed to China's economic slowdown. If corruption no longer assists entrepreneurs in slipping past bureaucratic barriers, it will put additional pressure on Xi to institute economic reforms that genuinely reduce those obstacles.

The Politics of Anticorruption

Since the anticorruption campaign is just one of a number of major changes taking place in the Xi era, it's difficult to forecast what path it might take. In a pessimistic scenario, the campaign would end in failure after strong resistance within the top party leadership and the bureaucratic system forces Xi to back down. That outcome would be a catastrophe. Corruption would likely rise to pre-2012 levels (at the very least), destabilizing the economy, reducing investor confidence, and seriously eroding Xi's authority, making it difficult for him to lead.

In a more optimistic scenario, Xi would manage to overcome internal resistance and move on to broader economic, legal, and political reforms. Ideally, the campaign will strengthen Xi's power base enough and win him the support necessary to reduce the party's tight grip on policy and regulatory and administrative power, creating a favorable environment for the growth of a more independent private sector. Xi has no interest in creating a Western-style democratic system, but he does think that China could produce a cleaner and more effective form of authoritarianism. To better serve that goal, Xi should consider adding a number of more ambitious elements to the anti-corruption crusade, including a step that both Transparency International and the G-20 have called for: improving public registers to clarify who owns and controls which companies and land, which would make it harder for corrupt officials and businesspeople to hide their illicit profits.

At the moment, there is more reason for optimism than pessimism. Xi has already consolidated a great deal of control over the state's power structures and is determined and able to remove anyone who might resist or challenge his authority or policies. So far, within the senior leadership and the wider bureaucratic system, resistance to the anticorruption campaign has been passive rather than active: some bureaucrats have reportedly slowed down their work in a rather limited form of silent protest. Meanwhile, the anticorruption campaign continues to enjoy strong public support, especially from low- and middle-income Chinese who resent the way that corruption makes the Chinese system even more unfair than it already is. Anti-corruption thus represents a way for the party to ease the social tensions and polarization that might otherwise emerge as the economy slows, even as dramatic economic inequalities persist. To maintain this public support, the trick for Xi will be calibrating the scope and intensity of the campaign: not so narrow or moderate as to seem halfhearted, but not so broad or severe as to seem like a form of abuse itself.

Critical Thinking

1. What lines is president Xi willing to cross in the battle against corruption?
2. Why is corruption so endemic in China?
3. What path will the anti-corruption campaign take?

Internet References

Embassy of the People's Republic of China in the United States
china-embassy.org

Freedom House
https://freedom house.org

Kleptocracy Initiative
kleptocracyinitiative.org

James Leung is a pseudonym for an economist with extensive experience in China, Europe, and the United States.

Article

Prepared by: Robert Weiner, *University of Massachusetts, Boston*

Latin Americans Stand Up to Corruption

The Silver Lining in a Spate of Scandals

JORGE G. CASTAÑEDA

Learning Outcomes

After reading this article, you will be able to:

- Define the problem of kleptocracy in Latin America.
- Define about the problem of kleptocracy in Brazil.

Just a few years ago, Latin America was on a roll. Its economies, riding on the back of the Chinese juggernaut, were flourishing. A boom in commodity prices and huge volumes of foreign direct investment in agriculture and natural resources generated a golden decade. Ambitious government programs began to reduce inequality. And relations with the United States, long a source of friction, were improving—even as they became less important to the region's success.

Today, the picture looks very different. Latin America's economies are grinding to a halt: in 2015, average GDP growth slipped below 1 percent. Inequality is still declining, but more slowly. And according to the annual Latinobarómetro poll, satisfaction with democracy in Latin America is lower than it is in any other region and is at its lowest point in almost a decade, at 37 percent. In Brazil and Mexico, it has descended to just 21 percent and 19 percent, respectively.

Yet on one count at least, the lands south of the Rio Grande are faring better than ever: Latin Americans are denouncing corruption as never before. In decades past, residents of the region seemed resigned to the problem, treating it as an ordinary, if lamentable, part of everyday life. In 1973, for example, Argentines elected Juan Perón to a third term as president despite his infamous criminality; as a popular saying put it, "Mujeriego y ladrón, pero queremos a Perón" (Philanderer and thief, we still want Perón).

Such tolerance is now a thing of the past. According to the same Latinobarómetro poll, the region's inhabitants identify corruption as their third most important problem, behind crime and unemployment but above inflation and poverty. Latin Americans have also started judging their politicians based on their perceived trustworthiness. Of the five most unpopular chief executives in Latin America today—Brazil's Dilma Rousseff, Mexico's Enrique Peña Nieto, Paraguay's Horacio Cartes, Peru's Ollanta Humala, and Venezuela's Nicolás Maduro—three come from the countries with the worst ratings for government transparency (Brazil, Mexico, and Peru).

Several factors explain this change in attitude. First, the economic growth of the last 15 years has created a large middle class (now estimated at almost a third of the region's population, according to the World Bank, up from around 20 percent a decade ago, although higher in Brazil, Chile, Mexico, and Uruguay) with high expectations. Second, the region has grown more democratic. As the recent economic downturn has highlighted the damage corruption causes, this newly enlarged middle class has used its new freedoms to vent its frustration with those in charge.

Of all the region's recent uprisings against corruption, the most dramatic have been in Brazil.

Foreigners have also played a role. As Latin America has become more integrated into the world economy, international media and civil society organizations have begun to direct intense opprobrium at corrupt leaders and to lavish praise on reformers. Outside forces have also helped the region's more independent judiciaries and media outlets expose official malfeasance.

Together, all these forces have created a combustible mix, and when cases of graft have come to light in recent years, they have sparked major scandals in one country after another. High-level Latin American officials and business leaders have found themselves denounced by the media. Prosecutors and courts have issued indictments, and protesters have taken to the streets. Although few of the governments implicated in the scandals have actually fallen—and few others are likely to—the sheer scale of the social and political protest has been astonishing and represents an important positive trend in a part of the world with an otherwise gloomy forecast.

Brazilian Bribes

Of all the region's recent uprisings against corruption, the most dramatic has unfolded in Brazil. The problems began in late 2013, a time when popular discontent with the government was already running high. The previous president, Luis Inácio Lula da Silva (known as Lula), had been tarnished by a corruption scandal years earlier. Now the economy was stagnating, and protests had begun to erupt over Brazil's lavish spending on the coming World Cup. Then, in late 2014, shortly after Rousseff narrowly won reelection, the so-called *petrolâo* scandal hit.

The scale of the revelations was unprecedented. In November 2014, federal police arrested 18 people, including senior executives of Petrobras, Brazil's state oil company, for corruption in the first raid of the investigation. Numerous firms had paid high-ranking government officials, including members of Rousseff's Workers' Party, enormous sums of money to obtain contracts from Petrobras. The bribes were thought to have totaled around $3 billion. Prosecutors charged executives from more than a dozen of the country's largest construction companies with corruption and money laundering. As several Petrobras executives turned state's witness, the police investigation, known as Operation Car Wash, continued to expand. Before long, many Brazilians concluded that Rousseff, who served as chair of Petrobras' board from 2003 to 2010, must be guilty herself. Although she has not been charged, Brazil's Supreme Court has ruled that her predecessor, Lula, can be called in for questioning, and on September 1, his former chief of staff was charged with racketeering, receiving bribes, fraud, and money laundering.

Lula was already under intense official scrutiny at the time: just a few months earlier, prosecutors had concluded that they had enough evidence to launch a full investigation into allegations that Brazil's biggest building company had paid Lula to lobby overseas on the firm's behalf. In yet another case, Rousseff has been accused by Brazil's Controller General's Office of illegally using funds earmarked for social programs and development to cover up budget deficits. Taken together, all these charges have helped push Rousseff's approval ratings down into the single digits; talk of her impeachment is now in the air. As demonstrations continue and the economy languishes—Brazil is now in its worst recession in decades—it's looking increasingly likely that Rousseff will not manage to serve out her term, which ends in 2018.

Bad as all these revelations have been for Brazil, the public reckoning that has followed can also be read as a sign of progress. The fact that the police, prosecutors, and judges have been willing to investigate the country's most powerful politicians and business leaders has highlighted the independence of Brazil's judicial system. And the unprecedented level of anger suggests that business as usual will no longer satisfy a Brazilian public increasingly intolerant of high-level corruption.

Another country north of Brazil has also recently turned a corner. In early 2015, after it emerged that officials had siphoned off millions of dollars in customs revenue, thousands of Guatemalans took to the streets, gathering every Saturday for weeks in the central square of the capital. In May, they forced the resignation of the country's vice president and several cabinet ministers. But the protests continued, and on September 1, legislators from President Otto Pérez Molina's own party stripped him of his immunity. On September 2, he resigned and within hours was jailed on corruption charges—an extraordinary event in a country where politicians have long enjoyed great impunity.

The fight against corruption in Guatemala has benefited from outside support. That help has come in the form of the International Commission against Impunity in Guatemala (CICIG), which was created in 2006 as part of a larger agreement between the UN and Guatemala. The body, which was initially intended to investigate crimes committed during the civil war, is financed by the European Union, supported by the U.S. embassy, and led by a Colombian; it now numbers more than 200 foreign officials. It has become one of the most powerful instruments in the campaign against corruption. As a high government official told me in August, "It hurts to admit that we are unable to clean up our own house, but it is better that someone else does it than that nobody does."

Guatemala's example has sparked similar protests across Central America, in neighboring Honduras and to a lesser extent in El Salvador. In Tegucigalpa, the capital of Honduras, thousands of protesters gather every Friday in a *marcha de las antorchas*, or "march of the torches," to demand an investigation into the defrauding of the Honduran Institute of Social Security by the governing party. President Juan Orlando Hernández has attempted to placate the demonstrators by creating a commission similar to the CICIG, although without prosecutorial powers, but so far, these attempts have failed. As long as it lacks the teeth of its Guatemalan counterpart, such a commission is unlikely to satisfy the protesters, and if more scandals come to light, calls for Hernández's resignation will mount. And in El Salvador, there are calls for an external investigation into Alba Petróleos, the subsidized energy venture set up by the then Venezuelan President Hugo Chávez several years ago, which is suspected of making financial contributions to El Salvador's ruling party.

No Más

Foreign influence has played a crucial role in Mexico as well. The American press is responsible for unveiling three of the most important corruption cases in recent history: Walmart's bribing of Mexican municipal officials (which *The New York Times* reported on in 2012); the revelation that Luis Videgaray, the country's finance minister, had acquired property under suspicious circumstances (a story *The Wall Street Journal* broke in 2014); and the concealed purchase of multimillion-dollar condos in Manhattan and elsewhere by a former governor (another story broken by the *Times*, this one in 2015).

Yet Mexico's domestic media have also done their part. The radio reporter Carmen Aristegui, the newsweekly *Proceso*, and the daily *Reforma* have helped expose numerous scandals. Aristegui revealed that the $7 million modernist mansion in Mexico City built for the Peña Nieto family was in fact owned by a company to which the president had awarded hundreds of millions of dollars in public contracts and that was headed by a personal friend of his. A government investigation has since exonerated Peña Nieto, but many in Mexico have dismissed the inquiry as a cover-up. Other accusations have been leveled at the interior minister and several governors. These scandals have all generated a great deal of anger and unhappiness in Mexican society. But so far, not much more has come of them.

Latin Americans are denouncing corruption as never before.

Yet, public opinion has come to matter more and more in today's democratic Mexico. Online social networks now provide the new middle class with an outlet halfway between public protest and private complaint: Mexicans use Facebook and Twitter to vent their anger and share information (not all of it accurate) about high-level corruption. This allows for a measure of catharsis but has yet to produce actual change: although many think that Peña Nieto's government is Mexico's most corrupt since the late 1980s, so far calls for the president's resignation have foundered.

A major corruption scandal is causing a similar reaction in Chile. It began in February 2015 with accusations of influence peddling against the son and daughter-in-law of President Michelle Bachelet. Other scandals soon emerged, involving tax fraud and campaign finance crimes on the part of opposition leaders and members of the governing coalition, several of whom were jailed. As of May, most had been released, but some were under house arrest. Even Marco Enríquez-Ominami, a former independent candidate for president and one of Chile's most popular politicians, has been caught up in a controversy regarding campaign finance, according to Chilean media reports. Large financial and mining conglomerates—one of them led by the ex-son-in-law of Chile's former dictator, Augusto Pinochet—have been accused of fraudulently contributing to electoral campaigns.

Yet as in Mexico, the scandals have prompted only muted protests so far. That may be a consequence of their relatively small scale. Historically, Chile's has ranked as one of the more honest governments in Latin America, and the amounts at stake in the country's recent scandals pale in comparison to those in Brazil and Mexico: the most serious charge against a Chilean official involves a loan of $10 million.

Yet, the allegations still represent the most serious challenge Bachelet has faced in her two terms in power. Although she tried to show that she takes the issue seriously by asking for the resignation of her entire cabinet in May and by calling for a new constitution, by August, her approval rating had dropped by 30 percentage points in one year. Popular protests are likely to become louder unless Chile's elites take genuine steps to reduce corruption, especially if the country's copper-dependent economy doesn't pick up soon. But once again, this represents good news as well as bad. Chile's independent and honest judiciary and its free press were central to uncovering the corruption scandals—an important sign of the growing effectiveness of Chile's democratic institutions.

Even in Venezuela, where flagrant corruption is still the norm, there are signs that the public's patience is running out. According to Latinobarómetro, Venezuela is the region's second least transparent country, and at the end of 2015, Maduro was its third most unpopular president. The United States recently leaked accusations that many of the country's leaders—including Diosdado Cabello, the head of the National Assembly and Maduro's closest aide—have used illegal means to enrich themselves immensely, partly through links with Colombian drug cartels. In June, the deterioration of Venezuela's economy and the increase in violence and human rights violations forced Maduro to call elections for December. Although candidates and voters mostly focused on the economy, violence, and repression, more and more of them raised corruption as well.

A Glass Half Full

Yet, there are exceptions to this trend. In Argentina, there are a few positive developments in the fight against corruption. The outgoing vice president, Amado Boudou, is awaiting trial for corruption, but many suspect the charges will be dismissed. Allegations also surround outgoing President Cristina Fernández de Kirchner, whose net wealth surged to a reported $6 million over the 13 years she and her late husband ruled the country, but the chances of a prosecution are slim. Outsiders have less influence in Argentina than in many of its neighbors—the country's tradition of Peronist nationalism makes it hostile to perceived meddling. And the public seems resigned to the status quo: although hundreds of thousands of Argentines joined demonstrations when the prosecutor Alberto Nisman died under mysterious circumstances—as he was investigating Kirchner in connection with the 1994 bombing of a Buenos Aires Jewish community center—they have remained stubbornly passive when it comes to corruption.

The people of Nicaragua, meanwhile, seem even more complacent. President Daniel Ortega is currently focused on an enormous undertaking: an attempt to build a second interoceanic canal just north of the existing Panama Canal. A Chinese businessman has agreed to underwrite the cost, which could reach up to $100 billion. Some Nicaraguans think that the businessman is working for the Chinese government, but given China's economic problems, it is in no position to foot the bill, and more than two years after the project was announced, excavation has yet to begin. Many Nicaraguans believe that the whole venture is nothing more than an elaborate scheme designed to enrich the Ortega family and that no canal will ever be built. Yet, Nicaraguans have done little to register their displeasure: there have been no massive protests, for example.

As all these stories suggest, corruption remains deeply embedded in Latin American political and social life. Some countries have seen little improvement from the bad old days decades ago. Yet, the outraged reactions to the wave of scandals currently sweeping the continent may be the first sign that Latin American publics are no longer prepared to tolerate systemic dishonesty in their governments. The region's new middle classes, aided by pressure from abroad and by increasingly confident and independent domestic institutions, have begun demanding better governance.

The outcome of all these movements is still uncertain. Some may generate new institutions: autonomous controller's offices, more powerful and independent judiciaries, greater transparency, and more active and conscious civil societies. Others may take a populist turn, as candidates for office run on antielite platforms. And in some countries, the movements will subside. But in all cases, something will have changed in Latin America, and much for the better.

Critical Thinking

1. What are the causes of kleptocracy in Latin America?
2. Why should the U.S. be concerned about corruption in Brazil?
3. What are the factors that have stimulated opposition to corruption in Latin America?

Internet References

Kletocracy Initiative
kleptocracy/initiative.org

Natural Resources Governance Initiative
www.resourcegovernance.org/

Transparency International
www.transparency.org/

Jorge G. Castañeda is a global distinguished professor of Politics and Latin American and Caribbean Studies at New York University. He served as Mexico's Foreign Minister from 2000 to 2003. Follow him on Twitter @JorgeGCastaneda. This article draws on columns he wrote for *Project Syndicate* last year.

Article

Prepared by: Robert Weiner, *University of Massachusetts, Boston*

The G-Word: The Armenian Massacre and the Politics of Genocide

THOMAS DE WAAL

Learning Outcomes

After reading this article, you will be able to:

- Understand the historical background of the Armenian genocide.
- Understand the position of the Armenian community on the question today.
- Discuss the current Turkish position on the question.

One hundred years ago this April, the Ottoman Empire began a brutal campaign of deporting and destroying its ethnic Armenian community, whom it accused of supporting Russia, a World War I enemy. More than a million Armenians died. As it commemorates the tragedy, the U.S. government, for its part, still finds itself wriggling on the nail on which it has hung for three decades: Should it use the term "genocide" to describe the Ottoman Empire's actions toward the Armenians, or should it heed the warnings of its ally, Turkey, which vehemently opposes using the term and has threatened to recall its ambassador or even deny U.S. access to its military bases if the word is applied in this way? The first course of action would fulfill the wishes of the one-million-strong Armenian American community, as well as many historians, who argue that Washington has a moral imperative to use the term. The second would satisfy the strategists and officials who contend that the history is complicated and advise against antagonizing Turkey, a loyal strategic partner.

No other historical issue causes such anguish in Washington. One former State Department official told me that in 1992, a group of top U.S. policymakers sat in the office of Brent Scowcroft, then national security adviser to President George H. W. Bush, and calculated that resolutions related to the topic were consuming more hours of their time with Congress than any other matter. Over the years, the debate has come to center on a single word, "genocide," a term that has acquired such power that some refuse to utter it aloud, calling it "the G-word" instead. For most Armenians, it seems that no other label could possibly describe the suffering of their people. For the Turkish government, almost any other word would be acceptable.

U.S. President Barack Obama has attempted to break this deadlock in statements he has made on April 24, the day when Armenians traditionally commemorate the tragedy, by evoking the Armenian language phrase Meds Yeghern, or "Great Catastrophe." In 2010, for example, he declared, "1.5 million Armenians were massacred or marched to their death in the final days of the Ottoman Empire. . . . The Meds Yeghern is a devastating chapter in the history of the Armenian people, and we must keep its memory alive in honor of those who were murdered and so that we do not repeat the grave mistakes of the past."

Armenian descendants seeking recognition of their grandparents' suffering could find everything they wanted to see there, except one thing: the word "genocide." That omission led a prominent lobbying group, the Armenian National Committee of America, to denounce the president's dignified statement as "yet another disgraceful capitulation to Turkey's threats," full of "euphemisms and evasive terminology."

In a sense, Obama had only himself to blame for this over-the-top rebuke. After all, during his presidential campaign, he had, like most candidates before him, promised Armenian American voters that he would use the word "genocide" if elected, but once in office, he had honored the relationship with Turkey and broken his vow. His 2010 address did go further than those of his predecessors and openly hinted that he had the G-word in mind when he stated, "My view of that history has not changed." But if he edged closer to the line, he stopped short of crossing it.

History as Battleground

Back in 1915, there was nothing controversial about the catastrophe suffered by ethnic Armenians in the Ottoman Empire. The Young Turkish government, headed by Mehmed Talat Pasha and two others, which ruled what was left of the empire, had entered World War I the year before on the side of Germany, fighting against its longtime foe Russia. The leadership accused Christian Armenians—a population of almost two million, most of whom lived in what is now eastern Turkey—of sympathizing with Russia and thus representing a potential fifth column. Talat ordered the deportation of almost the entire people to the arid deserts of Syria. In the process, at least half of the men were killed by Turkish security forces or marauding Kurdish tribesmen. Women and children survived in greater numbers but endured appalling depredation, abductions, and rape on the long marches.

Leading statesmen of the time regarded the deportation and massacre of the Armenians as the worst atrocity of World War I. One of them, former U.S. President Theodore Roosevelt, argued in a 1918 letter to the philanthropist Cleveland Dodge that the United States should go to war with the Ottoman Empire "because the Armenian massacre was the greatest crime of the war, and failure to act against Turkey is to condone it." Some of the best sources on the horrific events were American. Because the United States had remained neutral during the war's early years, dozens of its diplomatic officials and missionaries in the Ottoman Empire had stayed on the ground and witnessed what happened. In May 1915, Henry Morgenthau, the U.S. ambassador in Turkey, delivered a démarche from the Ottoman Empire's three main adversaries—France, Russia, and the United Kingdom—that denounced the deportation of the Armenians. The statement condemned the Ottoman government for "crimes against humanity," marking the first known official usage of that term. In July 1915, Morgenthau cabled to Washington, "Reports from widely scattered districts indicate systematic attempts to uproot peaceful Armenian populations." These actions, he wrote, involved arbitrary arrests, torture, and large-scale deportations of Armenians, "accompanied by frequent instances of rape, pillage, and murder, turning into massacre."

At the other corner of the Ottoman Empire, Jesse Jackson, the U.S. consul in Aleppo, watched as pitiful convoys of emaciated Armenians arrived in Syria. In September 1916, Jackson sent a cable to Washington that described the burial grounds of nearly 60,000 Armenians near Maskanah, a town in today's northern Syria: "As far as the eye can reach mounds are seen containing 200 to 300 corpses buried in the ground pele mele, women, children and old people belonging to different families."

By the end of World War I, according to most estimates of the time, around one million Armenians had died. Barely one-tenth of the original population remained in its native lands in the Ottoman Empire. The rest had mostly scattered to Armenia, France, Lebanon, and Syria. Many, in ever-greater numbers over the years, headed to the United States.

From the 1920s on, the events of the Great Catastrophe became more a matter of private grief than public record. Ordinary Armenians concentrated on building new lives for themselves. The main political party active in the Armenian diaspora, the Armenian Revolutionary Federation (which had briefly ruled an independent Armenia in 1918–20, before it became a Soviet republic), expended most of its efforts fighting the Soviet Union rather than Turkey. Only in the 1960s did Armenians seriously revive the memory of their grandparents' suffering as a public political issue. They drew inspiration from "Holocaust consciousness," the urge for collective remembrance and action that brought together the Jewish people after the 1961 trial of Adolf Eichmann for Nazi war crimes.

The Republic of Turkey, founded by Mustafa Kemal in 1923, was a state rooted in organized forgetting—not only of the crimes committed in the late Ottoman period against Armenians, Assyrians, and Greeks but also of the suffering of the Muslim population in a string of wars in Anatolia and the Balkans prior to 1923. As the new Turkish state developed, the vanishing of the Armenians became a political, historical, and economic fait accompli. In Turkey, only one substantial book addressing the issue was published between 1930 and the mid-1970s.

When Turkish historians finally returned to the topic in the late 1970s, they did so in response to a wave of terrorist attacks on Turkish diplomats in Western Europe, most of them carried out by Armenian militants based in Beirut. The campaign set off a war among nationalist historians. A simplistic Armenian narrative told of Turkish perpetrators, callous international bystanders, and innocent Armenian victims, downplaying the role that radical Armenian political parties had played in fueling the crackdown. Countering this story was an even cruder narrative spun by some pro-Turkish scholars, several of whom were receiving funding from the Turkish government. That story line portrayed the Armenians as traitors and Muslims as victims of scheming Christian great powers that sought to break up the Ottoman Empire.

The United States served as the main arena for these assertions and denials. In one book published in 1990, Heath Lowry, the head of the newly established Institute of Turkish Studies in Washington, D.C., pursued a common line of Turkish argument: casting doubt on the authenticity of Westerners' eyewitness testimonies. His account, *The Story Behind "Ambassador Morgenthau's Story,"* alleged that Morgenthau was an unreliable witness. Others argued that U.S. missionaries were untrustworthy sources because of their anti-Muslim bias. Over the years, efforts to discredit dozens of primary sources have grown increasingly tortuous. The U.S.-based Turkish website Tall Armenian Tale, for example, laboriously tries to cast doubt on every single one of the hundreds of eyewitness testimonies of the massacre.

A more legitimate line of historical inquiry has focused on the hitherto overlooked tribulations of Muslims in Anatolia and the Caucasus during World War I. These accounts have pointed out that the Armenians were not the only people to face persecution in eastern Turkey. The Kurdish and Turkish populations, too, suffered grievously at the hands of the Russian army, which contained several Armenian regiments, when these forces occupied swaths of eastern Turkey not long after the Armenian deportations. Later, in 1918–20, Muslim Azerbaijanis were deported from the briefly independent Republic of Armenia before it was conquered by the Bolsheviks.

The wartime context of the Armenian massacre and the multiple actors involved—in addition to Armenians and Turks: Assyrians, Azerbaijanis, Greeks, Kurds, British, Germans, and Russians—make it harder to tell the story in all its nuance. The history of the Armenian genocide lacks the devastating simplicity of the Holocaust's narrative. But a new generation of historians has finally taken up the challenge of explaining the full context of the tragedy. Some of them, such as Raymond Kevorkian, are Armenian, whereas others, including Donald Bloxham and Erik-Jan Zurcher, hail from Europe. Several come from Turkey, including Fikret Adanir, Taner Akcam, Halil Berktay, and Fuat Dundar.

At the heart of most of these histories lies a hard kernel of truth: although Muslims suffered enormously during World War I, in both Anatolia and the Caucasus, the Armenian experience was of a different order of pain. Along with the Assyrians, the Armenians were subjected to a campaign of destruction that was more terrible for being organized and systematic. And even though some Armenian nationalists helped precipitate the brutal Ottoman response, every single Armenian suffered as a result. As Bloxham has written, "Nowhere else during the First World War was the separatist nationalism of the few answered with the total destruction of the wider ethnic community from which the nationalists hailed. That is the crux of the issue."

Word as Weapon

If the issue of the experience of the Armenians in World War I were merely a matter of historical interpretation, a way forward would be clear. The huge volume of primary source material, combined with Armenian oral histories, authenticates the veracity of what Armenians recall—as does the plain fact that an entire people vanished from their historical homeland. All that historians have to do, it would seem, is fill out the context of the events and explain why the Young Turks treated the Armenians the way they did.

But what dominates the public discourse today is the word "genocide," which was devised almost three decades after the Armenian deportations to designate the destruction not just of people but also of an entire people. The term is closely associated with the man who invented it, the Polish-born Jewish lawyer Raphael Lemkin. Lemkin barely escaped the horror of the Holocaust, which wiped out most of his family in Poland after he immigrated to the United States. As he would later explain in a television interview, "I became interested in genocide because it happened so many times. It happened to the Armenians, and after the Armenians, Hitler took action."

Lemkin had a morally courageous vision: to get the concept of genocide enshrined in international law. His tireless lobbying soon paid off: in 1948, just four years after he invented the term, the United Nations adopted the Genocide Convention, a treaty that made the act an international crime. But Lemkin was a more problematic personality than the noble crusader depicted in modern accounts, such as Samantha Power's book *A Problem From Hell.* In his uncompromising pursuit of his goal, Lemkin allowed the term "genocide" to be bent by other political agendas. He opposed the Universal Declaration of Human Rights, adopted a week after the Genocide Convention, fearing that it would distract the international community from preventing future genocides—the goal that he thought should surpass all others in importance. And he won the Soviet Union's backing for the convention after "political groups" were excluded from the classes of people it protected.

The final definition of "genocide" adopted by the UN had several points of ambiguity, which gave countries and individuals accused of this crime legal ammunition to resist the charge. For example, Article 2 of the convention defines "genocide" as "acts committed with intent to destroy, in whole or in part, a national, ethnical, racial or religious group, as such." The meaning of the words "as such" is far from clear. And alleged perpetrators often deny that the destruction was "committed with intent"—an argument frequently made in Turkey.

Soon, however, only a careful few were bothering to refer to the UN convention in evoking the term. In the broader public's mind, the association with the Holocaust gave the word "genocide" totemic power, making it the equivalent of absolute evil. After 1948, the legal term that had initially been created to deter mass atrocities became an insult traded between nations and peoples accusing each other of past and present horrors. The United States and the Soviet Union each freely accused the other of genocide during the Cold War.

The Armenian diaspora saw the word as a perfect fit to describe what had happened to their parents and grandparents and began referring to the Meds Yeghern as "the Armenian genocide." The concept helped activate a new political movement. The year 1965 marked both the 50th anniversary of the massacre and the moment when the Armenian diaspora made seeking justice for the victims a political cause.

In the postwar United States, it was normal practice to put the words "Armenian" and "genocide" together in the same sentence. This usage came with the assumption that the UN convention—one of its first signatories was Turkey—had no retroactive force and therefore could not provide the basis for legal action related to abuses committed before 1948. For instance, in 1951, U.S. government lawyers submitted an advisory opinion on the Genocide Convention to the International Court of Justice, in The Hague, citing the Turkish massacre of the Armenians as an instance of genocide. In April 1981, in a proclamation on the Holocaust, U.S. President Ronald Reagan mentioned "the genocide of the Armenians before it, and the genocide of the Cambodians which followed it."

Political circumstances changed this thinking in the 1980s. Reagan himself performed an abrupt about-face following the 1982 assassination of Kemal Arikan, the Turkish consul general to the United States, by two young Armenian militants in Los Angeles. The death of a diplomat of a close NATO ally in Reagan's own home state enraged and embarrassed the president. He and his team concluded that on three of the foreign policy issues that concerned them the most—the Soviet Union, Israel, and terrorism—Turkey was staunchly on the U.S. side. Armenians, by contrast, were not.

Seven months after the killing of Arikan, the State Department's official bulletin published a special issue on terrorism, which included a piece titled "Armenian Terrorism: A Profile." A note at the end of the article said, "Because the historical record of the 1915 events in Asia Minor is ambiguous, the Department of State does not endorse allegations that the Turkish government committed a genocide against the Armenian people. Armenian terrorists use this allegation to justify in part their continuing attacks on Turkish diplomats and installations." In response to furious Armenian complaints, the bulletin ended up publishing not one but two clarifications of that statement. But from that point on, a new line had been drawn by the executive branch, and the term "Armenian genocide" was outlawed in the White House.

Deadlock on the Hill

Congress, meanwhile, was plowing its own furrow. By the 1970s, one million Armenians lived in the United States. Younger generations were no longer willing to limit the discussions of their ancestors' deaths to Sunday dinners, requiem services, and low-circulation newspapers. Many Armenian Americans who had political savvy and wealth, such as the Massachusetts businessman Stephen Mugar, began to lobby Congress. They found an ally in the Speaker of the House of Representatives, Tip O'Neill, whose congressional district included the de facto capital of the Armenian American community: Watertown, Massachusetts. In early 1975, urged on by Mugar and others, O'Neill managed to get the House to pass a resolution authorizing the president to designate April 24 of that year as the "National Day of Remembrance of Man's Inhumanity to Man" and observe it by honoring all victims of genocide, "especially those of Armenian ancestry who succumbed to the genocide perpetrated in 1915."

That occasion marked the only time Congress has passed any kind of resolution recognizing the Armenian genocide. In 1990, the Senate spent two days in fierce debate over whether April 24 should again be officially designated as a national day of remembrance, this time of the "Armenian Genocide of 1915–1923." Kansas Senator Bob Dole led the argument in favor of the motion, but opponents managed to block it. Ever since, with the White House opposed to officially recognizing the phrase "Armenian genocide," resolutions of this kind have failed. They have become an increasingly tired and predictable exercise: however much historical evidence the Armenian lobbyists produce to support their case, the Turks play the trump card of national security, lightly threatening that a yes vote would jeopardize the United States' continued use of the Incirlik Air Base, which is on Turkish territory, a key supply hub for U.S. military operations in the region. In 2007, when one genocide resolution appeared certain to pass the House, no fewer than eight former secretaries of state intervened with a joint letter advising Congress to drop the issue—which it ultimately did.

The fight for genocide recognition has now become the raison d'être for the two dominant Armenian American organizations, the Armenian Assembly of America and the Armenian National Committee of America. They do not conceal that the campaign helps them preserve a collective identity among the Armenian diaspora—an increasingly assimilated group that is losing other common bonds, such as the Armenian language and attendance at services of the Armenian Apostolic Church. But they do not like to admit that the campaign has also damaged their cause. For many Americans, the phrase "Armenian genocide" now evokes not a story of terrible human suffering but an exasperating, eye-roll-inducing tale of lobbying and congressional bargaining. Inevitably, the need to secure votes for any given resolution on the topic means that the memory of the Ottoman Armenians is cheapened by being tied to other items of congressional business. What results is routine horse-trading, as in, "You vote for the farm bill, and I'll back you on the genocide resolution."

A few thoughtful Armenians object to such genocide-recognition lobbying campaigns on the grounds that they turn the deaths of their grandparents into one big homicide case. They see that their fellow Armenians are less interested in grieving for the dead than in demonstrating outside the Turkish embassy with pictures of dead bodies—the more gruesome, the better—and struggling to prove something that they already know to be true. The obsession with genocide, argues the French Armenian philosopher Marc Nichanian, "forbids mourning." Armenian campaigners have a point when they contend that their pursuit of genocide recognition has had the benefit of focusing Turkey's mind on an issue that the country would rather have forgotten. But their campaign has also heightened Turkish passions, since their efforts have indirectly strengthened the Turkish nationalist story line of World War I. That partial, but not entirely inaccurate, account portrays the great powers of the time as conspirators plotting to undermine the Ottoman Empire. Consequently, any resolution passed by a modern great power condemning Turkey's historical crimes would only inflame a sore spot.

Fueling this paranoia, many Turkish policymakers have expressed their suspicion that a genocide resolution would pave the way for territorial concessions. These fears have little basis in reality. Although some radical groups, such as the Armenian Revolutionary Federation, continue to make territorial claims, the Republic of Armenia has all but officially recognized Turkey's current borders. Reestablishing full diplomatic relations between the two countries, which have been on hold since the Armenian-Azerbaijani war in the early 1990s, would make this recognition formal. No statements made by a political party that last ruled Armenia in 1920 can change that reality.

As for reparations, it is hard to see how Washington's adoption of the word "genocide" would make the case for them. Most international legal opinions are clear that the UN Genocide Convention carries no retroactive force and therefore could not be invoked to bring claims on dispossessed property. Such a scenario is all the more difficult to imagine because it would trigger a nightmarish relitigation of the whole of World War I, during which not only Armenians but also Azerbaijanis, Greeks, Kurds, and Turks were robbed of their possessions in Anatolia, the Balkans, and the Caucasus. Yet the invocation of the controversial word still fills Turkey with dread.

A Turkish Thaw

The only good news in this bleak historical tale comes from Turkey itself. Since the election in 2002 of the post-Kemalist government led by the Justice and Development Party (known as the AKP), in a process largely unconnected to outside pressure, Turkish society has begun to revisit some of the dark pages of its past, including the oppression of the non-Turkish populations of the late Ottoman Empire. This growing openness has allowed the descendants of forcibly Islamized Armenians to come out of the shadows, and a few Armenian churches and schools have reopened. Turkish historians have begun to write about the late Ottoman period without fear of retribution. And they have finally started to challenge the old dominant narrative, which the historian Berktay has called "the theory of the immaculate conception of the Turkish Republic."

From the Armenian standpoint, this opening has been too slow. But it could hardly have proceeded at a faster pace. As one of the key figures behind the thaw, the late Istanbul-based Armenian journalist Hrant Dink, pointed out, Turkey had been a closed society for three generations; it takes time and immense effort to change that. "The problem Turkey faces today is neither a problem of 'denial' or 'acknowledgement,'" Dink wrote in 2005. "Turkey's main problem is 'comprehension.' And for the process of comprehension, Turkey seriously needs an alternative study of history and for this, a democratic environment. . . . The society is defending the truth it knows."

In that spirit, Dink, a stalwart of the left and a confirmed antiimperialist, criticized genocide resolutions in foreign parliaments on the grounds that they merely replicated previous great-power bullying of Turkey. He saw his mission as helping Turks understand Armenians and the trauma they have passed down over generations, while helping Armenians recognize the sensitivities and legitimate interests of the Turks. Dink's stand broke both Turkish and Armenian taboos, and he paid the highest price for his courage: in 2007, he was assassinated by a young Turkish nationalist.

Dink's insights suggest that the word "genocide" may be the correct term but the wrong solution to the controversy. Simply put, the emotive power of the word has overpowered Armenian-Turkish dialogue. No one willingly admits to committing genocide. Faced with this accusation, many Turks (and others in their position) believe that they are being invited to compare their grandparents to the Nazis.

It may be that the word "genocide" has exhausted itself, and that the success of Lemkin's invention has also been its undoing. Lemkin probably never anticipated that coining a new standard of awfulness would set off an unfortunate global competition in which nations—from Armenia's neighbor Azerbaijan to Sudan and Tibet—vie to get the label applied to their own tragedies. As the philosopher Tzvetan Todorov has observed, even though no one wants to be a victim, the position does confer certain advantages. Groups that gain recognition as victims of past injustices obtain "a bottomless line of moral credit," he has written. "The greater the crime in the past, the more compelling the rights in the present—which are gained merely through membership in the wronged group." Conversely, the grandchildren of the alleged perpetrators aspire to absolve their ancestors of guilt and, by association, of a link to Adolf Hitler and the Holocaust.

In *A Problem From Hell,* Power chastised the international community for its timidity and failure to stop genocides even after this appalling phenomenon had been named and outlawed. But the problem can be posed the other way around: Could it be that international actors hide behind the ambiguities of genocide terminology in order to do nothing—and that the very power of the word "genocide" and the responsibilities it invokes deter action? It may be no coincidence that the first successful prosecution under the UN Genocide Convention, that of a Rwandan war criminal, came only in September 1998, nearly 50 years after the convention was adopted. In the Armenian case, the phrase "Armenian genocide" has become customary in the scholarly literature. Those who avoid it today risk putting themselves in the company of skeptics who minimize the tragedy or deny it outright. Many progressive Turkish intellectuals, too, now use the term. Among them are such brave voices as the journalist Hasan Cemal, grandson of Ahmed Cemal Pasha, one of the three Young Turkish leaders who ran the brutal Ottoman government in 1915.

But that does not mean that Meds Yeghern is an inferior and less expressive phrase. If it becomes more widely used, it might acquire the same resonance as the words "Holocaust" and "Shoah" have in describing the fate of the European Jews. There is also the legal term "crimes against humanity," first applied in 1915 specifically in reference to the Armenian massacre. This concept lacks the emotional charge and the definitional problems of the word "genocide" and covers mass atrocities not falling under its narrow definition—those in which the perpetrators may not have intended to eradicate an entire nation but have still killed an awful lot of innocent people.

The challenge for the United States, then, is not simply to find a way to once again use the term "Armenian genocide," a phrase it has employed before, but to do so while also accepting the limitations of a concept that has grown emotionally fraught and overly legalistic. The mere act of using the term, without a deeper engagement with the history of the Armenians and the Turks, would do little to resolve the bigger underlying question—namely, how to persuade Turkey to honor the losses of the Ottoman Armenians and other minorities a hundred years ago.

Having been a neutral power in 1915, the United States can assert that it bears no historical grudge against Turkey. Washington can therefore help bring about the rapprochement between the Armenians and the Turks that Dink advocated. The United States can urge Turkey to hasten the process of historical reckoning by taking steps to keep the small Armenian Turkish population from leaving the country, to conserve what little Armenian cultural heritage survives in Turkey, and to restore the place of Armenians and other ethnic minorities in Turkey's history books.

Armenians need to be able to finally bury their grandparents and receive an acknowledgment from the Turkish state of the terrible fate they suffered. These steps toward reconciliation will surely become more possible as a more open Turkey begins to confront its past as a whole. If that can be made to happen, everything else will follow.

Critical Thinking

1. Why won't the U.S. label the killing of the Armenians as genocide?
2. Why does the author argue that the term genocide is overly legalistic?
3. Why did the Ottoman Turkish government target the Armenians?

Internet References

Genocide Convention
preventgenocide.org

Turkish Embassy, Washington DC
vasington.be.mfa.gov.tr

U.S. Holocaust Museum
www.ushmm.org/

Thomas de Waal is a Senior Associate at the Carnegie Endowment for International Peace and the author of the forthcoming *Great Catastrophe: Armenians and Turks in the Shadow of Genocide.* Follow him on Twitter @TomdeWaalCEIP.

Article

Prepared by: Robert Weiner, *University of Massachusetts, Boston*

Race in the Modern World: The Problem of the Color Line

Kwame Anthony Appiah

Learning Outcomes

After reading this article, you will be able to:

- Explain the difficult efforts of scholars to come up with a definition of race.
- Discuss the importance of Pan-Africanism.

In 1900, in his "Address to the Nations of the World" at the first Pan-African Conference, in London, W. E. B. Du Bois proclaimed that the "problem of the twentieth century" was "the problem of the color-line, the question as to how far differences of race—which show themselves chiefly in the color of the skin and the texture of the hair—will hereafter be made the basis of denying to over half the world the right of sharing to their utmost ability the opportunities and privileges of modern civilization."

Du Bois had in mind not just race relations in the United States but also the role race played in the European colonial schemes that were then still reshaping Africa and Asia. The final British conquest of Kumasi, Ashanti's capital (and the town in Ghana where I grew up), had occurred just a week before the London conference began. The British did not defeat the Sokoto caliphate in northern Nigeria until 1903. Morocco did not become a French protectorate until 1912, Egypt did not become a British one until 1914, and Ethiopia did not lose its independence until 1936. Notions of race played a crucial role in all these events, and following the Congress of Berlin in 1878, during which the great powers began to devise a world order for the modern era, the status of the subject peoples in the Belgian, British, French, German, Spanish, and Portuguese colonies of Africa—as well as in independent South Africa—was defined explicitly in racial terms.

Du Bois was the beneficiary of the best education that North Atlantic civilization had to offer: he had studied at Fisk, one of the United States' finest black colleges; at Harvard; and at the University of Berlin. The year before his address, he had published *The Philadelphia Negro,* the first detailed sociological study of an American community. And like practically everybody else in his era, he had absorbed the notion, spread by a wide range of European and American intellectuals over the course of the nineteenth century, that race—the division of the world into distinct groups, identifiable by the new biological sciences—was central to social, cultural, and political life.

Even though he accepted the concept of race, however, Du Bois was a passionate critic of racism. He included anti-Semitism under that rubric, and after a visit to Nazi Germany in 1936, he wrote frankly in *The Pittsburgh Courier,* a leading black newspaper, that the Nazis' "campaign of race prejudice . . . surpasses in vindictive cruelty and public insult anything I have ever seen; and I have seen much." The European homeland had not been in his mind when he gave his speech on the color line, but the Holocaust certainly fit his thesis—as would many of the centuries' genocides, from the German campaign against the Hereros in Namibia in 1904 to the Hutu massacre of the Tutsis in Rwanda in 1994. Race might not necessarily have been *the* problem of the century—there were other contenders for the title—but its centrality would be hard to deny.

Violence and murder were not, of course, the only problems that Du Bois associated with the color line. Civic and economic inequality between races—whether produced by government policy, private discrimination, or complex interactions between the two—were pervasive when he spoke and remained so long after the conference was forgotten.

All around the world, people know about the civil rights movement in the United States and the antiapartheid struggle in

South Africa, but similar campaigns have been waged over the years in Australia, New Zealand, and most of the countries of the Americas, seeking justice for native peoples, or the descendants of African slaves, or East Asian or South Asian indentured laborers. As non-Europeans, including many former imperial citizens, have immigrated to Europe in increasing numbers in recent decades, questions of racial inequality there have come to the fore, too—in civic rights, education, employment, housing, and income. For Du Bois, Chinese, Japanese, and Koreans were on the same side of the color line as he was. But Japanese brutality toward Chinese and Koreans up through World War II was often racially motivated, as are the attitudes of many Chinese toward Africans and African Americans today. Racial discrimination and insult are a global phenomenon.

Of course, ethnoracial inequality is not the only social inequality that matters. In 2013, the nearly 20 million white people below the poverty line in the United States made up slightly more than 40 percent of the country's poor. Nor is racial prejudice the only significant motive for discrimination: ask Christians in Indonesia or Pakistan, Muslims in Europe, or LGBT people in Uganda. Ask women everywhere. But more than a century after his London address, Du Bois would find that when it comes to racial inequality, even as much has changed, much remains the same.

Us and Them

Du Bois' speech was an invitation to a global politics of race, one in which people of African descent could join with other people of color to end white supremacy, both in their various homelands and in the global system at large. That politics would ultimately shape the process of decolonization in Africa and the Caribbean and inform the creation of what became the African Union. It was politics that led Du Bois himself to become, by the end of his life, a citizen of a newly independent Ghana, led by Kwame Nkrumah.

But Du Bois was not simply an activist; he was even more a scholar and an intellectual, and his thinking reflected much of his age's obsession with race as a concept. In the decades preceding Du Bois' speech, thinkers throughout the academy—in classics, history, artistic and literary criticism, philology, and philosophy, as well as all the new life sciences and social sciences—had become convinced that biologists could identify, using scientific criteria, a small number of primary human races. Most would have begun the list with the black, white, and yellow races, and many would have included a Semitic race (including Jews and Arabs), an American Indian race, and more. People would have often spoken of various subgroups within these categories as races, too. Thus, the English poet Matthew Arnold considered the Anglo-Saxon and Celtic races to be the main components of the population of the United Kingdom; the French historian Hippolyte Taine thought the Gauls were the race at the core of French history and identity; and the U.S. politician John C. Calhoun discussed conflicts not only between whites and blacks but also between Anglo-Canadians and "the French race of Lower Canada."

People thought race was important not just because it allowed one to define human groups scientifically but also because they believed that racial groups shared inherited moral and psychological tendencies that helped explain their different histories and cultures. Of course, there were always skeptics. Charles Darwin, for example, believed that his evolutionary theory demonstrated that human beings were a single stock, with local varieties produced by differences in environment, through a process that was bound to result in groups with blurred edges. But many late-nineteenth-century European and American thinkers believed deeply in the biological reality of race and thought that the natural affinity among the members of each group made races the appropriate units for social and political organization.

Essentialism—the idea that human groups have core properties in common that explain not just their shared superficial appearances but also the deep tendencies of their moral and cultural lives—was not new. In fact, it is nearly universal, because the inclination to suppose that people who look alike have deep properties in common is built into human cognition, appearing early in life without much prompting. The psychologist Susan Gelman, for example, argues that "our essentializing bias is not directly taught," although it is shaped by language and cultural cues. It can be found as far back as Herodotus' *Histories* or the Hebrew Bible, which portrayed Ethiopians, Persians, and scores of other peoples as fundamentally other. "We" have always seen "our own" as more than superficially different from "them."

What was new in the nineteenth century was the combination of two logically unrelated propositions: that races were biological and so could be identified through the scientific study of the shared properties of the bodies of their members and that they were also political, having a central place in the lives of states. In the eighteenth century, the historian David Hume had written of "national character"; by the nineteenth century, using the new scientific language, Arnold was arguing that the "Germanic genius" of his own "Saxon" race had "steadiness as its main basis, with commonness and humdrum for its defect, fidelity to nature for its excellence."

If nationalism was the view that natural social groups should come together to form states, then the ideal form of nationalism would bring together people of a single race. The eighteenth-century French American writer J. Hector St. John de Crèvecoeur's notion that in the New World, all races could be "melted into a new race of man"—so that it was the nation that made the race, not the race the nation—belonged to an older way of thinking, which racial science eclipsed.

The Other Dismal Science

In the decade after Du Bois' address, however, a second stage of modern argumentation about human groups emerged, one that placed a much greater emphasis on culture. Many things contributed to this change, but a driving force was the development of the new social science of anthropology, whose German-born leader in the United States, Franz Boas, argued vigorously (and with copious evidence from studies in the field) that the key to understanding the significant differences between peoples lay not in biology—or, at least, not in biology alone—but in culture. Indeed, this tradition of thought, which Du Bois himself soon took up vigorously, argued not only that culture was the central issue but also that the races that mattered for social life were not, in fact, biological at all.

In the United States, for example, the belief that anyone with one black grandparent or, in some states, even one black great-grandparent was also black meant that a person could be socially black but have skin that was white, hair that was straight, and eyes that were blue. As Walter White, the midcentury leader of the National Association for the Advancement of Colored People, whose name was one of his many ironic inheritances, wrote in his autobiography, "I am a Negro. My skin is white, my eyes are blue, my hair is blond. The traits of my race are nowhere visible upon me."

Strict adherence to thinking of race as biological yielded anomalies in the colonial context as well. Treating all Africans in Nigeria as "Negroes," say, would combine together people with very different biological traits. If there were interesting traits of national character, they belonged not to races but to ethnic groups. And the people of one ethnic group—Arabs from Morocco to Oman, Jews in the Diaspora—could come in a wide range of colors and hair types.

In the second phase of discussion, therefore, both of the distinctive claims of the first phase came under attack. Natural scientists denied that the races observed in social life were natural biological groupings, and social scientists proposed that the human units of moral and political significance were those based on shared culture rather than shared biology. It helped that Darwin's point had been strengthened by the development of Mendelian population genetics, which showed that the differences found between the geographic populations of the human species were statistical differences in gene frequencies rather than differences in some putative racial essence.

In the aftermath of the Holocaust, moreover, it seemed particularly important to reject the central ideas of Nazi racial "science," and so, in 1950, in the first of a series of statements on race, unesco (whose founding director was the leading biologist Sir Julian Huxley) declared that national, religious, geographic, linguistic and cultural groups do not necessarily coincide with racial groups: and the cultural traits of such groups have no demonstrated genetic connection with racial traits. . . . The scientific material available to us at present does not justify the conclusion that inherited genetic differences are a major factor in producing the differences between the cultures and cultural achievements of different peoples or groups.

Race was still taken seriously, but it was regarded as an outgrowth of sociocultural groups that had been created by historical processes in which the biological differences between human beings mattered only when human beings decided that they did. Biological traits such as skin color, facial shape, and hair color and texture could define racial boundaries if people chose to use them for that purpose. But there was no scientific reason for doing so. As the unesco statement said in its final paragraph, "Racial prejudice and discrimination in the world today arise from historical and social phenomena and falsely claim the sanction of science."

Construction Work

In the 1960s, a third stage of discussion began, with the rise of "genetic geography." Natural scientists such as the geneticist Luigi Luca Cavalli-Sforza argued that the concept of race had no place in human biology, and social scientists increasingly considered the social groups previously called "races" to be social constructions. Since the word "race" risked misleading people on this point, they began to speak more often of "ethnic" or "ethnoracial" groups, in order to stress the point that they were not aiming to use a biological system of classification.

In recent years, some philosophers and biologists have sought to reintroduce the concept of race as biological using the techniques of cladistics, a method of classification that combines genetics with broader genealogical criteria in order to identify groups of people with shared biological heritages. But this work does not undermine the basic claim that the boundaries of the social groups called "races" have been drawn based on social, rather than biological, criteria; regardless, biology does not generate its own political or moral significance. Socially constructed groups can differ statistically in biological characteristics from one another (as rural whites in the United States differ in some health measures from urban whites), but that is not a reason to suppose that these differences are caused by different group biologies. And even if statistical differences between groups exist, that does not necessarily provide a rationale for treating individuals within those groups differently. So, as Du Bois was one of the first to argue, when questions arise about the salience of race in political life, it is usually not a good idea to bring biology into the discussion.

It was plausible to think that racial inequality would be easier to eliminate once it was recognized to be a product of sociology and politics rather than biology. But it turns out that all sorts of status differences between ethnoracial groups can persist long after governments stop trying to impose them. Recognizing that institutions and social processes are at work rather than innate qualities of the populations in question has not made it any less difficult to solve the problems.

Imagined Communities

One might have hoped to see signs that racial thinking and racial hostility were vanishing—hoped, that is, that the color line would not continue to be a major problem in the twenty-first century, as it was in the twentieth. But a belief in essential differences between "us" and "them" persists widely, and many continue to think of such differences as natural and inherited. And of course, differences between groups defined by common descent can be the basis of social identity, whether or not they are believed to be based in biology. As a result, ethnoracial categories continue to be politically significant, and racial identities still shape many people's political affiliations.

Once groups have been mobilized along ethnoracial lines, inequalities between them, whatever their causes, provide bases for further mobilization. Many people now know that we are all, in fact, one species, and think that biological differences along racial lines are either illusory or meaningless. But that has not made such perceived differences irrelevant.

Around the world, people have sought and won affirmative action for their ethnoracial groups. In the United States, in part because of affirmative action, public opinion polls consistently show wide divergences on many questions along racial lines. On American university campuses, where the claim that "race is a social construct" echoes like a mantra, black, white, and Asian identities continue to shape social experience. And many people around the world simply find the concept of socially constructed races hard to accept, because it seems so alien to their psychological instincts and life experiences.

Race also continues to play a central role in international politics, in part because the politics of racial solidarity that Du Bois helped inaugurate, in co-founding the tradition of pan-Africanism, has been so successful. African Americans are particularly interested in U.S. foreign policy in Africa, and Africans take note of racial unrest in the United States: as far away as Port Harcourt, Nigeria, people protested against the killing of Michael Brown, the unarmed black teenager shot to death by a police officer last year in Missouri. Meanwhile, many black Americans have special access to Ghanaian passports, Rastafarianism in the Caribbean celebrates Africa as the home of black people, and heritage tourism from North and South America and the Caribbean to West Africa has boomed.

Pan-Africanism is not the only movement in which a group defined by a common ancestry displays transnational solidarity. Jews around the world show an interest in Israeli politics. People in China follow the fate of the Chinese diaspora, the world's largest. Japanese follow goings-on in São Paulo, Brazil, which is home to more than 600,000 people of Japanese descent—as well as to a million people of Arab descent, who themselves follow events in the Middle East. And Russian President Vladimir Putin has put his supposed concern for ethnic Russians in neighboring countries at the center of his foreign policy.

Identities rooted in the reality or the fantasy of shared ancestry, in short, remain central in politics, both within and between nations. In this new century, as in the last, the color line and its cousins are still going strong.

Wouldn't It Be Nice?

The pan-Africanism that Du Bois helped invent created, as it was meant to, a new kind of transnational solidarity. That solidarity was put to good use in the process of decolonization, and it was one of the forces that helped bring an end to Jim Crow in the United States and apartheid in South Africa. So racial solidarity has been used not just for pernicious purposes but for righteous ones as well. A world without race consciousness, or without ethnoracial identity more broadly, would lack such positive mobilizations, as well as the negative ones. It was in this spirit, I think, that Du Bois wrote, back in 1897, that it was "the duty of the Americans of Negro descent, as a body, to maintain their race identity until . . . the ideal of human brotherhood has become a practical possibility."

But at this point, the price of trying to move beyond ethnoracial identities is worth paying, not only for moral reasons but also for the sake of intellectual hygiene. It would allow us to live and work together more harmoniously and productively, in offices, neighborhoods, towns, states, and nations. Why, after all, should we tie our fates to groups whose existence seems always to involve misunderstandings about the facts of human difference? Why rely on imaginary natural commonalities rather than build cohesion through intentional communities? Wouldn't it be better to organize our solidarities around citizenship and the shared commitments that bind political society?

Still, given the psychological difficulty of avoiding essentialism and the evident continuing power of ethnoracial identities, it would take a massive and focused effort of education, in schools and in public culture, to move into a postracial world. The dream of a world beyond race, unfortunately, is likely to be long deferred.

Critical Thinking

1. What are W. E. B. Dubois' major ideas about race?
2. What is meant by an imagined community?

Internet References

Brennan Center for Justice
www.brennancenter.org

W. E. B. Dubois papers
credo.library.umass.edu

Kwame Anthony Appiah is Professor of Philosophy and Law at New York University. His most recent book is *Lines of Descent: W. E. B. Du Bois and the Emergence of Identity*. Born to a British mother and Ghanaian father, Kwame Anthony Appiah grew up traveling between his two homelands, an experience that has shaped his wideranging writing on ethnicity, identity, and culture. He is the author of numerous books, including *Lines of Descent,* and has won scores of prizes, among them the National Humanities Medal and the Arthur Ross Book Award. Now a professor at New York University, Appiah explores the past, present, and future of thinking about race in "Race in the Modern World" (page 1).

Article

Prepared by: Robert Weiner, *University of Massachusetts, Boston*

Democracy and Its Discontents

JOHN SHATTUCK

Learning Outcomes

After reading this article, you will be able to:

- Explain why liberal democracy has been replaced by illiberal democracy in Eastern Europe.
- Define what is meant by illiberal democracy.

What's happening to democracy in Eastern Europe? A new authoritarianism, "illiberal governance," has taken over in Hungary and Poland. It's been boosted by the Paris and Brussels attacks and the fear of terrorism. Hungary and Poland are not isolated cases. A trend away from democratic pluralism is also sweeping through Western Europe. Where will it lead?

In trying to answer this question, I'll follow the advice of Václav Havel, to "keep the company of those who seek the truth, but run from those who claim to have found it." I promise to make no such claim, but I will seek the truth about "illiberal governance"—the new threat to democracy that is prominent in the headlines these days. My tentative answer is that this modern form of "soft authoritarianism" may not prove sustainable.

Competing forces were unleashed by the fall of the Berlin Wall. Forces of integration broke down barriers, promoted democratic development, created economic interdependence, and facilitated the digital revolution. Forces of disintegration tore apart failed states, stimulated ethnic and religious violence, and spurred nationalist leaders to challenge transnational entities like the European Union. There are conflicting scenarios about how these forces will play out. An optimistic view envisions slow and steady progress toward the universal realization of democracy. A negative, almost dystopian, perspective sees the increasing clash of cultures and civilizations—a steady regress toward ongoing conflict among cultures, religions, and societies. These two visions have been caricatured as alternative realities of the post–Cold War world, but they provide a useful starting point for understanding what's happening today to democracy in Europe and the United States.

I

Discontent with democracy is widespread. In 2014, a European Commission poll revealed that 68 percent of Europeans distrusted their national governments, and 82 percent distrusted the political parties that had produced these governments.[1] In the United States, a Gallup poll in the same year found that 65 percent of Americans were dissatisfied with their system of government and how it works—a striking increase from only 23 percent in 2002.[2]

One reason for this discontent may be a growing sense that the world is spinning out of control, and that democracy is only making matters worse. A deeper reason may be that people today are confused about the meaning of democracy—demanding both greater participation in their own governance and greater efficiency in the way government operates. The very idea of democracy may be at war with itself; people look to democratic governments to solve their problems but are unwilling to recognize their own responsibility for keeping democracy healthy.

Digging deeper, the roots of discontent can be found in four democratic revolutions of the last 50 years. As my colleague Ivan Krastev has written, these four upheavals have simultaneously strengthened and weakened democracy in Europe and the United States.

The Cultural Revolution of the 1960s gave birth to a modern world of individual rights and freedoms. At the same time, the rights revolution reduced the sense of collective purpose essential to democratic governance. It transformed democratic society, but a counterrevolution pushed back, turning the struggle for human rights and civil liberties into an endlessly divisive political battleground.

The Market Revolution of the 1980s released the power of the market economy to produce economic growth. It also cut way back on the role of government in regulating the economy, destroying the Keynesian consensus about the social benefits of a mixed economy and a welfare state. It paved the way for the rise of new economic elites, globalization, and inequality, while breeding political resentment among the overwhelming majority left behind.

The Political Revolution of 1989 marked the end of Communism and the Cold War, the opening of borders, and the beginning of a transition to democracy and market freedom in Eastern Europe. But it also marked the collapse of longstanding social support systems in the East, and in the West an end to the informal social contract between economic elites and the people.

The most recent democratic upheaval, the Internet Revolution, opened the floodgates of information, creating unlimited opportunities for peer-to-peer communication and horizontal grassroots pressure for change. At the same time, it spawned vast echo chambers and ghettoes of communication, reducing discourse across political divides and increasing the polarization of democratic societies.

II

Democratic discontent is especially acute in Eastern Europe, where the roots of democratic governance are shallow. Eastern Europeans were ruled for centuries by successive empires of Ottoman, Russian, Hapsburg, fascist, and communist authoritarian regimes. A long-suppressed hunger for national identity and honor among the peoples of the region constantly fueled their anger against outside oppressors—the Hapsburgs, who executed the first elected Hungarian prime minister in 1849; the Russians, who dominated Poland throughout the 19th century; and the Turks, who defeated the Serbs in the Battle of Kosovo Polje at the end of the 14th century. The collective memory of this ancient defeat in Serbia was so powerful that Slobodan Milosevic was able to invoke it 600 years later, when he launched his notorious ethnic cleansing campaign against the Kosovar Muslims.

In the 20th century, communism destroyed civil society in Eastern Europe by limiting civic engagement to activities relating to or mandated by the state. It also destroyed the sense of personal responsibility to the community that is essential for the growth of democracy. In Prague in the 1990s "volunteering" still meant collaborating with the regime. In Budapest today common spaces in apartment buildings are still rarely cared for by the residents. Communism's alternative to civil society was state employment and social security, but of course these were dismantled after the fall of the Berlin Wall.

After 1989, hopes in Eastern Europe that democracy would bring immediate economic benefits went unfulfilled. Standards of living failed to keep pace with popular expectations, especially after the financial crisis hit the region in 2009. In this neuralgic environment, Eastern Europeans found themselves attracted to political leaders who claimed they could defend the people against outsiders, like the foreign banks that had called in their mortgages when the financial markets collapsed.

These festering resentments were the building blocks of a new nationalism. Two basic elements went into its construction.

First was the politics of national identity. The longing for national identity had been largely ignored by the proponents of post–Cold War European integration, but it was taken up with a vengeance by nationalist leaders who developed new narratives to appeal to a resentful and confused populace.

In Hungary, which had been on the losing side of both world wars as an ally of Germany, the new nationalist narrative depicted Hungarians as victims, stripped of ⅔ of their lands and separated from their compatriots by the Treaty of Trianon after the First World War, then occupied by Germany and allegedly forced to participate in the Holocaust at the end of the Second World War. A particularly dangerous charge in this twisted national narrative was that "Brussels is the new Moscow."[3] After decades of being dictated to by a distant Soviet regime, Hungarians were susceptible to this claim. Casting the European Union as a hostile foreign power served the interests of nationalist politicians like Viktor Orban whose popularity was bolstered whenever EU authorities questioned the quality of Hungarian democracy.

A second building block of nationalism is the politics of fear. Today, leaders are linking the threat of terrorism in their countries to the refugees fleeing the violence in the Middle East. In Hungary, Slovakia, and Poland, the governing parties have called Muslim refugees "a threat to Christian civilization." Not to be outdone, the Hungarian government has warned that refugees in Europe are all potential terrorists, and is now preparing to enact an antiterror law to give the government emergency powers to declare "a state of terror threat" and suspend the constitution for 60 days, subject to continuous extension.

III

Once an Eastern European nationalist state was fully constructed, its form of government was given a new name—illiberal democracy. The term was coined in July 2014 by Prime Minister Viktor Orban of Hungary for the Hungarian government. He asserted that Hungary and its Eastern European neighbors had rejected the liberal values of individual rights and were returning to the traditional collective values of their nation-states. To emphasize his point, he asserted that "the Hungarian nation is not a pile of individuals" like people in the West after the rights revolution of the 1960s.[4] Orban claimed that liberal democracy was a failure, pointing to political division and economic inequality in the United States, and dysfunction in the EU on issues of financial policy and migration. In his view, countries that are "capable of making us competitive" in the global economy "are not Western, not liberal democracies, maybe not even democracies," citing as models the governments of Russia, China, Turkey, and Singapore.[5]

What are the elements of an "illiberal democracy?" The entry point is an election, to establish its claim—however tenuous—to be a democracy. Beyond that, the critical feature is majoritarian rule, implemented by a parliamentary supermajority that guarantees total control by the ruling party. In Hungary, this supermajority has opened the door to constitutional changes abolishing checks and balances and other key distinguishing features of a pluralist democracy.

The central claims of the new illiberal system are its promises of efficiency, collective purpose, and national pride. The trade-offs to achieve these goals are the centralization of power and the curtailment of individual rights. A question mark hanging over the system is whether it is sustainable, especially when it is inside a larger transnational system like the European Union. In his 2014 speech, Viktor Orban challenged the EU, claiming, "I don't think our EU membership precludes building an illiberal new state based on a national foundation."[6]

The Hungarian government has rejected the values and structures of a liberal democratic order. These values and institutional structures are intended to maximize accountability and liberty within a framework of democratic governance—checks and balances; freedoms of expression and assembly; due process of law; independence of the judiciary and the media; the protection of minorities; a pluralist civil society; and the rule of law.

The European Union was built on these values. They are at the heart of the political culture that has promoted the integration of Europe, but the new illiberal regimes of Eastern Europe are alien to this culture, and their neo-authoritarian leaders are rejecting it. Forces of disintegration unleashed by the refugee crisis and the Eurocrisis, combined with Viktor Orban's challenge to European values, are threatening the very concept of European integration.

Last fall, this new model of illiberal democracy galvanized nationalists across Europe when Hungary constructed razor wire fences on its borders and stationed its army and police to keep out refugees. The result was a huge boost to the governing party's flagging popularity at home, and the Hungarian Prime Minister's emergence on the European stage as a challenger to German Chancellor Angela Merkel, whose response to the refugee crisis was based on the liberal values of the EU. The refugee crisis provided a golden opportunity for Viktor Orban to burnish his illiberal credentials without having to make the kinds of sober compromises that a liberal leader like Merkel has had to do to support both European values and European security. To paraphrase the Polish sociologist Zygmunt Bauman, illiberal leaders use chaos to create the opportunity for imposing order.

The new Polish government is now emulating the Hungarian model. It made the refugee issue a central feature of its election campaign last fall, promising that religious and ethnic nationalism would protect Poles from an invasion of Muslims into Poland's homogeneous Catholic society. The government took a page out of the Hungarian playbook by attacking the Polish Constitutional Court and the independence of the Polish judiciary.

A pitched battle is now shaping up in Europe between liberal and illiberal democracy. At stake are the values that safeguard Europe against a repeat of its catastrophic experience with 20th century fascism and communism. These values are challenged not only by the proponents of illiberal democracy but also from within liberal democracies in Europe and the United States. Disturbing signs are everywhere about the health of Western democracies—their steady decline in voter participation, their broad distrust of political leaders, their alienation from distant decision-makers, their susceptibility to the influence of money in politics, their inability to make decisions on urgent issues like the Eurocrisis, refugees and immigration, and their increasing polarization and gridlock. Out of this discontent, new nationalists and demagogues on both sides of the Atlantic like Marine Le Pen and Donald Trump are gaining popularity.

IV

This is why the winter of our discontent will not be ending soon. But if we step back and ask some questions, I think some surprising answers may indicate the state of democracy may not be as bleak in the long run as it may seem today.

Can the EU Survive the Challenge Posed to it from within by Illiberal Governance?

The EU is clearly vulnerable. Without major structural reforms, EU institutions make easy targets for nationalist movements. The Brussels bureaucracy is remote and voters have no real connection to it. Only the member states participate directly in EU governance, and so far, their leaders have shown little inclination to discipline a member state like Hungary or Poland that defies EU rules and principles—probably, because they may want to do so themselves one day, as many British leaders are doing now by promoting "Brexit," or British exit from the EU.

Paradoxically, Eastern European illiberal states may not be as big a threat to the EU as they appear because the benefits they receive are far greater than their costs of staying in. Two basic factors tie Hungary and its Eastern European neighbors to the EU—money and politics.

The money is plentiful, and flows freely in the form of structural funds with few strings attached. Over the next five years, Hungary is guaranteed to receive EUR 22 billion from the EU. Many of the country's major capital projects, public investment opportunities, and employment strategies are connected to this beneficent and benign funding source.

The second factor is politics. The EU provides an attractive political target for Eastern European politicians who benefit from biting the hand that feeds them with their rallying cry that "Brussels is the new Moscow." And despite their assault on the

EU's liberal values, Eastern European countries benefit substantially from the Schengen rules on freedom of movement within the EU that guarantee employment mobility for their citizens. Without the EU, Hungary and its neighbors would be cast adrift in a chaotic environment. They have no natural resources, and would become economic vassals of the two big illiberal states to the East—Russia and Turkey—whose economic and security situation is far more uncertain even than that of the EU. This is why Viktor Orban is trying to prevent the EU from detaching Eastern Europe from the Schengen zone, and also why he is seeking to maintain social benefits for Hungarian workers in the UK. These may be losing battles for him, especially if he continues to resist the EU quota rules on accepting refugees, but they show how much he and his neighbors need the EU.

Are the New Illiberal Democracies in Eastern Europe Sustainable?

If an illiberal government can be changed by democratic means, then the system may be sustainable. But if the centralization of power is so successful that the government can fend off any democratic challenge, then, paradoxically, an illiberal system may not be sustainable in the long run. There are four key weaknesses in the system.

First, the legacy of state control over the economy and its eventual collapse under communism show that it may be difficult for centralized illiberal regimes to deliver economically to their citizens without liberalizing their political institutions. This is particularly true for countries like Hungary and Poland that have been incorporated into a much larger interconnected market economy like the EU. Russia and China, the two main countries cited by Viktor Orban as models of illiberal governance, are both faltering economically because of the way they are governed politically.

Second, illiberal governance tends to lead to systemic corruption, which is a drag on economic growth and a source of instability, as the situation in Russia shows. Eastern European countries have unfavorable ratings compared to other EU member states on Transparency International's European Corruption Index.

Third, illiberal governance is vulnerable to the digital revolution, which allows increased peer-to-peer flows of information and creates horizontal pressures for change. Traditional media may have fallen under the control of illiberal regimes, but digital media have not. In Hungary, over 100,000 people took to the streets in 2014 when the government threatened to tax the use of the internet, and the government had to back down.

Fourth, as the Internet tax controversy shows, illiberal regimes have few institutional safety valves for citizen discontent. When popular pressures build, the regime must either back down or resort to coercion. The Euromaidan protests in Ukraine demonstrated that the use of violence by an illiberal regime can lead to greater public discontent and pressure for more radical change.

A far greater challenge to the EU than illiberal governance in Eastern Europe is coming from one of the world's oldest democracies in the West—the United Kingdom. Now that the EU has given Prime Minister David Cameron what he has been asking for, it would be devastating for both sides of the Channel if the Brexit referendum were to pass.

Is Liberal Democracy in Recession, or a State of Permanent Decline?

This question can be answered in different ways. If one looks at the increasing popular demands for participation in governance and engagement in decision-making—as demonstrated by democracy movements around the world from Euromaidan to Taksim Gezi Park, to Tahrir Square, to Hong Kong, to Black Lives Matter in the United States—the ideas of democracy have greater appeal today than ever, even as the supply of healthy democratic governance may be diminishing.

On the other hand, if one looks at the popular appeal of the politics of national identity and security, and the demand for stability and efficiency in governance, as the opinion polls in Europe and the United States seem to show, then liberal democracy with its aging pluralist institutions and short-term election perspectives may be in decline.

In the end, it will depend on democracy's capacity to reform itself—to use the tools of the digital revolution to stimulate participation while leveling the playing field and curtailing the economic power of the top 1 percent to exercise disproportionate influence over decision-making. It will also depend on liberal democracy giving more recognition to national identity and security, and creating new channels for national participation in supranational structures like the EU.

What about the United States—Will They Elect a Nationalist, Populist, Unilateralist, Illiberal President?

There are certainly threats to liberal values in the United States from the far right—on immigration, racial issues, and women's rights, to name a few. But there's also plenty of energy, especially on the left, for economic and political reforms to strengthen liberal democracy. On foreign policy, no one should mistake populist discontent for support for foreign intervention. Military deployment is deeply unpopular in the wake of the disastrous 2003 intervention in Iraq. If anything, I'm concerned that the United States is being swept up in a wave of neo-isolationism that may keep it from engaging as a leader in the world, and particularly from working with Europe and Russia to address the crises in Ukraine and Syria, and manage the global refugee crisis.

My prediction is that the United States will not elect a nationalist, populist, unilateralist, illiberal president, but that gridlock and polarization will continue to plague American politics unless one party wins both the presidency and the Congress, especially now that the Supreme Court is up for grabs. This is a sorry commentary on the state of democracy in America. Democratic politics are about compromise and negotiation between opposing viewpoints, not about zero-sum scorched-earth attacks on anyone who does not follow the orthodoxy of one political group. The Tea Party movement was the harbinger of contemporary anticompromise, antidemocracy politics in America, and Donald Trump is its apotheosis. Trump may not succeed in capturing the presidency, but what he represents is a more dangerous American version of the nationalist illiberal democracy movements in Europe.

V

The rise of illiberal governance in Eastern Europe is rooted in a long legacy of authoritarianism. Democratic solutions must come from within and will take time to develop. These regimes do not pose an existential threat to the European Union—in fact, the benefits the EU provides them may make them stronger EU supporters than liberal democracies in the West like the UK. Illiberal democracies stimulate and feed on popular fears and anxieties, but without an institutional safety valve for popular discontent, they may not be sustainable in the long run.

The popular demand for democratic participation is growing, but it needs new language and new structures beyond those of traditional liberal democracy.

Democracy always sparks discontent, but discontent can also spark change. While democracy offers a path for change, illiberal governance is a dead end: its proponents are determined to control all the levers of power, and block all the avenues for change. In the end, democracy, as Winston Churchill famously pointed out, is the worst form of government, apart from all the others.

To return to Václav Havel, his words sum up very well the challenge of democracy and its discontents: "I'm not an optimist because I don't believe all ends well. I'm not a pessimist because I don't believe all ends badly. Instead, I'm a realist who carries hope, and hope is the belief that democracy has meaning, and is worth the struggle."

Legal Topics

For related research and practice materials, see the following legal topics:

Communications Law, Related Legal Issues, Taxation, Immigration Law, Refugees Eligibility, International Trade Law, Trade Agreements General Overview.

Footnotes

1. European Commission, Directorate-General for Communication. *Standard Eurobarometer 82 "Public Opinion in the European Union, First Results."* December 2014, http://ec.europa.eu/public_opinion/archives/eb/eb82/eb82_first_en.pdf (accessed April 11, 2016).
2. McCarthy, Justin. "In U.S., 65% Dissatisfied with How Gov't System Works." Gallup. January 22, 2014, http://www.gallup.com/poll/166985/dissatisfied-gov-system-works.aspx (accessed April 19, 2016).
3. "Rechtsruck in Ungarn," DW, January 1, 2013, http://www.dw.com/de/rechtsruck-in-ungarn/a-16561346 (accessed April 19, 2016).
4. "Viktor Orban's illiberal world,"*Financial Times*, July 30, 2014, http://blogs.ft.com/the-world/2014/07/viktor-orbans-illiberal-world/ (accessed April 19, 2016).
5. "Orban Wants to Build Illiberal State,"*EU Observer*, July 28, 2014, https://euob-server.com/political/125128 (accessed April 19, 2016).
6. "Orban Says He Seeks to End Liberal Democracy in Hungary,"*Bloomberg*, July 28, 2014, http://www.bloomberg.com/news/articles/2014-07-28/orban-says-he-seeks-to-end-liberal-democracy-in-hungary (accessed April 19, 2016).

Critical Thinking

1. Why is illiberal democracy threatening European integration?
2. Why has illiberal democracy developed in Poland and Hungary?
3. Why can illiberal democracy lead to either fascism or communism?

Internet References

Democracy and Rule of Law Program, Carnegie Endowment for International Peace
www.carnegieendowment.org

Freedom House
www.freedomhouse.org/

Organization for Security and Cooperation in Europe, Office for Democratic Institutions and Human Rights
www.osce.org/odihr

UN Electoral Assistance Unit
www.un.org/undp/en/elections